T0317414

SMALL UNMANNED FIXED-WING AIRCRAFT DESIGN

Related titles
Aerospace Series

Small Unmanned Fixed-wing Aircraft Design: A Practical Approach	Keane	October 2017
Performance of the Jet Transport Airplane: Analysis Methods, Flight Operations, and Regulations	Young	August 2017
Differential Game Theory with Applications to Missiles and Autonomous Systems Guidance	Faruqi	May 2017
Advanced UAV Aerodynamics, Flight Stability and Control: Novel Concepts, Theory and Applications	Marques and Da Ronch	April 2017
Introduction to Nonlinear Aeroelasticity	Dimitriadis	April 2017
Introduction to Aerospace Engineering with a Flight Test Perspective	Corda	March 2017
Aircraft Control Allocation	Durham, Bordignon and Beck	January 2017
Remotely Piloted Aircraft Systems: A Human Systems Integration Perspective	Cooke, Rowe, Bennett. Jr. and Joralmon	October 2016
Adaptive Aeroservoelastic Control	Tewari	March 2016
Theory and Practice of Aircraft Performance	Kundu, Price and Riordan	November 2015
The Global Airline Industry, Second Edition	Belobaba, Odoni and Barnhart	July 2015
Modeling the Effect of Damage in Composite Structures: Simplified Approaches	Kassapoglou	March 2015
Introduction to Aircraft Aeroelasticity and Loads, 2nd Edition	Wright and Cooper	December 2014
Aircraft Aerodynamic Design: Geometry and Optimization	Sóbester and Forrester	October 2014
Theoretical and Computational Aerodynamics	Sengupta	September 2014
Aerospace Propulsion	Lee	October 2013
Aircraft Flight Dynamics and Control	Durham	August 2013
Civil Avionics Systems, 2nd Edition	Moir, Seabridge and Jukes	August 2013
Modelling and Managing Airport Performance	Zografos, Andreatta and Odoni	July 2013
Advanced Aircraft Design: Conceptual Design, Analysis and Optimization of Subsonic Civil Airplanes	Torenbeek	June 2013
Design and Analysis of Composite Structures: With Applications to Aerospace Structures, 2nd Edition	Kassapoglou	April 2013
Aircraft Systems Integration of Air-Launched Weapons	Rigby	April 2013
Design and Development of Aircraft Systems, 2nd Edition	Moir and Seabridge	November 2012
Understanding Aerodynamics: Arguing from the Real Physics	McLean	November 2012
Aircraft Design: A Systems Engineering Approach	Sadraey	October 2012
Introduction to UAV Systems 4e	Fahlstrom and Gleason	August 2012
Theory of Lift: Introductory Computational Aerodynamics with MATLAB and Octave	McBain	August 2012
Sense and Avoid in UAS: Research and Applications	Angelov	April 2012
Morphing Aerospace Vehicles and Structures	Valasek	April 2012
Gas Turbine Propulsion Systems	MacIsaac and Langton	July 2011
Basic Helicopter Aerodynamics, 3rd Edition	Seddon and Newman	July 2011
Advanced Control of Aircraft, Spacecraft and Rockets	Tewari	July 2011
Cooperative Path Planning of Unmanned Aerial Vehicles	Tsourdos et al	November 2010
Principles of Flight for Pilots	Swatton	October 2010
Air Travel and Health: A Systems Perspective	Seabridge et al	September 2010
Design and Analysis of Composite Structures: With applications to aerospace Structures	Kassapoglou	September 2010
Unmanned Aircraft Systems: UAVS Design, Development and Deployment	Austin	April 2010
Introduction to Antenna Placement & Installations	Macnamara	April 2010

Principles of Flight Simulation	Allerton	October 2009
Aircraft Fuel Systems	Langton et al	May 2009
The Global Airline Industry	Belobaba	April 2009
Computational Modelling and Simulation of Aircraft and the Environment: Volume 1 - Platform Kinematics and Synthetic Environment	Diston	April 2009
Handbook of Space Technology	Ley, Wittmann Hallmann	April 2009
Aircraft Performance Theory and Practice for Pilots	Swatton	August 2008
Aircraft Systems, 3rd Edition	Moir & Seabridge	March 2008
Introduction to Aircraft Aeroelasticity And Loads	Wright & Cooper	December 2007
Stability and Control of Aircraft Systems	Langton	September 2006
Military Avionics Systems	Moir & Seabridge	February 2006
Design and Development of Aircraft Systems	Moir & Seabridge	June 2004
Aircraft Loading and Structural Layout	Howe	May 2004
Aircraft Display Systems	Jukes	December 2003
Civil Avionics Systems	Moir & Seabridge	December 2002

SMALL UNMANNED FIXED-WING AIRCRAFT DESIGN

A Practical Approach

Andrew J. Keane

University of Southampton
UK

András Sóbester

University of Southampton
UK

James P. Scanlan

University of Southampton
UK

Library of Congress Cataloging-in-Publication Data
Names: Keane, Andrew J., author. | Sóbester, András, author. | Scanlan, James P., author.
Title: Small unmanned fixed-wing aircraft design : a practical approach /
    Andrew J. Keane, University of Southampton, UK, András Sóbester,
    University of Southampton, UK, James P. Scanlan, University of
    Southampton, UK.
Description: First edition. | Hoboken, NJ, USA : John Wiley & Sons, Inc.,
    [2017] | Series: Aerospace series | Includes bibliographical references and index. |
Identifiers: LCCN 2017024962 (print) | LCCN 2017027876 (ebook) | ISBN 9781119406327 (pdf) |
    ISBN 9781119406310 (epub) | ISBN 9781119406297 (cloth)
Subjects: LCSH: Drone aircraft–Design and construction. | Airplanes–Design and construction.
Classification: LCC TL685.35 (ebook) | LCC TL685.35 .K43 2017 (print) | DDC 629.133/39–dc23
LC record available at https://lccn.loc.gov/2017024962

Cover image: Courtesy of the authors
Cover design by Wiley

Set in 10/12pt, TimesLTStd by SPi Global, Chennai, India

10 9 8 7 6 5 4 3 2 1

This book is dedicated to the students of the University of Southampton who have designed, built and flown many UAVs over the last decade and who have been great fun to work with.

Contents

List of Figures   xvii

List of Tables   xxxiii

Foreword   xxxv

Series Preface   xxxvii

Preface   xxxix

Acknowledgments   xli

PART I   INTRODUCING FIXED-WING UAVS

1   Preliminaries   **3**
1.1   Externally Sourced Components   4
1.2   Manufacturing Methods   5
1.3   Project DECODE   6
1.4   The Stages of Design   6
     1.4.1   *Concept Design*   8
     1.4.2   *Preliminary Design*   10
     1.4.3   *Detail Design*   11
     1.4.4   *Manufacturing Design*   12
     1.4.5   *In-service Design and Decommissioning*   13
1.5   Summary   13

2   Unmanned Air Vehicles   **15**
2.1   A Brief Taxonomy of UAVs   15
2.2   The Morphology of a UAV   19
     2.2.1   *Lifting Surfaces*   21
     2.2.2   *Control Surfaces*   22
     2.2.3   *Fuselage and Internal Structure*   23
     2.2.4   *Propulsion Systems*   24

	2.2.5 *Fuel Tanks*	24
	2.2.6 *Control Systems*	24
	2.2.7 *Payloads*	27
	2.2.8 *Take-off and Landing Gear*	27
2.3	Main Design Drivers	29

PART II   THE AIRCRAFT IN MORE DETAIL

3	**Wings**	**33**
3.1	Simple Wing Theory and Aerodynamic Shape	33
3.2	Spars	37
3.3	Covers	37
3.4	Ribs	38
3.5	Fuselage Attachments	38
3.6	Ailerons/Roll Control	40
3.7	Flaps	41
3.8	Wing Tips	42
3.9	Wing-housed Retractable Undercarriage	42
3.10	Integral Fuel Tanks	44

4	**Fuselages and Tails (Empennage)**	**45**
4.1	Main Fuselage/Nacelle Structure	45
4.2	Wing Attachment	47
4.3	Engine and Motor Mountings	48
4.4	Avionics Trays	50
4.5	Payloads – Camera Mountings	51
4.6	Integral Fuel Tanks	52
4.7	Assembly Mechanisms and Access Hatches	54
4.8	Undercarriage Attachment	55
4.9	Tails (Empennage)	57

5	**Propulsion**	**59**
5.1	Liquid-Fueled IC Engines	59
	5.1.1 *Glow-plug IC Engines*	62
	5.1.2 *Spark Ignition Gasoline IC Engines*	62
	5.1.3 *IC Engine Testing*	65
5.2	Rare-earth Brushless Electric Motors	66
5.3	Propellers	68
5.4	Engine/Motor Control	70
5.5	Fuel Systems	70
5.6	Batteries and Generators	71

6	**Airframe Avionics and Systems**	**73**
6.1	Primary Control Transmitter and Receivers	73
6.2	Avionics Power Supplies	76

6.3	Servos	78
6.4	Wiring, Buses, and Boards	82
6.5	Autopilots	86
6.6	Payload Communications Systems	87
6.7	Ancillaries	88
6.8	Resilience and Redundancy	90
7	**Undercarriages**	**93**
7.1	Wheels	93
7.2	Suspension	95
7.3	Steering	95
7.4	Retractable Systems	97

PART III   DESIGNING UAVS

8	**The Process of Design**	**101**
8.1	Goals and Constraints	101
8.2	Airworthiness	103
8.3	Likely Failure Modes	104
	8.3.1 *Aerodynamic and Stability Failure*	105
	8.3.2 *Structural Failure*	106
	8.3.3 *Engine/Motor Failure*	107
	8.3.4 *Control System Failure*	107
8.4	Systems Engineering	110
	8.4.1 *Work-breakdown Structure*	110
	8.4.2 *Interface Definitions*	112
	8.4.3 *Allocation of Responsibility*	112
	8.4.4 *Requirements Flowdown*	112
	8.4.5 *Compliance Testing*	113
	8.4.6 *Cost and Weight Management*	114
	8.4.7 *Design "Checklist"*	117
9	**Tool Selection**	**119**
9.1	Geometry/CAD Codes	120
9.2	Concept Design	123
9.3	Operational Simulation and Mission Planning	125
9.4	Aerodynamic and Structural Analysis Codes	125
9.5	Design and Decision Viewing	125
9.6	Supporting Databases	126
10	**Concept Design: Initial Constraint Analysis**	**127**
10.1	The Design Brief	127
	10.1.1 *Drawing up a Good Design Brief*	127
	10.1.2 *Environment and Mission*	128
	10.1.3 *Constraints*	129

10.2   Airframe Topology   130
       10.2.1 Unmanned versus Manned – Rethinking Topology   130
       10.2.2 Searching the Space of Topologies   133
       10.2.3 Systematic "invention" of UAV Concepts   136
       10.2.4 Managing the Concept Design Process   144
10.3   Airframe and Powerplant Scaling via Constraint Analysis   144
       10.3.1 The Role of Constraint Analysis   144
       10.3.2 The Impact of Customer Requirements   145
       10.3.3 Concept Constraint Analysis – A Proposed Computational
                  Implementation   145
       10.3.4 The Constraint Space   146
10.4   A Parametric Constraint Analysis Report   146
       10.4.1 About This Document   146
       10.4.2 Design Brief   147
       10.4.3 Unit Conversions   149
       10.4.4 Basic Geometry and Initial Guesses   151
       10.4.5 Preamble   151
       10.4.6 Preliminary Calculations   152
       10.4.7 Constraints   154
10.5   The Combined Constraint Diagram and Its Place in the Design Process   162

11   Spreadsheet-Based Concept Design and Examples   **165**
11.1   Concept Design Algorithm   166
11.2   Range   169
11.3   Structural Loading Calculations   169
11.4   Weight and CoG Estimation   170
11.5   Longitudinal Stability   170
11.6   Powering and Propeller Sizing   171
11.7   Resulting Design: Decode-1   174
11.8   A Bigger Single Engine Design: Decode-2   177
11.9   A Twin Tractor Design: SPOTTER   182

12   Preliminary Geometry Design   **189**
12.1   Preliminary Airframe Geometry and CAD   190
12.2   Designing Decode-1 with AirCONICS   192

13   Preliminary Aerodynamic and Stability Analysis   **195**
13.1   Panel Method Solvers – XFoil and XFLR5   196
13.2   RANS Solvers – Fluent   200
       13.2.1 Meshing, Turbulence Model Choice, and y+   204
13.3   Example Two-dimensional Airfoil Analysis   208
13.4   Example Three-dimensional Airfoil Analysis   210
13.5   3D Models of Simple Wings   212
13.6   Example Airframe Aerodynamics   214
       13.6.1 Analyzing Decode-1 with XFLR5: Aerodynamics   215
       13.6.2 Analyzing Decode-1 with XFLR5: Control Surfaces   221

13.6.3 *Analyzing Decode-1 with XFLR5: Stability* 223
13.6.4 *Flight Simulators* 227
13.6.5 *Analyzing Decode-1 with Fluent* 228

14 Preliminary Structural Analysis **237**
14.1 Structural Modeling Using AirCONICS 240
14.2 Structural Analysis Using Simple Beam Theory 243
14.3 Finite Element Analysis (FEA) 245
14.3.1 *FEA Model Preparation* 246
14.3.2 *FEA Complete Spar and Boom Model* 250
14.3.3 *FEA Analysis of 3D Printed and Fiber- or Mylar-clad Foam Parts* 255
14.4 Structural Dynamics and Aeroelasticity 265
14.4.1 *Estimating Wing Divergence, Control Reversal, and Flutter Onset Speeds* 266
14.5 Summary of Preliminary Structural Analysis 272

15 Weight and Center of Gravity Control **273**
15.1 Weight Control 273
15.2 Longitudinal Center of Gravity Control 279

16 Experimental Testing and Validation **281**
16.1 Wind Tunnels Tests 282
16.1.1 *Mounting the Model* 282
16.1.2 *Calibrating the Test* 284
16.1.3 *Blockage Effects* 284
16.1.4 *Typical Results* 287
16.2 Airframe Load Tests 290
16.2.1 *Structural Test Instruments* 290
16.2.2 *Structural Mounting and Loading* 293
16.2.3 *Static Structural Testing* 294
16.2.4 *Dynamic Structural Testing* 296
16.3 Avionics Testing 300

17 Detail Design: Constructing Explicit Design Geometry **303**
17.1 The Generation of Geometry 303
17.2 Fuselage 306
17.3 An Example UAV Assembly 309
17.3.1 *Hand Sketches* 311
17.3.2 *Master Sketches* 311
17.4 3D Printed Parts 313
17.4.1 *Decode-1: The Development of a Parametric Geometry for the SLS Nylon Wing Spar/Boom "Scaffold Clamp"* 313
17.4.2 *Approach* 314
17.4.3 *Inputs* 314
17.4.4 *Breakdown of Part* 315
17.4.5 *Parametric Capability* 316

	17.4.6	More Detailed Model	317
	17.4.7	Manufacture	318
17.5	Wings		318
	17.5.1	Wing Section Profile	320
	17.5.2	Three-dimensional Wing	323

PART IV   MANUFACTURE AND FLIGHT

18	**Manufacture**		**331**
18.1	Externally Sourced Components		331
18.2	Three-Dimensional Printing		332
	18.2.1	Selective Laser Sintering (SLS)	332
	18.2.2	Fused Deposition Modeling (FDM)	335
	18.2.3	Sealing Components	335
18.3	Hot-wire Foam Cutting		337
	18.3.1	Fiber and Mylar Foam Cladding	339
18.4	Laser Cutting		339
18.5	Wiring Looms		342
18.6	Assembly Mechanisms		342
	18.6.1	Bayonets and Locking Pins	345
	18.6.2	Clamps	346
	18.6.3	Conventional Bolts and Screws	346
18.7	Storage and Transport Cases		347

19	**Regulatory Approval and Documentation**		**349**
19.1	Aviation Authority Requirements		349
19.2	System Description		351
	19.2.1	Airframe	352
	19.2.2	Performance	355
	19.2.3	Avionics and Ground Control System	356
	19.2.4	Acceptance Flight Data	358
19.3	Operations Manual		358
	19.3.1	Organization, Team Roles, and Communications	359
	19.3.2	Brief Technical Description	359
	19.3.3	Operating Limits, Conditions, and Control	359
	19.3.4	Operational Area and Flight Plans	360
	19.3.5	Operational and Emergency Procedures	360
	19.3.6	Maintenance Schedule	360
19.4	Safety Case		361
	19.4.1	Risk Assessment Process	362
	19.4.2	Failure Modes and Effects	362
	19.4.3	Operational Hazards	363
	19.4.4	Accident List	364

19.4.5	Mitigation List	364
19.4.6	Accident Sequences and Mitigation	366
19.5	Flight Planning Manual	368

20	**Test Flights and Maintenance**	**369**
20.1	Test Flight Planning	369
20.1.1	Exploration of Flight Envelope	369
20.1.2	Ranking of Flight Tests by Risk	370
20.1.3	Instrumentation and Recording of Flight Test Data	370
20.1.4	Pre-flight Inspection and Checklists	371
20.1.5	Atmospheric Conditions	371
20.1.6	Incident and Crash Contingency Planning, Post Crash Safety, Recording, and Management of Crash Site	371
20.2	Test Flight Examples	375
20.2.1	UAS Performance Flight Test (MANUAL Mode)	375
20.2.2	UAS CoG Flight Test (MANUAL Mode)	377
20.2.3	Fuel Consumption Tests	377
20.2.4	Engine Failure, Idle, and Throttle Change Tests	377
20.2.5	Autonomous Flight Control	378
20.2.6	Auto-Takeoff Test	380
20.2.7	Auto-Landing Test	380
20.2.8	Operational and Safety Flight Scenarios	381
20.3	Maintenance	381
20.3.1	Overall Airframe Maintenance	382
20.3.2	Time and Flight Expired Items	382
20.3.3	Batteries	383
20.3.4	Flight Control Software	383
20.3.5	Maintenance Record Keeping	384

| **21** | **Lessons Learned** | **385** |
| 21.1 | Things that Have Gone Wrong and Why | 388 |

| **PART V** | **APPENDICES, BIBLIOGRAPHY, AND INDEX** | |

| **A** | **Generic Aircraft Design Flowchart** | **395** |

| **B** | **Example AirCONICS Code for Decode-1** | **399** |

C	**Worked (Manned Aircraft) Detail Design Example**	**425**
C.1	Stage 1: Concept Sketches	425
C.2	Stage 2: Part Definition	429
C.3	Stage 3: "Flying Surfaces"	434
C.4	Stage 4: Other Items	435
C.5	Stage 5: Detail Definition	435

| **Bibliography** | | **439** |

| **Index** | | **441** |

List of Figures

Figure 1.1   The University of Southampton UAV team with eight of our aircraft, March 2015. See also https://www.youtube.com/c/SotonUAV and https://www.sotonuav.uk/.   4

Figure 1.2   The design spiral.   7

Figure 2.1   The Southampton University SPOTTER aircraft at the 2016 Farnborough International Airshow.   19

Figure 2.2   University of Southampton SPOTTER UAV with under-slung payload pod.   20

Figure 2.3   Integral fuel tank with trailing edge flap and main spars.   21

Figure 2.4   A typical carbon spar and foam wing with SLS nylon ribs at key locations (note the separate aileron and flap with associated servo linkages).   22

Figure 2.5   A typical SLS structural component.   23

Figure 2.6   A typical integral fuel tank.   25

Figure 2.7   Typical telemetry data recorded by an autopilot. Note occasional loss of contact with the ground station recording the data, which causes the signals to drop to zero.   26

Figure 2.8   Flight tracks of the SPOTTER aircraft while carrying out automated take-off and landing tests. A total of 23 fully automated flights totaling 55 km of flying is shown.   26

Figure 2.9   A typical UAV wiring diagram.   28

Figure 2.10   The SkyCircuits SC2 autopilot (removed from its case).   29

Figure 2.11   University of Southampton SPOTTER UAV with under-slung maritime flight releasable AUV.   29

Figure 3.1   Variation of airfoil section drag at zero lift with section Reynolds number and thickness-to-chord ratio. After Hoerner [9].   35

Figure 3.2   A UAV with significant FDM ABS winglets (this aircraft also has Custer ducted fans).   36

Figure 3.3   Wing foam core prior to covering or rib insertion −note strengthened section in way of main wing spar.   36

Figure 3.4   Covered wing with spar and rib – in this case, the rib just acts to transfer the wing twisting moment while the spar is bonded directly to the foam without additional strengthening.   37

Figure 3.5   A SPOTTER UAV wing spar under static sandbag test.   38

Figure 3.6   SLS nylon wing rib with spar hole – note the extended load transfer elements that are bonded to the main foam parts of the wing and also flap hinge point.   39

Figure 3.7   Two wing foam cores with end rib and spar inserted – note in this case the rib does not extend to the rear of the section, as a separate wing morphing mechanism will be fitted to the rear of the wing.   39

Figure 3.8   SPOTTER UAV wing under construction showing the two-part aileron plus flap, all hinged off a common rear wing spar – note also the nylon torque peg on the rib nearest the camera.   40

Figure 3.9   UAV that uses wing warping for roll control.   41

Figure 3.10   UAV that uses tiperons for roll control.   41

Figure 3.11   Fowler flap – note the complex mechanism required to deploy the flap.   42

Figure 3.12   Simple FDM-printed wing tip incorporated into the outermost wing rib.   43

Figure 3.13   UAV with pneumatically retractable undercarriage – the main wheels retract into the wings while the nose wheel tucks up under the fuselage (wing cut-out shown prior to undercarriage installation).   43

Figure 3.14   Integral fuel tank in central wing section for SPOTTER UAV.   44

Figure 4.1   SPOTTER SLS nylon engine nacelle/fuselage and interior structure.   46

Figure 4.2   Bayonet system for access to internal avionics (a) and fuselage-mounted switch and voltage indicators (b).   46

Figure 4.3   Load spreader plate on Mylar-clad foam core aileron.   46

Figure 4.4   Commercially produced model aircraft with foam fuselage (and wings).   47

Figure 4.5   Space frame structure made of CFRP tubes with SLS nylon joints and foam cladding.   47

Figure 4.6   DECODE aircraft with modular fuselage elements.   48

Figure 4.7   Wing attachment on SPOTTER fuselage. Note the recess for square torque peg with locking pin between main and rear spar holes.   48

Figure 4.8   Typical engine and motor mounts for SLS nylon fuselages and nacelles. Note the steel engine bearer in first view, engine hours meter in second image, and vibration isolation in third setup.   49

Figure 4.9   Frustratingly small fuselage access hatch.   50

Figure 4.10   Typical plywood avionics boards with equipment mounted. Note dual layer system with antivibration mounts in last image.   50

Figure 4.11   SULSA forward-looking video camera.   51

Figure 4.12   SPOTTER payload pods with fixed aperture for video camera (a) and downward and sideways cameras (b and c).   51

Figure 4.13   Simple two axis gimbal system and Hero2 video camera mounted in front of nose wheel.   52

Figure 4.14   Three-axis gimbal system and Sony video camera mounted in front of the nose wheel. Note the video receiver system on the bench that links to the camera via a dedicated radio channel.   53

Figure 4.15   SPOTTER integral fuel tank. Note internal baffle and very small breather port (top left) in the close-up view of the filler neck.   53

Figure 4.16   Aircraft with SLS nylon fuselage formed in three parts: front camera section attached by bayonet to rear two sections joined by tension rods. Note the steel tension rod inside the hull just behind bayonet in the right-hand image.   54

Figure 4.17   Example hatches in SLS nylon fuselages. Note the locking pins and location tabs on the right-hand hatch.   55

Figure 4.18   Metal-reinforced nose wheel attachment with steering and retract hinge in aluminum frame attached to SLS nylon fuselage. Note the nose wheel leg sized to protect the antenna.   56

Figure 4.19   Nylon nose wheel attachment. Note the significant reinforcement around the lower and upper strut bearings.   56

Figure 4.20   Tails attached directly to the fuselage. The right-hand aircraft is a heavily modified commercial kit used for piggy-back launches of gliders.   57

Figure 4.21   Tails attached using CFRP booms, both circular and square in cross-section.   57

Figure 4.22   All-moving horizontal stabilizer with port/starboard split to augment roll control and provide redundancy.   57

Figure 4.23   SLS nylon part to attach tail surfaces to a CFRP tail boom.   58

Figure 5.1   UAV engine/electric motor/propeller test cell. Note the starter generator on the engine behind the four-bladed propeller.   60

Figure 5.2   UAV engine dynamometer.   61

Figure 5.3   Typical maximum powers, weights, and estimated peak static thrusts of engines for UAVs in the 2–150 kg MTOW range.   61

Figure 5.4   OS Gemini FT-160 glow-plug engine in pusher configuration. Note the permanent wiring for glow-plugs.   62

Figure 5.5   OS 30 cc GF30 four-stroke engine installed in a hybrid powered UAV. Note the significant size of the exhaust system.   63

Figure 5.6   Saito 57 cc twin four-stroke engine in pusher configuration. Note the pancake starter generator fitted to this engine.   64

Figure 5.7   Twin 3W-28i CS single-cylinder two-stroke engines fitted to 2Seas UAV. Note again the significant size of the exhaust systems.   64

Figure 5.8   Twin OS 40 cc GF40 four-stroke engines installed in SPOTTER UAV, with and without engine cowlings.   64

Figure 5.9   Raw performance data taken from an engine under test in our dynamometer.   66

Figure 5.10   Hacker brushless electric motor.   67

Figure 5.11   Outputs from JavaProp "multi analysis" for a propeller operating at fixed torque. Note the differing horizontal scales. Note the design point: here a typical cruise speed of 25 m/s is shown by the small circle and is slightly below the peak efficiency for the design. Peak thrust occurs at about 4 m/s, ensuring a good ability to start the aircraft rolling on a grass field.   69

Figure 5.12   Large UAV fuel tanks. Note clunks and fuel level sensor fitting at rear left-hand corner of one tank.   70

Figure 5.13   SPOTTER fuel tank level sensors. One sensor lies behind the central flap in the upper wider part of the tank (just visible in the right-hand image), while the second one lies at the bottom just above the payload interface.   71

Figure 5.14   Engine-powered brushless generators driven directly or by toothed belt.   72

Figure 6.1   Outline avionics diagram for SPOTTER UAV.   74

Figure 6.2   Outline avionics diagram for SPOTTER UAV (detail) – note switch-over unit linking dual receivers and dual autopilots.   75

Figure 6.3   Typical avionics boards. Note the use of MilSpec connectors (the Futaba receivers are marked 1, the switch-over unit 2, the SC2 autopilot and GPS antenna 3, and the avionics and ignition batteries 4 and 5, respectively).   76

Figure 6.4   Fuselage with externally visible LED voltage monitor strips. Here, one is for the avionics system and the second for the ignition system.   77

Figure 6.5   Aircraft with twin on-board, belt-driven generators as supplied by the UAV Factory and a close-up of UAV Factory system.   77

Figure 6.6   On-board, belt-driven brushless motor used as generators.   78

Figure 6.7   Aircraft with a Sullivan pancake starter–generator system.   78

Figure 6.8   A selection of aircraft servos from three different manufacturers.   79

Figure 6.9   Variation of servo torque with weight for various manuafcturers' servos.   80

Figure 6.10   SPOTTER aircraft showing multiple redundant ailerons and elevators.   81

Figure 6.11   Servo cut-out in wing with SLS nylon reinforcement box.   81

Figure 6.12   Typical servo linkage. Note the servo arm, linkage, and servo horn (with reinforcing pad).   82

Figure 6.13   SPOTTER "iron bird" test harness layout. Note the full-size airframe drawing placed under the wiring.   83

Figure 6.14   Generator and drive motor for "iron bird" testing.   83

Figure 6.15   SPOTTER "iron bird" with resulting professionally built harness in place.   84

Figure 6.16   Decode-1 "iron bird" with harness that uses simple aero-modeler-based cable connections.   85

Figure 6.17   Baseboard with mil spec connections on left- and right-hand edges. Note SkyCircuits SC2 autopilot fitted top right with GPS antenna on top and switch-over unit in the center with very many wiring connections.   85

Figure 6.18   Laser-cut plywood baseboards.   86

Figure 6.19   Components located directly into 3D SLS nylon printed structure. The servo is screwed to a clip-in SLS part, while the motor is bolted directly to the fuselage.   86

Figure 6.20   Basic Arduino Uno autopilot components including GPS module on extension board, and accelerometer, barometer and three-axis gyro on daughter boards.   87

Figure 6.21   Pixhawk autopilot.   87

Figure 6.22   The SkyCircuits SC2 autopilot (removed from its case (a), and with attached aerials and servo connection daughter board (b). See also Figure 6.3, where the SC2 is fitted with its case and a GPS aerial on top).   88

Figure 6.23   A selection of professional-grade 5.8 GHz video radiolink equipment: (front) transmitter with omnidirectional antenna in ruggedized case, and (rear left to right) receiver, directional antenna, and combined receiver/high intensity screen unit.   89

Figure 6.24   A hobby-grade 5.8 GHz video radiolink: (a) receiver with omni-directional antenna and (b) transmitter with similar unclad antenna and attached mini-camera.   89

Figure 6.25   Futaba s-bus telemetry modules:(clockwise from left) temperature sensor, rpm sensor, and GPS receiver.   90

Figure 7.1   Some typical small UAV undercarriages.   94

Figure 7.2   An aircraft with spats fitted to its main wheels to reduce drag.   94

Figure 7.3   Nose wheel and strut showing suspension elements, main bearings, control servo, and caster.   95

Figure 7.4   Tail wheel showing suspension spring.   96

Figure 7.5   Nose wheel mechanism with combined spring-coupled steering and vertical suspension spring.   96

Figure 7.6   UAV with a pneumatic, fully retractable undercarriage system. Note also the nose camera that has been added to the aircraft shown in the image with undercarriage retracted.   97

Figure 7.7   Details of fully retractable undercarriage system.   97

Figure 8.1   Explosion of information content as design progresses.   102

Figure 8.2   2SEAS aircraft with redundant ailerons and elevators.   109

Figure 8.3   Autopilot system on vibration test.   109

Figure 8.4   Treble isolated engine mounting.   110

Figure 8.5   Typical military UAV work-breakdown structure interface definitions, from MIL-HDBK-881C for UAVs.   111

Figure 8.6   Example military system requirements flowdown [13]. Defence Acquisition University.   114

Figure 8.7   Systems engineering "V" model.   114

Figure 8.8   Weight prediction of SPOTTER UAV.   115

Figure 8.9   Pie chart plots of SPOTTER weight.   116

Figure 8.10   Example weight and cost breakdown.   117

Figure 9.1   Outline design workflow.   121

Figure 9.2   Analysis tool logic.   122

Figure 9.3   Mission analysis using the AnyLogic event-driven simulation
                   environment.   126

Figure 10.1   On May 14, 1954, Boeing officially rolled out the Dash-80, the proto-
                   type of the company's 707 jet transport. Source: This photo, by John
                   M. 'Hack' Miller, was taken during the rollout (Image courtesy of
                   the Museum of History & Industry, Seattle https://creativecommons.
                   org/licenses/by-sa/2.0/ – no copyright is asserted by the inclusion of
                   this image).   131

Figure 10.2   Four semi-randomly chosen points in an immense space of unmanned
                   aircraft topologies: (starting at the top) the Scaled Composites Proteus,
                   the NASA Prandtl-D research aircraft, the AeroVironment RQ-11 Raven,
                   and the NASA Helios (images courtesy of NASA and the USAF).   134

Figure 10.3   Minimum mass cantilever designed to carry a point load.   135

Figure 10.4   NASA oblique-wing research aircraft (images courtesy of NASA). Could
                   your design benefit from asymmetry?   138

Figure 10.5   Multifunctionality: the 3D-printed fuel tank (highlighted) of the SPOT-
                   TER unmanned aircraft does not only hold the fuel (with integral baffles)
                   but also generates lift and it has a structural role too, see also Figures 2.3,
                   2.6, and 3.14.   140

Figure 10.6   Typical constraint diagram. Each constraint "bites" a chunk out of the
                   P versus W/S space; whatever is left is the feasible region, wherein the
                   design will have to be positioned.   163

Figure 11.1   Decode-1 in the R.J. Mitchell wind tunnel with wheels and wing tips
                   removed and electric motor for propeller drive.   175

Figure 11.2   Decode-1 in flight with nose camera fitted.   175

Figure 11.3   Decode-1 spreadsheet snapshot – inputs page.   178

Figure 11.4   Decode-1 spreadsheet snapshot – results summary page.   179

Figure 11.5   Decode-1 spreadsheet snapshot – geometry page.   180

Figure 11.6   Decode-1 outer geometry as generated with the AirCONICS tool suite. 180

Figure 11.7   Decode-2 spreadsheet snapshot.   183

Figure 11.8   Decode-2 in flight with nose camera fitted.   183

Figure 11.9   Decode-2 outer geometry as generated with the AirCONICS tool suite. 184

Figure 11.10   SPOTTER spreadsheet snapshot.   186

Figure 11.11   SPOTTER in flight with payload pod fitted.   187

Figure 11.12   SPOTTER outer geometry as generated with the AirCONICS tool suite.   187

Figure 12.1   Basic AirCONICS airframe geometry for a single tractor engine,
                   twin-boom, H-tail design.   191

Figure 12.2 AirCONICS model of complete Decode-1 airframe with control surfaces, undercarriage, and propeller disk. 192

Figure 13.1 C_p and streamline plot for the NACA0012 foil at 16° angle of attack as computed with XFoil. 197

Figure 13.2 Results of XFoil analysis sweep for the NACA 64−201 foil at Mach 0.17 as computed with XFLR5. 199

Figure 13.3 Results of XFLR5 analysis sweep for a wing generated from the NACA 64−201 foil sections at Reynold's number of 4.4 million and Mach 0.17. 201

Figure 13.4 Convergence plot of two-dimensional k-ω SST RANS-based CFD analysis. 203

Figure 13.5 Pathlines and surface static pressure plot from Fluent RANS based CFD solution. 203

Figure 13.6 Section through a coarse-grained 3D Harpoon mesh for typical Spalart−Allmaras UAV wing model and close-up showing a boundary layer mesh. 206

Figure 13.7 Histogram of $y+$ parameter for typical boundary layer mesh using the Spalart−Allmaras one-parameter turbulence model. 207

Figure 13.8 Histogram of $y+$ parameter for typical boundary layer mesh using the $k - \omega$ SST turbulence model. 207

Figure 13.9 Lift versus drag polar for NAC0012 airfoil from XFoil and experiments. Note that when plotted in this way, both lift and drag coefficients may be found at a given angle of attack or, for a given lift coefficient, drag coefficient and angle of attack may easily be read off. 208

Figure 13.10 Low-resolution NASA Langley 2D mesh around the NACA0012 foil. 209

Figure 13.11 Middle-resolution NASA Langley 2D mesh around the NACA0012 foil. 209

Figure 13.12 ICEM 2D mesh around the NACA0012 foil (courtesy of Dr D.J.J. Toal). 210

Figure 13.13 Experimental and 2D computational lift and drag data for the NACA0012 airfoil (using the $k - \omega$ SST turbulence model). Adapted from Abbott and von Doenhoff 1959. 210

Figure 13.14 Computed two-dimensional flow past the NACA0012 foil when almost fully stalled. 211

Figure 13.15 Section through a fine-grained Harpoon 3D mesh around the NACA0012 foil suitable for the $k - \omega$ SST turbulence model. Note the wake mesh extending from the trailing edge. 211

Figure 13.16 Experimental and 3D computational lift and drag data for the NACA0012 airfoil (using the Spalart−Allmaras and $k - \omega$ SST turbulence models). NASA. 212

Figure 13.17 Experimental [20, 22] and computational lift and drag data for the NACA 64−210 section. 213

Figure 13.18 Pathlines from a RANS $k - \omega$ solution for the NACA 64−210 airfoil at 12° angle of attack. Note the reversed flow and large separation bubble on the upper surface. 214

Figure 13.19 Experimental and computational lift and drag data for the Sivells and Spooner [21] wing and $y+$ for the $k - \omega$ SST Harpoon mesh. NASA.   215

Figure 13.20 XFLR5 model of the Sivells and Spooner wing.   216

Figure 13.21 Experimental and computational lift and drag data for the Sivells and Spooner [21] wing with enhanced $k - \omega$ SST Harpoon mesh of 76 million cells and $y+$ for the enhanced mesh.   217

Figure 13.22 Pathlines and static pressure around the Sivells and Spooner [21] wing with enhanced $k - \omega$ SST Harpoon mesh at $11°$ angle of attack. NASA.   218

Figure 13.23 XFLR5 model of Decode-1 airframe as generated by AirCONICS with main wing setting angle of $0°$ and elevator setting angle of $-2.85°$, at an angle of attack of $2.6°$ and 30 m/s. Note the use of cambered sections for the main wing and symmetrical profiles for the elevator and fins. The green bars indicate the section lift, with the tail producing downforce to ensure pitch stability.   218

Figure 13.24 XFLR5-generated polar plot for Decode-1 airframe as generated by Air-CONICS with main wing setting angle of $0°$ and elevator setting angle of $-2.85°$, showing speed variations from 15 to 30 m/s. The black circles indicate flight at an angle of attack of $2.53°$ at which Cm is zero.   219

Figure 13.25 XFLR5-generated polar plot for Decode-1 airframe as generated by Air-CONICS with main wing setting angle of $2.53°$ and elevator setting angle of $-0.34°$, showing speed variations from 15 to 30 m/s. Note that Cl is 0.28 and Cm is zero at an angle of attack of $0°$ as required in the cruise condition.   220

Figure 13.26 XFLR5-generated polar plot for Decode-1 airframe at 30 m/s with main wing setting angle of $0°$, showing variations in center of gravity position by 100 mm, reduction in tail length by 300 mm, and elevator set at an angle of $0°$.   222

Figure 13.27 Time-domain simulation for XFLR5-generated eigenvalues at 30 m/s taken from Table 13.2 showing τ_R for the roll mode and T_2 for the spiral mode.   228

Figure 13.28 University of Southampton flight simulator.   229

Figure 13.29 Decode-1 mesh shown inside Harpoon along with wake surfaces and refinement zones.   230

Figure 13.30 Fluent mesh on the center plane for the Decode-1 airframe $k - \omega$ SST analysis at 30 m/s, together with resulting $y+$ histogram.   231

Figure 13.31 Fluent convergence plot for Decode-1 whole aircraft model at 30 m/s.   232

Figure 13.32 Polar plot for Decode-1 airframe at 30 m/s showing both Fluent and XLFR5 results for lift and drag. Those for Fluent include results for just the lifting surfaces and with the complete airframe fuselage, control surfaces, and undercarriage gear; those for XFLR5 show also the impact of adding a fixed parasitic drag coefficient of 0.0375.   232

Figure 13.33 AirCONICS model of Decode-1 lifting surfaces.   233

Figure 13.34 AirCONICS model of complete Decode-1 airframe with control surfaces, undercarriage, and propeller disk.   233

Figure 13.35 Streamlines colored by velocity magnitude around the complete Decode-1 airframe with deflected ailerons.   234

Figure 14.1   Typical composite Vn diagram for gust and maneuver loads on a small UAV (here for Decode-1 assuming maneuver load factors of +5 and −2, 9.1 m/s gust velocity, and a dive speed of 160% of the cruise speed).   238

Figure 14.2   Breakdown of Decode-1 outer mold line model into individual components for structural modeling.   240

Figure 14.3   Decode-1 components that will be produced by 3D printing or made from laser-cut ply.   242

Figure 14.4   Deflection and slope variations for the Decode-1 main spar when flying at 30 m/s and an angle of attack of 2.53° using loading taken from XFLR5, a load factor of 4, and simple beam theory analysis. The spar is assumed to be made from a circular CFRP section of outer diameter 20 mm, wall thickness 2 mm, Young's modulus of 70 GPa, and extending the full span of the aircraft, being clamped on the center plane.   245

Figure 14.5   Preliminary spar layout for Decode-1. Here the linking parts are taken directly from AirCONICS without being reduced to either thick-walled or thin-walled rib-reinforced structures.   246

Figure 14.6   Simplified Abaqus® main spar model with solid SLS nylon supports for Decode-1, showing subdivided spar and boundary conditions for a 4g maneuver loading.   247

Figure 14.7   Deformed shape and von Mises stress plot for Decode-1 main spar under 4g flight loads using a uniform spar load. The tip deflection is 189.7 mm.   248

Figure 14.8   Abaqus loading for full Decode-1 spar model under wing flight loads taken from XFLR5 together with a load factor of 4 plus elevator and fin loading based on Cl values of unity.   251

Figure 14.9   Deformed shape and von Mises stress plot for full Decode-1 spar model under wing flight loads taken from XFLR5 together with a load factor of 4 plus elevator and fin loading based on Cl values of unity. The main spar tip deflections are 143.9 mm, the elevator spar tip deflections are 10.8 mm, and the fin spar tip deflections are 11.1 mm.   252

Figure 14.10 Further details of the deformed shape and von Mises stress plot for full Decode-1 spar model under wing flight loads taken from XFLR5 together with a load factor of 4 plus elevator and fin loading based on Cl values of unity.   253

Figure 14.11 Deformed shape and von Mises stress plot for nylon support part in full Decode-1 spar model under wing flight loads taken from XFLR5 together with a load factor of 4 plus elevator and fin loading based on Cl values of unity.   255

Figure 14.12 Deformed shape and von Mises stress plot for full Decode-1 spar model with locally refined mesh under wing flight loads taken from XFLR5 together with a load factor of 4 plus elevator and fin loading based on Cl values of unity.   256

Figure 14.13 Deformed shape and von Mises stress plot for full Decode-1 spar model with fully refined mesh and reduced boundary conditions under wing flight loads taken from XFLR5 together with a load factor of 4 plus elevator and fin loading based on Cl values of unity.   257

Figure 14.14 Simplified Abaqus thick-walled structural model for Decode-1 SLS nylon part. The mesh for this part contains 25 000 elements.   257

Figure 14.15 Deformed shape and von Mises stress plot for thick-walled nylon part in full Decode-1 spar model with fully refined mesh and reduced boundary conditions under wing flight loads taken from XFLR5 together with a load factor of 4 plus elevator and fin loading based on Cl values of unity.   258

Figure 14.16 Deformed shape and von Mises stress plot for 2 mm thick-walled nylon part in full Decode-1 spar model with fully refined mesh and reduced boundary conditions under wing flight loads taken from XFLR5 together with a load factor of 4 plus elevator and fin loading based on Cl values of unity.   259

Figure 14.17 Abaqus model of foam core created with CAD shell and fillet commands and meshed with brick hex elements.   260

Figure 14.18 Abaqus model of glass-fiber wing cover created with CAD shell commands and meshed with continuum shell hex elements. Note the wedge elements used for the sharp trailing edge.   260

Figure 14.19 Abaqus assembly with foam parts added, highlighting the tie constraint between the foam and the SLS nylon support.   261

Figure 14.20 Pressure map on Decode-1 foam part under wing flight loads taken from XFLR5.   261

Figure 14.21 Resulting deflections and stresses in foam core and cover for wing under flight conditions.   263

Figure 14.22 Resulting deflections and stresses in SLS nylon part with foam mounting lug for wing under flight conditions.   264

Figure 14.23 Two-degrees-of-freedom model of wing aeroelasticity.   267

Figure 14.24 Truncated Abaqus contour plot of a first twist mode revealing the nodal line and hence the elastic axis.   270

Figure 14.25 Abaqus plots of first flap and twist modes for Decode-1 wing.   271

Figure 15.1 Channel wing aircraft being weighed after final assembly.   278

Figure 16.1 Decode-1 and channel wings on wind tunnel mounting rig. Note the circular boundary plate that stands in for the absent fuselage.   283

Figure 16.2 AirCONICS model of Decode-1 airframe in a representation of the R.J. Mitchell 11' × 8' wind tunnel working section at Southampton University, illustrating degree of blockage.   285

Figure 16.3   AirCONICS half-model of Decode-1 airframe in the R.J. Mitchell 11' ×
8' wind tunnel prior to mesh preparation.   286

Figure 16.4   Section through Fluent velocity magnitude results and Harpoon mesh for
Decode-1 airframe in the R.J. Mitchell 11' × 8' wind tunnel.
Note the extent of the boundary layer on the tunnel walls and the fine
boundary layer mesh needed to resolve this, along with the refinement
zone near the wing tip.   286

Figure 16.5   Decode-1 baseline wind tunnel results (control surfaces in neutral positions) under varying wind speed. (a) Lift coefficient. (b) Drag coefficient.
(c) Side coefficient. (d) Pitch coefficient. (e) Roll coefficient. (f) Yaw
coefficient.   288

Figure 16.6   Decode-1 elevator effectiveness with varying deflection angles and wind
speed. (a) Lift coefficient at 15 m/s. (b) Drag coefficient at 15 m/s. (c)
Pitch coefficient at 10 m/s. (d) Pitch coefficient at 15 m/s. (e) Pitch coefficient at 24 m/s.   289

Figure 16.7   Decode-1 rudder effectiveness with varying deflection angle. (a) Lift
coefficient at 24 m/s. (b) Drag coefficient at 24 m/s. (c) Side coefficient
at 24 m/s. (d) Pitch coefficient at 24 m/s. (e) Roll coefficient at 24 m/s.
(f) Yaw coefficient at 24 m/s.   291

Figure 16.8   Dial gauge in use to measure aiframe deflection during static test in the
lab.   292

Figure 16.9   Lab-quality force transducer, piezoelectric accelerometers, and electromagnetic shakers.   293

Figure 16.10 Flight-capable piezoelectric accelerometer and data-capture system.   293

Figure 16.11 Mounting system for wing and main spar assembly under sandbag load
test.   294

Figure 16.12 Clamping system for main spar.   294

Figure 16.13 Wing assembly under sandbag load test.   295

Figure 16.14 Partial failure of SLS nylon structural component during sandbag load
test. Note the significant cracks and large deformations.   296

Figure 16.15 Load testing of an undercarriage leg and associated SLS nylon mounting
structure. Note the dummy carbon-fiber tubes present to allow the SLS
structure to be correctly set up.   296

Figure 16.16 Ground vibration test of a Decode-1 wing showing support and mounting
arrangements.   297

Figure 16.17 Ground vibration test of a Decode-1 wing ((a) accelerometer on starboard
wing tip: (b) shaker and force transducer near wing root).   298

Figure 16.18 Frequency response from ground vibration test of a Decode-1 wing:
accelerometer on port wing tip and cursors on first flap mode.   298

Figure 16.19 Frequency response from ground vibration test of a Decode-1 wing: flapping mode accelerometer placement (upper) and twisting mode place-
ment (lower), cursors on first twist mode.   299

Figure 16.20 SPOTTER iron-bird being used to test a complete avionics build-up: note motors to spin generators in a realistic manner.   301

Figure 16.21 Avionics board under vibration test. Note the free-free mounting simulated by elastic band supports. In this case, a force transducer has been placed between the shaker and the long connecting rod that stimulates the board. The in-built accelerometer in the flight controller is used to register motions.   301

Figure 16.22 Typical Servo test equipment: (front left to right) simple low-cost tester, large servo, motor speed tester with in-built power meter, and servo control output; (rear) avionics battery and standard primary receiver.   302

Figure 17.1 Detail design process flow.   304

Figure 17.2 The structure of well-partitioned concept design models.   305

Figure 17.3 Example configuration studies.   306

Figure 17.4 Example 3D models of Rotax aircraft engine and RCV UAV engine. Courtesy of Chris Bill and RCV Engines Ltd.   307

Figure 17.5 Example of images used to create realistic looking 3D Solidworks geometry model.   308

Figure 17.6 2D side elevations of Rotax aircraft engine and RCV UAV engine. Courtesy of Chris Bill and RCV Engines Ltd.   308

Figure 17.7 Scaling dimension added to drawing (mm). Courtesy of Chris Bill.   308

Figure 17.8 "Spaceframe" aircraft structure.   309

Figure 17.9 Illustrative student UAV assembly.   310

Figure 17.10 UAV assembly model can be modified by changing design table parameters.   310

Figure 17.11 Plan and side view hand sketches.   311

Figure 17.12 Hand sketch scaled and positioned orthogonally in Solidworks.   312

Figure 17.13 Exact, dimensioned sketch being created on hand-sketch outline.   312

Figure 17.14 The "master" driving sketches in the assembly.   313

Figure 17.15 Design table for example UAV.   313

Figure 17.16 Input reference geometry.   314

Figure 17.17 The input geometry modeled as partitioned parts.   315

Figure 17.18 The assembly generated from reference geometry.   315

Figure 17.19 "Debugging" the detailed model.   316

Figure 17.20 Trimming of boom tube fairing.   317

Figure 17.21 Final detailed model.   318

Figure 17.22 Multipanel wing of PA-28. Photo courtesy Bob Adams https://creativecommons.org/licenses/by-sa/2.0/ – no copyright is asserted by the inclusion of this image.   319

Figure 17.23 NACA four-digit section coordinate spreadsheet.   320

Figure 17.24 Curve importing in Solidworks.   321

Figure 17.25 Use of "convert entities" in Solidworks.   321

Figure 17.26 Closing the 2D aerofoil shape.   322

Figure 17.27 Deleting sketch relationship with reference geometry.   322

Figure 17.28 Reference geometry.   323

Figure 17.29 Constraining curve to reference "scaffold" geometry.   323

Figure 17.30 "3D" scaffold to define the relative positions in space of two independently scalable wing sections.   324

Figure 17.31 Wing surface with span, twist, taper, and sweep variables.   324

Figure 17.32 Multipanel wing.   325

Figure 17.33 Example of a double-curvature composite wing.   325

Figure 17.34 Fabricated wing structures.   326

Figure 17.35 Simple wooden rib and alloy spar structure.   326

Figure 17.36 Parametric wing structure.   327

Figure 18.1 3D SLS nylon parts as supplied from the manufacturer.   332

Figure 18.2 3D SLS stainless steel gasoline engine bearer after printing and in situ.   333

Figure 18.3 3D SLS nylon manufacturing and depowdering.   333

Figure 18.4 Small office-based FDM printer. Parts as they appear on the platten and after removal of support material.   335

Figure 18.5 FDM-printed ABS fuselage parts.   336

Figure 18.6 Aircraft with FDM-printed fuselage and wing tips.   336

Figure 18.7 In-house manufactured hot-wire foam cutting machine. This cuts blocks of foam up to 1400 mm × 590 mm × 320 mm.   338

Figure 18.8 Large hot-wire foam cutting machine.   338

Figure 18.9 Hot-wire-cut foam wing parts: (Left) The original material blocks with and without cores removed; (right) with FDM-manufactured ABS joining parts.   339

Figure 18.10 Foam wings after cladding: glass fiber, Mylar, and filled glass fiber.   340

Figure 18.11 Aircraft with wings fabricated from laser-cut plywood covered with aero-modeler film.   341

Figure 18.12 Avionics base board and servo horn reinforcement made from laser cut plywood.   341

Figure 18.13 Foam reinforcement ribs made from laser-cut plywood.   341

Figure 18.14 Logical wiring diagram (detail).   343

Figure 18.15 Iron bird for building wiring looms.   344

Figure 18.16 Soldering station (note the clamps, heat-resistant mat, and good illumination).   344

Figure 18.17 Female and male bayonet produced in SLS nylon with quick-release locking pin.   345

Figure 18.18 Quick-release pin fitting used to retain a wing to a fuselage (note lug on wing rib).   345

Figure 18.19 SLS nylon clamping mechanisms.   346

Figure 18.20 Cap-screws and embedded retained nuts, here on an undercarriage fixing point.   347

Figure 18.21 Transport and storage cases.   347

Figure 19.1 Typical take-off performance.   356

Figure 19.2 Typical wiring schematic.   357

Figure 20.1 Typical flight log.   370

Figure 20.2 Typical pre-flight checklist.   372

Figure 20.3 Typical flight procedures checklist.   373

Figure 21.1 Our first student-designed UAV.   386

Figure 21.2 Not all test flights end successfully!   386

Figure 21.3 Aircraft with variable length fuselage. (a) Fuselage split open. (b) Spare fuselage section.   387

Figure 21.4 Student-designed flying boat with large hull volume forward and insufficient vertical tail volume aft.   389

Figure 21.5 Aircraft with split all-moving elevator. (a) Without dividing fence. (b) With fence.   389

Figure 21.6 Autopilot on vibration test.   390

Figure 21.7 Student UAV with undersized wings. The open payload bay also added to stability issues.   390

Figure 21.8 2SEAS aircraft after failure of main wheel axle.   391

Figure A.1 Generic aircraft design flowchart.   396

Figure C.1 Vans RV7 Aircraft. Cropped image courtesy Daniel Betts https://creativecommons.org/licenses/by-sa/2.0/ – no copyright is asserted by the inclusion of this image.   426

Figure C.2 Concept sketches of an aircraft.   426

Figure C.3 Side elevation hand sketch imported and scaled.   426

Figure C.4 Plan, side, and front view imported and scaled.   427

Figure C.5 Tracing the outline of the hand sketch to capture the "essential" geometry.   427

Figure C.6 Dimensioned parametric geometry sketch.   428

Figure C.7 View of all three of the dimensioned parametric geometry sketches.   428

Figure C.8 Center fuselage part, with side elevation parametric geometry sketch in the background.   429

Figure C.9 Underlying geometry for the center fuselage.   429

Figure C.10 Completion of center fuselage.   430

Figure C.11 Rear fuselage synchronized with center fuselage at shared interface.   430

Figure C.12 Fully realized fuselage geometry.   431

Figure C.13 Two-panel wing and wing incidence and location line in side elevation. 431

Figure C.14 All the major airframe surface parts added. 431

Figure C.15 Checking against original sketch. 432

Figure C.16 Addition of propeller disk and spinner so that ground clearance can be checked. 432

Figure C.17 Engine installation checking cowling clearance and cooling (note: lightweight decal engine geometry). 433

Figure C.18 Checking the instrument panel fit (again use of decal for instruments). 433

Figure C.19 Checking the ergonomics of crew seating and canopy clearance/view. 433

Figure C.20 Hand sketches–to parameterized sketches–to solid assembly. 434

Figure C.21 Whole aircraft parametric variables. 436

Figure C.22 Wing geometry used to calculate lift centers for static margin calculations. 437

Figure C.23 Final parametric aircraft design with all major masses added. 437

Figure C.24 Final detailed geometry. Courtesy of Vans Aircraft, Inc. 438

List of Tables

Table 1.1	Design system maturity.	9
Table 2.1	Different levels of UAV autonomy classified using the Wright–Patterson air force base scheme.	16
Table 5.1	Typical liquid-fueled IC engine test recording table (maximum rpms are of course engine-dependent).	66
Table 5.2	Typical IC engine BMEP values taken from various sources.	67
Table 6.1	Typical primary transmitter/receiver channel assignments.	76
Table 6.2	Typical servo properties.	79
Table 8.1	Example responsibility allocation matrix for a maintenance team.	113
Table 9.1	Concept design requirements.	124
Table 11.1	Typical fixed parameters in concept design.	167
Table 11.2	Typical limits on variables in concept design.	168
Table 11.3	Estimated secondary airframe dimensions.	168
Table 11.4	Variables that might be used to estimate UAV weights.	171
Table 11.5	Items for which weight estimates may be required and possible dependencies.	172
Table 11.6	Other items for which weight estimates may be required.	173
Table 11.7	Design brief for Decode-1.	176
Table 11.8	Resulting concept design from spreadsheet analysis for Decode-1.	176
Table 11.9	Design geometry from spreadsheet analysis for Decode-1 (in units of mm and to be read in conjunction with Tables 11.3 and 11.8).	177
Table 11.10	Design brief for Decode-2.	181
Table 11.11	Estimated secondary airframe dimensions for Decode-2.	181
Table 11.12	Resulting concept design from spreadsheet analysis for Decode-2.	181
Table 11.13	Design geometry from spreadsheet analysis for Decode-2 (in units of mm and to be read in conjunction with Tables 11.11 and 11.12).	184
Table 11.14	Design brief for SPOTTER.	184
Table 11.15	Estimated secondary airframe dimensions for SPOTTER.	185

Table 11.16 Resulting concept design from spreadsheet analysis for SPOTTER.   185

Table 11.17 Design geometry from spreadsheet analysis for SPOTTER (in units of mm and to be read in conjunction with Tables 11.15 and 11.16).   186

Table 13.1 A summary of some of the Fluent turbulence models based on information provided in Ansys training materials.   202

Table 13.2 Decode-1 eigenvalues as calculated from XFLR5 stability derivatives using the formulae provided by Phillips [24, 25] and the estimated inertia properties for a flight speed of 30 m/s and MTOW of 15 kg.   227

Table 14.1 Shear forces (Q), bending moments (M), slopes (θ, in radians), and deflections (δ) for Euler–Bernoulli analysis of uniform encastre cantilever beams.   243

Table 14.2 A selection of results from various Abaqus models of the Decode-1 airframe.   254

Table 14.3 Natural frequency results (Hz) using Abaqus modal analysis for the Decode-1 airframe.   271

Table 15.1 Typical weight and LCoG control table (LCoG is mm forward of the main spar).   274

Table 18.1 Typical properties of carbon-fiber-reinforced plastic (CFRP) tubes.   332

Table 18.2 Typical properties of SLS nylon 12.   334

Table 18.3 Typical properties of closed-cell polyurethane floor insulation foam.   337

Table 18.4 Typical properties of glass-fiber-reinforced plastics.   340

Table 19.1 Typical small UAS operations manual template part Ai.   351

Table 19.2 Typical small UAS operations manual template part Aii.   352

Table 19.3 Typical small UAS operations manual template part Bi.   353

Table 19.4 Typical small UAS operations manual template parts Bii, C, and D.   354

Table 19.5 Typical summary airframe description.   355

Table 19.6 Typical engine characteristics.   355

Table 19.7 Typical aircraft performance summary in still air.   356

Table 19.8 Radio control channel assignments.   358

Table 19.9 Risk probability definitions (figures refer to flight hours).   361

Table 19.10 Accident severity definitions.   361

Table 19.11 Risk classification matrix.   362

Table 19.12 Risk class definitions.   362

Table 19.13 Typical failure effects list (partial).   363

Table 19.14 Typical hazard list (partial).   364

Table 19.15 Typical accident list (partial).   365

Table 19.16 Typical mitigation list (partial).   366

Table 19.17 Typical accident sequences and mitigation list (partial).   367

Foreword

As a proud University of Southampton alumnus, I am delighted to have the privilege of writing a few introductory words. My own journey through the ranks of the Royal Navy, and especially in naval aviation, leadership, and the exploitation of technology, has taught me the advantage and fun to be had in challenging convention.

Meanwhile, it is the sheer scale of the emerging new technologies that makes today such an exciting time. And the Internet has unlocked access to diverse technical knowledge. So, now no one has an excuse for inhabiting a warm and comfortable technology stovepipe! The true strength in technical creativity now involves the willingness to mix knowledge, without fear or favor.

The design of drones explores this genuinely new frontier. Why? Because, as this guide book makes clear, the approach is a subtle mix of skills, based of course on aerodynamics and airplane design. But also law, regulation, autonomy, disposability, low cost, unorthodoxy, as well as novel construction, automation, integration, and artificial intelligence. In other words, the chance to think very differently, across numerous domains. For example, mix biology and 3D printing, and you have drone biodegradability options.

Conventional aviation industries and aircraft manufacturers are not best suited to this exploratory approach, because they do not have the freedoms of behavior built into their leadership and management, or their business plans. Nor are their shareholders interested, until the firm is going bust.

So, exciting ideas and courage will come from "left field," and this is your chance to think differently and be part of that.

Source: Courtesy of Sir George Zambellas.

Admiral Sir George Zambellas GCB DSC DL FRAeS

Series Preface

Unmanned air vehicles can now be seen in many applications from domestic, industrial, government/official to military. The range of configurations includes fixed wing, multi-rotorcraft, adaptive wing, and space re-entry vehicles, in both remotely piloted and autonomous modes of operation. As a result there are many classes of unmanned air vehicles in existence, and many types within each class, developed by many manufacturers. They are all capable of carrying some form of payload, including sensors, and of relaying sensor information to the ground – their primary use. They should all be designed and tested to meet the accepted airworthiness requirements for certification, although perhaps not all are.

This book is a welcome addition to the literature of unmanned air vehicles concentrating as it does on a particular class, that of small, fixed wing subsonic vehicles capable of carrying significant payloads – a class with little associated literature and a class that is likely to expand in the future. There are configurations in this class that are readily available to members of the public and small businesses who use them as observation or surveillance platforms to complement their business activities. The book has been written by authors with long experience of the development of this class from concept through design, build and test and operation in a teaching environment. This experience shows in the clear explanations assisted by many relevant diagrams. The book stresses the need for a robust design process for the airframe, the systems and the software tool set used to support designers. The completeness of the text results in a handbook on how to design, build and fly small fixed wing Unmanned Air Vehicles.

The *Aerospace Series* has continued to provide practical, topical and relevant information for people working in the field of aerospace design and development, including engineering professionals and operators, allied professions such as commercial and legal executives, and also engineers in academia. In this instance the book is especially suitable for final year graduates and those entering the industry and intending to start a career in the field of unmanned vehicles.

Peter Belobaba, Jonathan Cooper and Allan Seabridge

Preface

Unmanned air vehicles (sometimes uninhabited air vehicles or even systems, UAVs or UASs) are becoming an increasingly common sight across the globe. Originally the preserve of very secretive military organizations, they are now in routine use by film crews, farmers, search and rescue teams, hobbyists, and so on. Most of the technological difficulties in building a system that can start, take off, fly a mission, and return without human intervention have been overcome, and the wider adoption of these technologies is now mostly a matter of cost, public acceptance, and regulatory approval. The only remaining technological challenges essentially concern the degree of on-board autonomy and decision making such vehicles can provide. If secure and robust communications to a ground-based pilot can be maintained to provide decision-making capabilities, very ambitious missions can be quite readily accomplished. On-board decision making is less well advanced but developments continue apace.

The origins of our interest in UAVs stem from the many years we have spent in the business of design, both practical and academic, teaching, and research. This has exposed us to a great deal of related activity in the aerospace and marine sectors, whose processes have changed considerably over the time we have been involved. A reoccurring theme throughout has been rapid evolution in the software toolset used to support designers, and it is in this area we have been principally engaged. Central to our views is a way of looking at engineering design that distinguishes between synthesis (the business of generating new or changed descriptions of artifacts) and analysis (where one uses the laws of physics, experiments, and past experience to assess the likely or actual performance of the designed artifact). It is by the use of formal analysis and experimentation to ascribe value to an artifact that engineering design distinguishes itself from other forms of design. Thus, to be useful in the world of engineering design, tools must either help describe the product or process being designed, analyze it, or support the delivery and integration of these processes – all else is just bureaucracy: design should always be seen as a decision-making process.

In this book we focus on one particular aspect of the rapidly growing area of UAV technology: the design, construction, and operation of low-cost, fixed-wing UAVs in the 2–150 kg maximum take-off weight (MTOW) class flying at low subsonic speeds. Such vehicles can offer long-endurance, robust platforms capable of operating for 10 h or more on budgets well below $100 000, often less than $10 000. They can carry significant payloads and operate from relatively simple ground facilities. In what follows, an approach to designing and building such UAVs, developed over many years at the University of Southampton, is set out. While there are, no doubt, many other valid ways of producing UAVs, the one described here works for us, providing effective low-cost platforms for teaching, research, and commercial exploitation.

Andrew J. Keane, András Sóbester and James P. Scanlan
Southampton, UK, 2017

Acknowledgments

The support and commitment of colleagues in the Computational Engineering and Design Group at the University of Southampton are gratefully acknowledged. The University is a great place to work and a continued source of ideas and renewal. In particular, our thanks are due to a number of academic, post-doctoral, and post-graduate researchers, engineers, and pilots who have contributed to our understanding in this area, including Marc Bolinches, Mantas Brazinskas, Bob Entwistle, Mehmet Erbil, Mario Ferraro, Dirk Gorissen, Paul Heckles, Andrew Lock, Stephen Prior, Erika Quaranta, Jeroen van Schaik, Ben Schumann, Kenji Takeda, David Toal, and Keith Towell.

Part I

Introducing Fixed-Wing UAVs

1

Preliminaries

Fixed-wing aircraft have now been successfully designed and flown for over 100 years. Aero-modelers have been flying quite large aircraft at low subsonic speeds for decades, sometimes at scales as large as one-third the full size. Given the accumulated experience, it is therefore a relatively straightforward task to design, build, and fly a workable fixed-wing unmanned air vehicle (UAV) platform, armed with one of the many textbooks available on aircraft design (perhaps the most famous of these being that by Torenbeek [1], though there are many others). Even a cursory search of the Web will reveal hundreds of UAVs, many of them fixed-wing, and a number being offered for sale commercially. What is much less simple is to quickly make robust and reliable airframes in a repeatable manner at low cost, tailored to specific missions and suitable for commercial-grade operations.

If one has to rely on the craft skills of a highly gifted model-maker to construct an aircraft, costs rapidly rise, timescales lengthen, and repeatability becomes difficult to ensure. The use of bespoke molds and various forms of composites allows a much higher standard of airframe, but the initial production costs become then high and the ability to alter designs becomes very limited. Conversely, by using commodity off-the-shelf components combined with computer-aided design (CAD)-based digital manufacture, craft skills can be eliminated, costs lowered, and repeatability guaranteed. Clearly, if one has always to manually adapt an existing design to come up with a specification for a new aircraft, much design flexibility is lost; if, instead, lightweight decision support tools are linked to sophisticated parametric CAD models, high-quality design concepts can be rapidly developed to specific needs.

This is the fundamental design philosophy adopted by the UAV team at the University of Southampton (Figure 1.1) and forms the guiding approach of this book. The basic idea is to work in a digital, online world, buying parts where possible and manufacturing custom items only where absolutely necessary – essentially the aim is to source a kit of components either from part suppliers or companies offering online CAD-based manufacture, which then simply requires assembly to produce the finished aircraft. This means that the resulting UAVs are of a high and repeatable quality with as much emphasis on smart design as possible. This philosophy has become possible largely because of a revolution in bespoke digital manufacturing capabilities afforded by advanced CAD, Internet-based sourcing, low-cost computer numerical controlled (CNC) machining, and the widespread availability of 3D printing of functional components. In particular, the use of SLS nylon and metal has transformed the way in which main

Small Unmanned Fixed-wing Aircraft Design: A Practical Approach, First Edition.
Andrew J. Keane, András Sóbester and James P. Scanlan.

Figure 1.1 The University of Southampton UAV team with eight of our aircraft, March 2015. See also https://www.youtube.com/c/SotonUAV and https://www.sotonuav.uk/.

fuselage components and bespoke aircraft fittings can now be made. The core aims throughout our work have been to seek

1. low costs with highly repeatable and robust products,
2. rapid conversion of design changes into flying aircraft to meet new requirements, and
3. flexible payload systems

combined with

4. duplication of all flight critical systems,
5. sufficiently sophisticated avionics to allow fully autonomous takeoff, flight, and landing,
6. large and strong fixtures and joints to provide tolerance of uneven landing sites and day-to-day ground handling, and
7. low take-off and landing speeds to minimize risks of damage during operations.

These aims ensure long-lived and robust commercial-grade aircraft, which can survive hundreds of flight cycles and thousands of flight hours – something that model aircraft never see.

1.1 Externally Sourced Components

To test our evolving design environments and build capabilities, a range of aircraft types have been considered. In all cases, these started with the knowledge that some of the major airframe components have to be externally sourced and that one has therefore to work with what is readily available in appropriate sizes. The following list of such components forms a key starting point for what follows. To maintain low costs, some things simply have to be sourced *off the shelf*:

1. *Engines*. Either petrol or glow-plug internal combustion engines ranging from 10 up to 200 cc (cm^3) in single-, twin-, three-, and four-cylinder configurations;
2. *Electric motors*. Usually rare-earth permanent magnet motors with digital speed controllers – which are available in a wide range of sizes;
3. *Starters and generators*. External or in-built starters, direct drive or coupled via drive belts;
4. *Propellers*. Pusher and tractor propellers available in wood, nylon, and carbon-fiber-reinforced plastic (CFRP) with between two and six blades;
5. *Batteries*. NiMH, LiFe, or LiPo aircraft-grade batteries;
6. *Receiver/transmitter systems for primary flight control*. High end aero-modeler systems from companies such as Futaba (which now support two-way transmission of data including rpm, temperature, and geographical positioning system (GPS) sensors on the aircraft);
7. *Autopilots*. Many are available, but we use Arduino and SkyCircuits[1] systems (including ground stations and software environments);
8. *Servos and actuators*. High-quality, high-torque, metal-geared aero-modeler items;
9. *Undercarriages and wheels*. High-quality aero-modeler items, typically including suspension and sometimes a retract capability.

At larger take-off weights, items such as propellers and undercarriages are more difficult to source, and then it is sometimes necessary to have bespoke items made by specialist suppliers – even so, it is desirable to use companies with sufficient turnover and expertise so that costs can be controlled and quality maintained. Given a ready supply of such items and the intention to build a conventional fixed-wing monoplane, the primary layout choices available to the designer then concern the number and positioning of engines/motors and the choice of fuselage/empennage type.

1.2 Manufacturing Methods

As already noted, a key requirement for the manufacture of the UAVs being considered here is that ideally no craft skills be needed in construction. Thus the focus is on

- advanced parametric CAD-based geometry design;
- logical *and* CAD-based design of wiring looms including all plug/socket physical details with manufacture by dedicated specialists;
- numerically controlled digital manufacture involving

  - 3D printing – selective laser-sintered (SLS) nylon or metal and fused deposition modeling (FDM) ABS,
  - laser-cut wood and plastic,
  - hot-wire-cut foam (foam parts sometimes being covered by outsourcing to specialists);
- use of stock-sized materials such as off-the-shelf CFRP tubular sections.

---

[1] See http://www.skycircuits.com/.

None of these involves a novel approach, but it is certainly the case that typical aero-modelers do *not* use such methods. Moreover, the use of SLS nylon and FDM ABS in aircraft fuselage design is relatively new, being an approach championed at the University of Southampton (who flew the world's first all- SLS printed aircraft in August 2011[2]). Most UAVs currently rely on bespoke-molded CFRP fuselage sections which, although offering good strength-to-weight ratios, increase the cost and reduce the speed with which design changes can be implemented. Three-dimensional printing allows designers to continue to refine their work to within 48 h of flight trials with impunity.

1.3 Project DECODE

Although universities across the world conduct a huge amount of design-related research, it is relatively rare for academics to actually undertake the design and manufacture of complete, real, complex working systems. A considerable amount of the experience relayed in this book, and most of the examples used, are taken from one such project and its successors, which have given birth to a series of new aircraft and the attendant design systems and methodologies used to create them: the UK Engineering and Physical Sciences Research Council project DECODE (Decision Environment for COmplex DEsigns). A fundamental aim of DECODE has been to research our ideas about how aerospace design should be tackled and to see what kind of small, light-weight, low-cost, high-performance aircraft can be built using the latest software and manufacturing tools. In particular, our focus has been on the so-called value-driven design – where as many aspects as possible of the final system are explicitly analyzed and balanced against each other. To do this, we require that, wherever we can, design decisions be supported by documented rational processes that can be clearly justified rather than simply accepting perceived wisdom, van Schaik *et al.* [2], Gorissen *et al.* [3].

Most commercial design activity is a race against time and takes place against a backdrop of limited resources. DECODE faced similar issues: the initial team consisted of just five full-time engineers plus support from various academic staff; it also worked against an ambitious set of fixed milestones and design review points. When under time pressure, fast or even arbitrary design decisions often have to be made with little knowledge as to the effect of this uncertainty. One of the purposes of DECODE was to provide designers with an understanding of such risks. This necessitates a clear customer mission and the assembly of a significant quantity of design and analysis software. The mission adopted at the outset of DECODE was an air–sea surveillance task associated with the UK Coastguard. During the process, best-of-breed software tools were assembled to tackle concept and geometry definition (CAD), aerodynamics (CFD), structural analysis (FEA), weight, manufacture, and cost analysis and fleet/mission matching, Gorissen *et al.* [4]. A major deliverable of DECODE has been developing an integrated tool suite that allows the rapid redesign of UAVs against new or changing missions.[3]

1.4 The Stages of Design

Despite all the advances that have been made in computational analysis, it must be remembered that the fundamental hallmark of design is not analysis but synthesis – the choice of

---

[2] See https://www.youtube.com/watch?v=ffrJ0l2ETaU.
[3] See www.soton.ac.uk/~decode.

appropriate mechanism types, power sources, setting of dimensions, choice of features and subcomponents, selection of materials, and manufacturing processes – all these are acts of synthesis, and it is the skillful making of decisions in these areas that is the hallmark of the good designer. Of course, designers use analysis all the time, but design is about decision making, and analysis is, by contrast, an act of gaining understating but not of making decisions. Moreover, to be a good designer, the most often cited personal prerequisite is experience – this view is backed up by many observational studies in engineering design offices.

So, even though design decision making is commonly preceded by a great deal of information gathering and analysis, and although the gathering of such information, often using computational models, may be a very skilled and time-consuming activity, it should be made clear that whatever the cost, this remains just a precursor to the decisions that lie at the heart of design. Our researches have primarily addressed the tools that help support this decision-making process. The process may be thought of as being built from four fundamental components:

1. *Taxonomy*. The identification of the fundamental elements that may be used, be they gears, servos, airfoils, and so on;
2. *Morphology*. The identification of the steps and their order in the design process;
3. *Creativity or synthesis*. The creation of new taxonomies, morphologies, or (more rarely) fundamental elements (such as the linear induction motor);
4. *Decision making*. The selection of the best taxonomy, morphology, and design configurations, often based on the results of much analysis.

Perhaps the simplest and most traditional way of representing the design process is as a spiral, see Figure 1.2. The idea behind this view is that design is also iterative in nature with every aspect being reconsidered in turn and in ever-more detail as the design progresses. It begins with an initial concept and constraint review that attempts to meet a (perceived) customer need

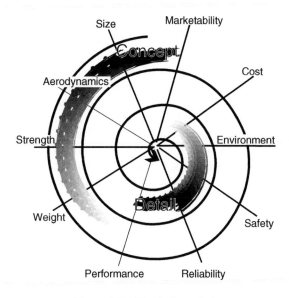

Figure 1.2   The design spiral.

specified in relatively few major requirements. For an aircraft, this might cover payload, range, speed, and anticipated cost; for an engine, it might be thrust, weight, fuel efficiency, and cost. This phase of the design process is often called *concept design*. It is then traditionally followed by preliminary design, detailed design, and the generation and verification of manufacturing specifications and tooling designs before production commences. For major aerospace systems, once the product is operational, a continuing "in-service" design team takes over to correct any emerging problems and deal with any desired through-life enhancements. Finally, decommissioning and waste disposal/reuse must be considered. In Rolls-Royce, for example, this is called the Derwent process, and is characterized by significant business decision gates at each stage.

In our UAV designs, mainly because the products being designed are nothing so complex as an airliner or a jet engine, the preliminary design phase is not separated from the concept stage – all design variables are considered to lie in the concept phase unless they live only within the CAD-based detail definition process, in which case they are considered detailed design variables. Similarly, considerations of manufacturing methods are not separated from the detailed design work – rather they are closely integrated so that the designs take maximum advantage from the manufacturing systems employed – here the focus has been to use modern rapid prototyping and numerically controlled machining so as to reduce the amount of manual input during manufacture to the absolute minimum. As we shall see, these approaches reduce, but do not eliminate, the dislocations that arise when a design moves from one such stage to the next. To support a truly seamless design process, the designer must be able to move between concept thinking and manufacturing details with ease whenever needed, without worrying that any changes made will make masses of existing design work redundant. Parametric, CAD-based detail modeling can help in this respect, but it is difficult to entirely remove them. The approaches described here represent our best attempt to mitigate such problems.

Nowadays, even the most traditional manual approach to design will probably make use of some computing facilities, but it will generally not draw on modern search and optimization strategies, knowledge-based systems, or grid-/cloud-based computing[4]. It will, in all probability, make heavy use of extremely experienced design staff and detailed experimental activities. At the other end of the spectrum, as much of the design process is automated as possible, high-level value metrics are used to balance competing concepts, formal optimization is used to generate new combinations of parameters, and data/knowledge stores are maintained electronically and centrally shared. Table 1.1 illustrates the kind of variation that can be encountered in organizations with different levels of design system maturity. Our aim is to operate at the highest maturity level in this spectrum. Before proceeding, however, it is useful to briefly step through the more commonly referred to stages of design in a little more depth. As just noted, in our UAV work the concept and preliminary stages are combined, as are the detailed and manufacturing stages. Formal decommissioning, over and above simply using the University's normal recycling systems, is not considered at all.

1.4.1 Concept Design

When embarking on a concept design (sometimes termed preliminary concept design), the designers must first decide which aspects of the product must be considered from the outset, and which tools will be used to make a start. Often, these are based on considerations of

---

[4] Widely distributed computing systems networked together to form a single, often remote, resource.

Table 1.1 Design system maturity.

Maturity level	Characteristics	Capabilities	Consequences
1	Cost implications of design ignored	Product performance focus	Product underperformance, massive redesign, and cost reduction activities. Large number of design concessions
2	Major acquisition cost implications of design recognized	"Design for manufacture" knowledge and processes implemented	Cost of ownership and longer term cost trade-offs not understood Customers initially attracted by purchase price become disillusioned by high cost of ownership Poor product/brand loyalty?
3	Excellent acquisition cost modeling capability; some life cycle cost awareness in design	Deeply embedded process capability knowledge; good relationship with supply chain and sharing of acquisition cost of knowledge	Better trade-offs and ability to consider long-term product contracts and reduced business risk
4	Designers are required to report on life cycle cost at design reviews Designers are skilled at formulating value-based objective functions	Holistic cost modeling; sophisticated logistics modeling and stochastic econometric forecasting embedded in the design process	Reputation for long-term capabilities and reliable performance Extended business opportunities in long-term product support
5	The organization systematically employs value-driven design metrics across multiple projects and uses these to identify optimal trade-offs both within and between projects	Very strong integrated optimization and modeling capability Routinely models stochastic behavior of product and its environment	World-class reputation for complex systems New products introduced with minimal risk to the business

previous designs: it is rare to design from a blank sheet of paper – even Whittle knew that attempts had been made to design and build jet engines before he filed his first patent; the Wright brothers were by no means the first to attempt powered, heavier-than-air flight. Given a few previous designs, it is usually possible to construct some basic design rules that will allow a designer to make outline predictions on a "back-of-the-envelope" basis, see, for example, the many approximate formulae and design curves in the books by Torenbeek [1] or Stinton [5]. If a great deal of data is available or the product very complex, such design rules may well be encoded into statistically based concept tools, such as those developed by Airbus [6]. Even today, however, it is not possible to provide first-principles analysis for all of the topics given in the list. In a traditional design process, the best that can be hoped for, particularly during the early concept stages, are empirical relationships and simple constraint analyses. The initial framework used to design our airframes is based on a linked series of highly structured Excel spreadsheet pages. Each page deals with a separate aspect of the design and is set up in a highly formalized way with input, calculation, and output sections, definitions, units, notes, and so on, all set in fixed locations. These are supported by static tables of key information for items such as propellers or engines.

The designs produced during concept work can usually be characterized by relatively few quantities. A typical aircraft design may be summarized at this stage by less than 100 numbers. Such designs are commonly produced fairly rapidly, in timescales measured in days and weeks. In the work described by Keane [7], a wing is characterized by just 11 variables, and in that by Walsh *et al.* at NASA Langley [8] a little under 30. The designs produced during concept design will normally be used to decide whether to proceed further in the design process. As Torenbeek [1] notes, the "aim is to obtain the information required in order to decide whether the concept will be technically feasible and possess satisfactory economic possibilities". Thus traditional concept design may be seen as part of forming a business case, as opposed to an engineering process. The designs considered following the decision to proceed may well differ radically from those produced by the concept team. The concept designers should ideally allow for this dislocation, so that performance specifications can be delivered in practice by those charged with preliminary and detailed design. The merging of the concept and preliminary design phases within DECODE eliminates these difficulties to a large extent (though it requires a greater effort to set up and develop).

It is in the area of concept design where formal optimization has perhaps had its greatest effect so far. Even so, the decision-making process, where many competing design topologies are traded off against each other, is still normally carried out manually. This aspect of design is beginning to change with the advent of multiobjective optimization and the use of game theory and search methods in decision making. The researchers at the Computational Engineering and Design Research Group at the University of Southampton have built a series of state-of-the-art tools for use in this phase of design.

1.4.2 Preliminary Design

Once the (usually economic/business) decision to proceed further with the design is taken, the preliminary design stage can begin (this is sometimes termed full concept or development definition/design). At this point, it is traditional for the different aspects or components of the design to be considered by dedicated teams. For aircraft, this might consist of an aerodynamics division, a structures division, a control systems division, costing teams, and so on, or it might consist of a team considering the fuselage, another considering wing design, others looking at the tail plane, propulsion systems, and so on. For aeroengines, it is common to break the design teams up into divisions based on the various components of the engine, such as fans, compressors, turbines, combustors, and so on. An alternative and more modern approach to managing design is via the use of integrated project teams (IPTs). Such teams are normally formed specifically for the product being designed and grow in size progressively throughout the preliminary design phase. If an IPT-based approach is used, it is then usually supported by specialist divisions who are charged with providing technical input across a range of project teams. These specialist divisions are responsible for the retention and development of core technologies and capabilities. Managing the inter-relationships between such teams becomes a key part of managing the design process, especially if they are geographically widely dispersed, as is now often the case.

Whatever form the design teams take, the tools used by them will be much more sophisticated than during concept work. For example, the designers considering structures will, as a matter of routine, make use of quite detailed stress analysis, normally by means of finite element analysis (FEA). Those considering the wings will pay close attention to predictions of the airflow. This may well involve significant experimental programs as well as the extensive use of computational fluid dynamics (CFD) methods. A key observation about how computational tools are used in the traditional approach to design is that they tend to have no *direct* impact on the geometry of the design being produced. They are used, instead, for analysis, and it is left to the designer to make decisions on how to change the design as a result of studying the resulting outputs. Of course, this means the designers must be highly experienced if they are to make effective use of the results obtained. It is often by no means obvious how to change a design to improve its performance or reduce its cost simply by studying the results of computational analysis. Trading performance improvement for cost saving is even more difficult to carry out in a rational manner. Even so, it is common to find that design teams spend enormous amounts of time during the preliminary design stage preparing the input to CFD and FEA runs and studying the results. The amount of such effort that can be afforded for a new UAV design will very much depend on the overall project budget. For low-cost systems, teams are generally replaced by individuals, and the sophistication of any analysis or experimentation that can be carried out is strictly limited, especially if little by way of automation of the analysis process has been set up.

At the present time, preparing the input for computational analysis is commonly very far from automated. If manual input is required, it may well take the designers 2 or 3 weeks to prepare and run the meshes needed for accurate CFD analysis of a complete aircraft configuration and to assimilate the results. This severely restricts the number of different configurations that can be considered during preliminary design. The result is that the design is frozen early on during the design process, so that even when the design team knows that the design suffers from some shortcomings, it is often very difficult to go back and change it. Developments in advanced parametric CAD systems are beginning to make some inroads into this problem. A major thrust in our research has been to automate this process as much as possible and merge it into the concept stage – this requires a good deal of systems architecting and is perhaps possible only because the UAV systems being considered are much less complicated than large aircraft or gas turbine engines – nonetheless, these UAVs include the full range of aircraft technologies from aerodynamics, through structural analysis, to control systems and manufacturing processes – and they thus pose a not insignificant design challenge.

1.4.3   Detail Design

Once the preliminary design is complete, detail, production, or embodiment design takes over. This stage will also focus on design verification and formal approval or acceptance of the designs. The verification and acceptance process may well involve prototype manufacture and testing. Such manufacturing needs to fully reflect likely production processes if potentially costly re-engineering during production is to be avoided. Apart from concerns over product performance, issues such as robustness, reliability, safety, and maintainability will be major concerns at this stage. The objective of detailed design is a completely specified product that meets both customer and business needs. At this stage in the process, relatively little design work is routinely automated or parameterized and so the staff effort involved in the design will tend to increase greatly. This is just as true in small UAV programs as large civil aerospace ones – detailing CAD representations with all the information needed for manufacturing remains an intensely manual task.

In all aerospace engineering, detailed design is dominated by the CAD system in use. Moreover, the capabilities of that CAD system can significantly affect the way that design is carried forward. If the system in use is little more than a drafting tool, then drafting becomes the fundamental design process. Increasingly, however, more advanced CAD tools are becoming the norm in aerospace companies. These generally allow parametric descriptions of the features to be produced, which permit more rapid changes to be carried through. They may also allow for information other than simple geometry to be captured alongside the drawings. Such information can address manufacturing processes, costing information, supplier details, and so on. The geometric capabilities of the system may also influence the way that complex surfaces are described: it is very difficult to capture the subtleties of modern aerodynamic surfaces with simplistic spline systems, for example.

In the traditional approach, detailed design revolves around the drafting process, albeit one that uses an electronic description of the product. It is, additionally, quite usual for the CAD system description and the analysis models used in preliminary design to be quite separate from each other. Even in detailed design, these two worlds often continue to run in parallel. Consequently, when any fresh analysis is required, the component geometries must be exported from the CAD system into the analysis codes for study. This conversion process, which commonly makes use of standards such as STEP or IGES, is often very far from straightforward. Moreover, even though such standards are continuously being updated, it is almost inevitable that they will never be capable of reflecting all of the complexity in the most modern CAD systems, since these are continually evolving themselves.

The effort required to convert full geometries into descriptions capable of being analyzed by CFD or FEA codes is often so great that such analyses are carried out less often than might otherwise be desirable. Current developments are increasing the ability of knowledge-based systems to control CAD engines so that any redrafting can be carried out automatically. Nonetheless, the analyst is often faced with the choice of using either an analysis discretization level that is far too fine for preference, or of manually stripping out a great deal of local detail from a CAD model to enable a coarser mesh to be used. The further that the design is into the detailed design process, the greater this problem becomes. In many cases, it leads to a parallel analysis geometry being maintained alongside the CAD geometry. In most small-scale UAV programs, it can mean that full stress analysis is rarely carried out, with reliance instead being placed on prototype testing.

1.4.4   Manufacturing Design

Once a detailed design is completed, manufacturing can begin; sometimes, if one is sufficiently confident in the design process, some areas of manufacture can begin even before the full design is completed. However, a completed description of the final product is rarely enough to answer all questions concerning manufacture – sometimes the desired material properties will mandate a certain manufacturing process, which require specific geometric properties; sometimes special tooling is required, which itself must be designed. This leads to the idea of manufacturing design – this encompasses both tooling design and the adaptation of product design to facilitate manufacture. Ideally, the manufacturing processes to be used are familiar and well understood so that the detail designer has allowed for all aspects of manufacture and no further adaptation is required; then just tooling design is considered at this stage. In our UAV work, all manufacturing considerations are dealt with during detail design, with heavy emphasis on fully automated, numerically controlled manufacturing, so that only final assembly requires significant manual intervention, and no airframe-specific tooling is required. Because rapid prototyping systems have been central to our work, detailed knowledge of what such systems can, and cannot, do has impacted noticeably on the design process. The one exception to this focus on automated digital manufacture has been the UAV wiring looms, where such manufacture would be prohibitively expensive for the small production runs involved. Even so, full geometric detailing of the looms is carried out in the CAD environment so that, when manufactured, they fit the airframe correctly.

1.4.5 In-service Design and Decommissioning

Very often, when a product reaches manufacture, the design process does not stop. Few products prove to be completely free of design faults when first produced, and, equally, it is common for operators to seek enhancements to designs throughout the life of a product. The lifetime of current aerospace systems can be extremely long, sometimes as much as 50 or even 60 years, and so this phase of design is commonly the longest, even if it is carried through with a relatively small team of engineers. Even when a product reaches the end of its life, it must be decommissioned and any waste disposed of or recycled. This process itself requires the input of designers, who may need to design specialist facilities to deal with waste or with stripping-down equipment. In our studies, because a series of UAVs have been designed and built, the in-service design function has been subsumed into redesign during the development of the next mark of airframe. Here no design effort has been given to decommissioning, though ease of maintainability and repair during operation has been a significant concern in the making of design decisions.

1.5 Summary

The rest of this book aims to set our experiences in designing, building, and flying a range of small fixed-wing aircraft, following the philosophy set out in this chapter. While it is perhaps inevitable that single-engine aircraft are where we started, they inherently prevent any form of single-point-of-failure tolerance. So, although we consider both single- and twin-engine designs, we firmly believe that only the latter designs offer the degree of resilience that regulators will inevitably demand for commercial activities. With regard to empennage, we have concluded that canard-based designs offer no significant advantages in the roles being considered, but beyond that, a range of tail types can be effective – we have successfully tried several. In all cases, we make use of rapid prototyping (generally SLS) along with carbon-fiber tubing to build the bulk of our fuselage and empennage structural components. Lifting surfaces have been made from carbon tube reinforced foam, covered in either Mylar film or glass/carbon fiber to increase strength and provide chemical and handling resistance. Our design work has been aided by a growing suite of computational tools that span from spreadsheets to full Reynolds-averaged Navier–Stokes fluid dynamics solvers. These have been augmented by the large amount of experimental data we have accumulated from wind tunnel tests, structures labs, and in-flight data recording.

2

Unmanned Air Vehicles

The term "unmanned air vehicle" is broadly used to describe any flight- capable vehicle that has nobody on board during its flight – sometimes UAV is taken to mean "uninhabited air vehicle." This encompasses an extremely broad range of systems, but generally does not include missiles or other guided weapons (though for cruise missiles the distinctions can be very blurred). Even so, it is useful to next set out a brief taxonomy of UAVs, indicating where in this space of potential designs the work described in this book primarily lies (noting that it was, of course, the aim of the research undertaken during the DECODE project to address wide-ranging issues that would apply to many complex design tasks).

2.1 A Brief Taxonomy of UAVs

Here we use five axes to categorize UAVs:

1. *Size*. The maximum take-off weight (MTOW) is used to distinguish between micro: <2 kg, small: 2–20 kg, medium: 20–150 kg, and large >150 kg UAVs; this distinction also maps to typical Aviation Authority definitions and quite closely to cost, since air vehicle costs correlate quite strongly with MTOW.
2. *Mission*. Six basic mission types are considered: surveillance, transport, combat, communications relay, support, and target.
3. *Capability*. The performance of the UAV in terms of endurance, range, speed, payload mass, and operational ceiling.
4. *Degree of autonomy*. Here this is based on the chart developed at the US Wright–Patterson Air Force Base, which uses an 11-point scale from simple remotely piloted vehicles through to those with complete autonomy, which are essentially indistinguishable from piloted vehicles to the outside observer (see Table 2.1) – note that most current systems operate at level 3 or lower on this scale.
5. *Aero-structural configuration*. The absence of the need to accommodate crew on UAVs has lead to a range of unconventional configurations being considered, so this axis ranges from conventional fuselage/wings with traditional control surfaces though morphing and deformable structures and blended wing-bodies, to aircraft using things such as Coanda effect controls.

Small Unmanned Fixed-wing Aircraft Design: A Practical Approach, First Edition.
Andrew J. Keane, András Sóbester and James P. Scanlan.
© 2017 John Wiley & Sons Ltd. Published 2017 by John Wiley & Sons Ltd.

Table 2.1 Different levels of UAV autonomy classified using the Wright–Patterson air force base scheme.

Level	Level descriptor	Observe	Orient	Decide	Act
10	Fully autonomous	Cognizant of all within the battlespace	Coordinates as necessary	Capable of total independence	Requires little guidance to do job
9	Battlespace swarm cognizance	Battlespace inference – intent of self and others (allies and foes); complex/intense environment – on-board tracking	Strategic group goals assigned; enemy strategy inferred	Distributed tactical group planning; individual determination of tactical goal; individual task planning/execution; choose tactical targets	Group accomplishment of strategic goal with no supervisory assistance
8	Battlespace cognizance	Proximity inference – intent of self and others (allies and foes); reduced dependence upon off-board data	Strategic group goals assigned; enemy tactics inferred	Coordinated tactical group planning; individual task planning/execution; choose targets of opportunity	Group accomplishment of strategic goal with minimal supervisory (example: go SCUD hunting)
7	Battlespace knowledge	Short track awareness – history and predictive battlespace data in limited range, timeframe, and numbers; limited inference supplemented by off-board data	Tactical group goals assigned; enemy trajectory estimated	Individual task planning/execution to meet goals	Group accomplishment of tactical goal with minimal supervisory assistance

6	Real-time multivehicle cooperation	Ranged awareness – on-board sensing for long range, supplemented by off-board data	Tactical group goals assigned; enemy location sensed/estimated	Individual task planning/execution to meet goals	Group accomplishment of tactical goal with minimal supervisory assistance
5	Real-time multivehicle coordination	Sensed awareness – local sensors to detect others, fused with off-board data	Tactical group plan assigned; real-time health diagnosis; ability to compensate for most failures and flight conditions; ability to predict onset of failures (e.g., prognostic health management); group diagnosis and resource management	On-board trajectory replanning – optimizes for current and predictive conditions; collision avoidance	Group accomplishment of tactical plan as externally assigned; air collision avoidance; possible close air space separation (1–100 yds) for automated aerial refueling (AAR), formation in non-threat conditions
4	Fault/ event-adaptive vehicle	Deliberate awareness – allies communicate data	Tactical plan assigned; assigned rules of engagement; real-time health diagnosis; ability to compensate for most failures and flight conditions - inner-loop changes reflected in outer-loop performance	On-board trajectory replanning - event-driven self-resource management; deconfliction	Self-accomplishment of tactical plan as externally assigned; medium vehicle airspace separation (100s of yds)

(continued)

Table 2.1 (Continued)

Level	Level descriptor	Observe	Orient	Decide	Act
3	Robust response to real-time faults/events	Health/status history and models	Tactical plan assigned; real-time health diagnostic (what is the extent of the problem?); ability to compensate for most control failure and flight conditions (i.e., adaptive inner-loop control)	Evaluate status versus required mission capabilities; abort/return to base if insufficient	Self-accomplishment of tactical plan as externally assigned
2	Changeable missions	Health/status sensors	Real-time health diagnosis (do I have problems?); off-board replan (as required)	Execute preprogrammed or uploaded plans in response to mission and health conditions	Self-accomplishment of tactical plan as externally assigned
1	Execute preplanned mission	Preloaded mission data; flight control and navigation sensing	Pre/post-flight bit (built-in test); report status	Preprogrammed mission and abort plans	Wide airspace separation requirements (miles)
0	Remotely piloted vehicle	Flight control (altitude, rates) sensing; nose camera	Telemetered data; remote pilot commands	N/A	Control by remote pilot

Figure 2.1 The Southampton University SPOTTER aircraft at the 2016 Farnborough International Airshow.

At Southampton, the design process has been focused on small and medium, 0.5–8 h endurance, medium range, low-speed surveillance- and transport-based designs, which operate changeable but essentially preplanned missions but can make use of unconventional control surfaces or morphing wings. The aim has been to keep MTOW below 150 kg to take advantage of the reduced certification requirements that are then typically applicable and also to keep the cost of a typical system below (often well below) $150 000 to make the entire program affordable while still allowing multiple airframes to be built and flown. A key aim has been to produce well engineered, rugged designs capable of many flights that could readily be used in a commercial context. Figure 2.1 shows our SPOTTER aircraft at the 2016 Farnborough International Airshow; Figure 2.2 provides a line drawing of the aircraft.

2.2 The Morphology of a UAV

All heavier-than-air UAVs contain many similar and well-understood components that the design team need to consider:

1. *Lifting surfaces*. Traditionally wings or rotors but this can include blended wing-bodies – certainly it is common for fuselages to generate lift. As this book is concerned with fixed-wing UAVs, rotorcraft are not considered further.
2. *Control surfaces or their equivalents*. Typically elevators, rudders, ailerons, and perhaps flaps and air-brakes (sometimes a single surface provides multiple functions).
3. *Fuselages*. To house systems, but these may be subsumed into a blended wing-body configuration or engine nacelles.
4. Internal structure to support all the loads seen by the vehicle and to connect the components together).

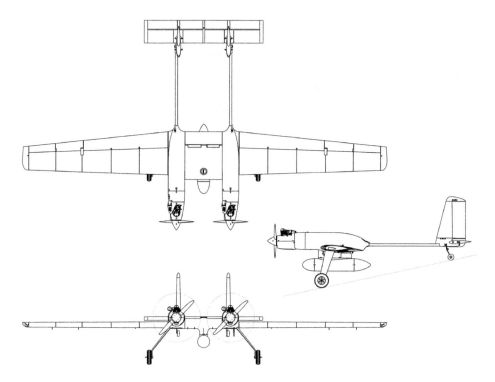

Figure 2.2 University of Southampton SPOTTER UAV with under-slung payload pod.

5. *Propulsion system.* Normally propeller-, turbo-fan-, or jet-based. Here, the focus is on propeller-driven aircraft that use piston engines or electric motors.
6. Fuel tanks or other energy sources for propulsion and possible on-board generation. Here, they are generally JP8 kerosene, gasoline, methanol, or LiPo batteries.
7. Command, communication, and control systems and associated on-board power system (generally supported by (LiFe) battery, generator, or the main engine).
8. *Payload.* Commonly sensors or munitions but sometimes emergency aid, medicines, or other lightweight high-value goods.
9. *Take-off and (normally) landing gear.* Generally wheels with suspension and steering, sometimes retractable (this can include catapult attachment points or landing hooks).

Figures 2.1 and 2.2 show the University of Southampton SPOTTER aircraft with most of these components including an external under-slung payload pod; Figure 2.3 shows the integral center fuel tank and main spars for this aircraft. It is noticeable that even in the most sophisticated vehicles, payload volume and mass are still a small percentage of the total, though they are a larger fraction than a manned aircraft would permit because all of the life-support systems and accommodation spaces can be dispensed with: on SPOTTER, the maximum payload mass is 5 kg excluding fuel, while the remaining airframe weight is 23 kg.

There are, of course, a considerable number of ways these basic elements can be laid out to form an aircraft, even if canard designs are not considered. First, the number and type of engines must be chosen – here we have considered only single-, twin-, and occasionally three-engine/propeller configurations. Second, we must choose whether to use tractor or pusher propellers, or perhaps a tractor/pusher combination. This naturally leads to a single

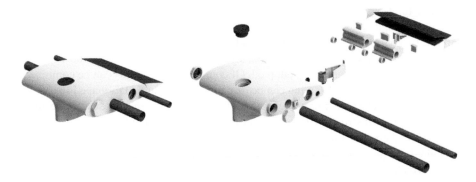

Figure 2.3 Integral fuel tank with trailing edge flap and main spars.

central fuselage, a central fuselage with twin wing-mounted engine nacelles, or twin fuselages that incorporate the nacelles. Third, the type of wing has to be chosen: generally low, middle, or high wing monoplane designs are adopted but biplane configurations can have some advantages. Next, one must consider the type of tail-plane from simple "T," inverted-"T," "V" or inverted-"V," and various forms of "U" or "H." It will immediately be clear that may tens of possible configurations can adopted, and any brief survey will show that almost all possible combinations have been tried at some time or other. Finally, one must choose the undercarriage layout assuming one is to be fitted – this is generally either a nose-wheel or tail-wheel configuration with the main undercarriage under the wings.

Ideally, one would systematically consider all possible combinations of these choices at the concept stage. However, to do so requires that the concept model can adequately distinguish between the various options, and unless one has a large quantity of prior data, this is unlikely to be possible – accurately estimating the impact of tail choice on structural weight is a nontrivial calculation, for example. At Southampton we have built and flown the following combinations:

1. Single-engine tractor configurations with conventional undercarriages and tails.
2. Single-engine pusher configurations with a range of tails, with and without undercarriage with low, middle, and high wing locations.
3. Twin-engine, dual-fuselage tractor designs with conventional "H tails," middle wings, and tail wheel undercarriage.
4. Hybrid designs with tractor gasoline and twin electric propulsion.

We next set out in general terms the approaches adopted at Southampton in the various areas of the aircraft before going on to more detailed descriptions in Part II. Clearly, there are many alternatives in each case that we have not considered – those described here reflect the overall approach already set out, their typical (low) costs, and their ready availability.

2.2.1 Lifting Surfaces

The design of the lifting surfaces (wings, tailplane, fins, etc.) for traditional aircraft can be a highly involved and complex activity. The aim is, of course, to use efficient aerodynamic sections together with planforms that balance the aerodynamicist's desire for high aspect ratios with the structural engineer's wish for small bending moments. However, before making decisions on planform, it is important to understand the manufacturing philosophy to be adopted. Essentially, two basic approaches can be considered: stresses can either be carried by a series of dedicated structural elements (spars, ribs, stringers, etc.), or the skin of the wing can be stressed and a monocoque approach adopted. Of course, these methods can be combined: for example, the front quarter of the wing can be made into a structural "D"-shaped spar to which is attached a nonstructural rear part to complete the airfoil shape. In all cases, the desire is for a light, torsionally stiff structure that can carry the required bending loads.

Since our fundamental aim is for simplicity of manufacture, so as to give low-cost, easily manufactured designs, we have adopted the use of standard carbon-fiber-reinforced circular tube spars which support selective laser-sintered (SLS) nylon ribs, with the main body of the wings being made from numerically (CNC)-cut closed cell foam which just transfers pressure loads to the other structural elements. The foam elements are bonded to the ribs, and these slide over the spars to form the wing. By using a trailing spar and another at the quarter chord point, we provide a structure suitable for attaching the control surfaces. The wing is then clad either with a very light weight glass-fiber-reinforced layer or a Mylar film. This covering increases bending and torsional strength, protects the foam from damage by fuel and oil, and also gives resistance to ground handling. Figure 2.4 shows a typical wing made in this way.

2.2.2   Control Surfaces

It is normal practice to fit control surfaces of some kind to all the lifting surfaces on the aircraft (or in the case of the tail, provide all steerable surfaces). Conventionally, these provide ailerons

Figure 2.4   A typical carbon spar and foam wing with SLS nylon ribs at key locations (note the separate aileron and flap with associated servo linkages).

and flaps on the main wings and elevators and rudders on the tail (Figure 2.4 shows a wing with an aileron and a flap). We mostly opt for simple hinged conventional surfaces but have experimented with Fowler flap mechanisms and main wing morphing for roll control. If "V" or inverted-"V" tails are used, the rudders and elevators become combined, and a suitably mixed control strategy is required to separate the control functions. To make our designs aerodynamically and structurally efficient, the tails adopt the same construction philosophy as the main surfaces: that is, we use a carbon spar onto which small ribs are placed to support a numerically cut foam body. By carrying the spar along the length of the wing, tail, or rudder, this can strengthen the main element against torsional loads while providing a convenient hinge point, as can also be seen in Figure 2.4 (note the pockets for control servos). The surfaces are operated by standard servos attached to the ribs with through-bolted horns with load spreaders where they meet the foam. If redundant systems are required, the functions of a single control surface can be provided with multiple elements each with its own servo, albeit at increased cost and weight.

2.2.3   Fuselage and Internal Structure

Having initially made fuselages and other complex structural components from pre-preg carbon fiber patches laid into molds or laser-cut plywood with various forms of covering, we have now switched almost exclusively to the use of 3D printed parts, either made from fused deposition modeling acrylonitrile butadiene styrene (FDM ABS) or, more commonly, SLS nylon. This approach allows for considerable structural sophistication, permitting various forms of stiffeners, hatches, bayonet joints, reinforcing in way of the main spars and landing gear, and so on. The parts thus produced represent the single biggest investment of design effort in the airframe, often involving hundreds of hours of computer-aided design (CAD) effort. However, once designed, they can be readily manufactured and offer many benefits, although they are somewhat heavier than equivalent fiber-reinforced molded structures. Perhaps the greatest advantage they offer is the rapid customization and modification that this approach lends to the design efforts. Each new part that is manufactured automatically reflects the latest design standard, and the turnaround time for any design change to having flight-ready parts is typically 48 h. Repeatability is very easily achieved, making replacement parts interchangeable with those on an airframe without further effort. Figure 2.5 shows a typical SLS structural

Figure 2.5   A typical SLS structural component.

component from the SPOTTER aircraft, illustrating the complexity that can be readily achieved by this approach to design.

2.2.4 Propulsion Systems

All our aircraft have been propeller-driven either using internal combustion engines or electric motors. Both tractor and pusher configurations have been used, and in some cases hybrid designs with both engines and motors have been employed. Currently, four-stroke gasoline or JP8 kerosene-fueled engines give the best range and endurance to aircraft, but they do need fuel supplies to be kept in mind; they can also cause vibration problems with on-board sensors unless careful consideration is given to antivibration mountings (of both engines and sensors).

Fitting an aircraft with electric motors for propulsion is relatively straightforward. A wide range of relatively inexpensive motors is readily available, and by adopting rare-earth magnets, very good power to weight ratios can be achieved. They tend to be very smooth running and extremely reliable since the main rotor and its bearings are usually the only moving parts. The Achilles heel of electric propulsion is the energy density of the batteries currently available.[1] LiPo batteries are readily available and reasonably priced, but using them to lift useful payloads typically restricts endurance to less than 2 h, often dramatically so. The advent of lithium–sulfur batteries may mitigate these problems to some extent, as would modestly priced fuel cells – currently neither technology is viable for low-cost UAV systems.

2.2.5 Fuel Tanks

When dealing with internal combustion engines, suitable fuel tanks are, of course, required. These are either sourced externally and mounted within the fuselage (typically injection molded nylon with simple screw-top fittings) or form part of the aircraft structure itself – we have tried both approaches. Currently, we tend to design integral fuel tanks built into the SLS nylon structure near the aerodynamic center of the aircraft. Tanks formed this way need to be thoroughly cleaned after manufacture and suitably sealed, but they work extremely well. The use of SLS nylon permits a host of desirable features such as internal baffling, fittings for caps, fuel supply lines and level sensors, and so on. Figure 2.6 shows a cut-away view of the integral SLS nylon fuel tank from the SPOTTER aircraft (note the corrugated internal baffle that also adds structural rigidity and support for the payload which is slung below the tank).

2.2.6 Control Systems

As UAV systems become more capable and need greater resilience, so the on-board command and control systems tend to grow in complexity. At the most basic level, a simple aero-modeler-grade receiver coupled to the control servos and supplied by a precharged battery is all that is needed. Even so, we find that the various makes of equipment vary in their robustness and capabilities: we tend to adopt Futaba systems and have had good experience

---

[1] Typical energy densities for LiPo batteries are currently around 0.36–0.95 MJ/kg, while gasoline has an energy density of 46.4 MJ/kg. Even allowing for the greater efficiencies of electric motors compared with gasoline engines, this is a massive disadvantage when range or endurance is important.

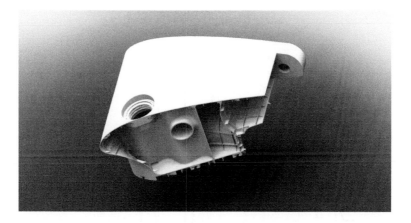

Figure 2.6   A typical integral fuel tank.

with them. They will allow control of the aircraft within an operational radius dictated by the range of the transmitter, the pilot's ability to monitor the aircraft, and any airspace regulations in force. Few aero-modeler systems have ranges beyond 1500 m, and even at this range it is extremely difficult for a pilot to observe the behavior of the aircraft as it is beyond visual line of sight (BVLOS). The alternative is long-range transmitters, but even these do not deal with the difficulty of feedback to the pilot. Air regulations typically prohibit normal aero-modeler flying beyond a radius of 500 m.

The next step up in complexity requires some form of on-board autopilot. At the most basic level, autopilots need to be able to fly an aircraft from place to place while maintaining speed and height. This necessarily requires that they can override any instructions coming from the receiver normally controlled by the pilot. The pilot does, of course, need some method of switching between manual control and autopilot control. Obviously, it is also necessary to be able to upload way-point instructions to the autopilot, either before takeoff or during flight. Any flights out to locations beyond the manual control transmitter range will rely solely on the autopilot to fly the aircraft, without any prospect of direct intervention. To enable the autopilot to decide where the aircraft is, it is normal to rely on global positioning systems (GPS), sometimes backed up by estimated positions using dead reckoning, given speed and compass heading. GPS vary in accuracy but even the most straightforward will give locations to within a few tens of meters. Provided the final way-point of a mission is within, say 500 m of the pilot, it is then possible for the pilot to take back direct control and land the aircraft. Most autopilots will also provide telemetry data via some form of radio downlink or store such data on board for subsequent analysis, see Figure 2.7.

Further to such basic operations, it is possible for the autopilot to conduct taxi-out, take-off, mission, and landing maneuvers without pilot intervention – the SkyCircuits system often used by the team at Southampton is capable of all these functions (we have achieved startlingly good repeatability in landings using this system coupled with a laser-based height sensor, see Figure 2.8). It also provides downlink telemetry of various health monitors on the aircraft such as fuel remaining, engine temperatures, and so on.[2]

---

[2] http://www.skycircuits.com/.

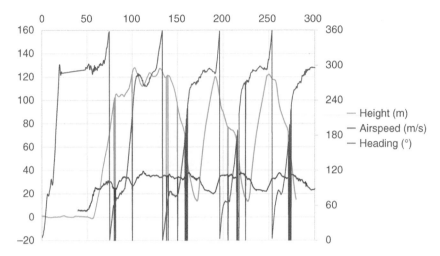

Figure 2.7 Typical telemetry data recorded by an autopilot. Note occasional loss of contact with the ground station recording the data, which causes the signals to drop to zero.

Figure 2.8 Flight tracks of the SPOTTER aircraft while carrying out automated takeoff and landing tests. A total of 23 fully automated flights totaling 55 km of flying is shown.

In addition to an autopilot, more complex UAVs may carry some means of trying to avoid collisions with other aircraft. At their simplest, these are transponders that allow aircraft to be alerted to the presence of each other and their intended direction, speed, and altitude. Mode C transponders transmit a four-digit code and pressure altitude. Mode S transponders transmit additional information such as the aircraft identity, direction, speed, and so on. In some areas, additional data channels, known as *automatic dependent surveillance – broadcast* (ADSB), allow a full picture of nearby aircraft including height and direction information to be plotted. The main problems with such equipment are size, weight, and cost – this tends to restrict their use to larger and more expensive UAVs. In addition to transponders, research is progressing on the so-called "sense-and-avoid" systems that take active steps to change the UAV's flight path when a possible collision is detected. Such systems are still in their infancy at the time of writing, but rapid developments can be expected, linked to cameras, radars, and transponder systems.

Beyond this, the degree of complexity on board is practically limitless given sufficient computing power. However, in our work the only other non-payload avionics we have seen necessary to fit are concerned with battery charging and health monitoring, so that mission endurance is not limited by finite battery capacity. Even so, the main wiring diagram for an aircraft with twin engines and multiple redundant receivers and charging circuits with a range of control surfaces can quickly become quite involved. Figure 2.9 shows the main wiring diagram from the SPOTTER aircraft (more detailed views are given in later chapters), and Figure 2.10 shows the SkyCircuits autopilot we typically use.

2.2.7 Payloads

Of course, the primary reason for operating a UAV, or indeed any vehicle, is to carry a payload, whether it be cargo, sensors, cameras, and so on. The approach adopted at Southampton has been to try and separate as much as possible the payload from the rest of the airframe design. Moreover, if the payload can be sited near the airframe's center of mass, it is then possible to swap payloads of varying weight without significant impact on the UAV. In our work, payloads have typically been cameras, often attached to dedicated radio downlink systems, but we have also carried flight-launchable maritime autonomous underwater vehicles (AUVs) for example, see Figure 2.11. From the airframe perspective, what is supplied is a structural mounting point together with (typically) 12 V electrical power supplies. Payload masses are usually around 10–25% of the take-off weight. Electrical powers of up to 150 W are readily available if generators supported by internal combustion engine are fitted, typically attached to the main engines. Greater power levels can be provided either by dedicated generators or by loading the main engine more heavily, but this can compromise take-off performance.

2.2.8 Take-off and Landing Gear

Take-off and landing gear has a surprisingly significant impact on the design of an aircraft. The total mass of such elements can be a noticeable fraction of overall weight, and the drag caused by unretracted wheels can be greater than that of the lifting surfaces combined. Also, when landing on grass runways, significant structural loads arise – we have measured accelerations of 20 g on undercarriage elements and have seen main landing wheel axles fail as a result

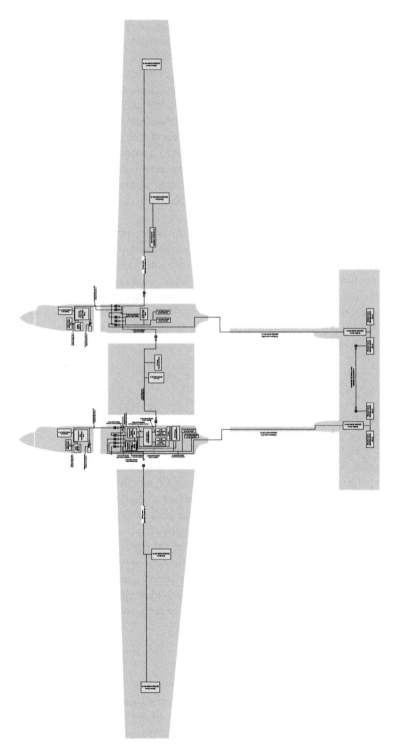

Figure 2.9   A typical UAV wiring diagram.

Figure 2.10   The SkyCircuits SC2 autopilot (removed from its case).

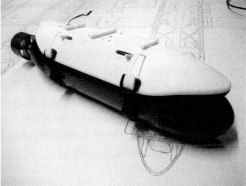

Figure 2.11   University of Southampton SPOTTER UAV with under-slung maritime flight releasable AUV.

of low-cycle fatigue. It is generally important that the undercarriage provides some form of compliance to cushion ground forces from passing unattenuated into the main airframe, either by the use of slightly flexible structural elements or through spring/damper systems. Some part of the system will also need to be steerable. If high flight speeds are important, some form of drag mitigation will be needed, either via wheel spats or by retract systems (typically electrical or pneumatic). We generally seek to buy off-the-shelf units for undercarriages for all but the very largest aircraft and mostly do not bother with the added complexity of retractable systems.

2.3   Main Design Drivers

Having set out an outline catalog of the component parts of our UAVs, and before going on to consider each aspect in detail, it is useful to consider the key design drivers that control our medium and small fixed-wing UAVs. Inevitably, the payload mass and the range/endurance/cruise speed of the aircraft are the fundamental inputs to any UAV platform design – in fact, they are the key inputs to *any* vehicle design whether they be airborne, seaborne, or landborne. The next key aspect to consider in fixed-wing UAVs turns out to be the landing speed – this controls the size and complexity of the wings, since low landing speeds require large wings and/or increasingly sophisticated high-lift devices. In small UAVs, the emphasis tends to be on large wings and simple flaps, while in medium-sized aircraft it becomes viable to consider more complex high-lift devices. Since we are concerned with UAVs under 150 kg MTOW and often have to operate from less than perfect landing fields, low landing speeds strongly impact on the likelihood of damage and wear and tear on the airframe. Ideally, landing speeds would be 12 m/s or less, but are often as high as 20 m/s. While this is still very low by comparison with large jet aircraft, it can still be frighteningly fast when things go wrong. High landing speeds can also lead to the need for long runways unless braking systems are used, which can be heavy, complicated, and expensive.

For low-cost designs, the next important area to consider is propulsion, since engines or motors have to be purchased off the shelf. When selecting electric motors, a very wide range of choices is available, but for internal combustion engines, particularly for larger airframes, the choice is much more restricted. It then often turns out that only one or two particular engines are suitable for the configuration under consideration, and this can significantly impact on the resulting design. One can be caught between selecting an engine that is slightly too small and risks giving an underpowered result, or one that is rather too large giving an over-heavy design. Sometimes we find that it is sensible to revisit payload and mission choices in the light of available engines rather than dogmatically insisting that these numbers are fixed. As ever in design, one is seeking a *balanced* result that acceptably trades a range of characteristics off into an harmonious whole. Given such information, it is possible to consider the boundaries of the wing loading (W/S) versus thrust-to-weight ratio (T/W) design domain. These boundaries are representations of the basic constraints that enforce the adherence of the design to the numbers specified in the design brief.

Armed with this information and a catalog of possible engines or motors, it is then relatively easy to start to build some form of concept model from which many of the principal aircraft parameters can be derived, starting with wing loading and principal dimensions and working through to a weight summary and control requirements to ensure that a balanced design can be achieved. Although such trade-off studies can be carried out using the proverbial back-of-an-envelope methods, engineers typically reach for their collection of computer-based tools at this stage – quite commonly the ubiquitous spreadsheet during initial concept definition: in Part III, we use Microsoft Excel. We will illustrate our approach with the concept design of several aircraft we have built and flown. Before heading in this direction, however, it is useful to consider the airframe components in more detail so as to understand the whole toolset that will be needed in the design process and how these various tools will be linked together. Having run through the airframe components and demonstrated a concept design process, we then follow this with CAD and physics-based analysis of one of our aircraft, together with details of its manufacture, regulatory approval, trials, and documentation before setting out how we operate and maintain our UAVs.

Part II

The Aircraft in More Detail

3

Wings

The most fundamental parts of any aircraft are the lifting surfaces that permit flight in the first place. For the UAVs being considered here, these are rigid wings, usually incorporating control surfaces such as ailerons and flaps. After more than 100 years of powered heavier-than-air flight, a great deal is known about the way the geometry of a wing impacts on its ability to generate lift with low drag. It is therefore useful to start by very briefly recapping the basics of wing theory before looking at how wings can be realized in practice at the size needed for UAVs of between 2 and 150 kg MTOW.

3.1 Simple Wing Theory and Aerodynamic Shape

The nonsymmetrical cross-section of a wing when presented to an oncoming airflow, caused either by using a cambered section or a symmetrical one inclined at some angle of incidence (attack) to the air, causes the air pressure on the upper surface of the wing to be lower than that on the lower one; the net imbalance in pressure generates the lift. For low-speed airfoils, the precise shape of the airfoils used does not make much difference to the amount of lift that can be produced with modest camber or low angles of incidence. In fact, a simple (two-dimensional – i.e., infinitely long span) flat plate inclined to the airflow will generate lift, and it is possible to show theoretically that if air was an inviscid fluid (i.e., it caused no losses due to friction, produced no boundary layer, and did not exhibit flow separation), the section lift coefficient for such a plate would be related to the angle of attack as $C_1 = 2\pi \sin \alpha$ (where α is expressed in radians), which means that for small angles of attack the slope of the lift coefficient curve with angle of attack is just $2\pi \, \mathrm{rad}^{-1}$ (or $\pi^2/90 \simeq 0.11 \, \mathrm{deg}^{-1}$). This is termed the classical lift slope; note that in this analysis the lift is not caused by the impact of the air on the surface of the plate, rather it is due to the air circulation around it.

Real airfoils, of course, have thickness and often camber – camber serves to increase the lift available at a given angle of incidence, while thickness allows for internal structure, and so on, and also permits the wing to operate over a range of angles of attack without separation. When suitable sections are used to form a three-dimensional wing, the issue of the flow around

Small Unmanned Fixed-wing Aircraft Design: A Practical Approach, First Edition.
Andrew J. Keane, András Sóbester and James P. Scanlan.
© 2017 John Wiley & Sons Ltd. Published 2017 by John Wiley & Sons Ltd.

the wingtip then arises – this is because the pressure difference between the upper and lower surfaces of a simple straight wingtip will give rise to rotating flow and a vortex stemming from the tip and trailing backward into the flow. This rotating flow acts to alter the effective angle of attack of the wing and leads to so-called induced drag, which would be present even in an ideal fluid. A great many variations in geometry can be used to limit such losses; the four most common approaches are as follows:

1. The use of large aspect ratios so that most of the lifting surface is distant from the vortex at the wing tip;
2. Reduction in section chord toward the tip to reduce the pressure difference near the tip and hence lessen the tip vortex (as seen in the classical elliptical shape of the Spitfire fighter aircraft);
3. Decambering or twisting of the wing to either reduce the local section lift coefficient or the local section angle of incidence near the tip, again reducing the strength of the tip vortex;
4. The use of winglets or other wingtip shape modifications to control and position the tip vortex (a feature increasingly common on commercial aircraft over the last 10 years).

When deciding on wing sections and planform shape, a number of key aerodynamic aspects must be considered. Assuming the wing is large enough to lift the aircraft at sensible angles of attack, the designer must first consider the trade-off between aspect ratio and weight – a high aspect ratio wing will, in general, be more aerodynamically efficient than a low aspect ratio one because of the reduction in induced drag. However, as a wing is essentially a cantilever beam, the greater the aspect ratio, the larger the bending moments and the heavier it is likely to be for a given planform area and structural design approach. We find aspect ratios between 6 and 9 to be a sensible range to consider. For shorter wings, some form of taper becomes increasingly important to control induced drag.

The other important driver of wing drag is the section thickness to chord ratio. As the data collected together by Hoerner [9], and sketched in Figure 3.1, show, section aerodynamic drag is fundamentally driven by the thickness to chord ratio – thin sections have lower zero-lift drag (also once the flow is turbulent, the drag is not strongly impacted by the section Reynolds number). However, thin sections suffer from two main drawbacks: first, they limit the internal room within the wing for structural elements and control systems; second, they tend to stall at lower angles of attack because the leading edge radius is necessarily limited.

A key problem in aircraft design is achieving adequate lift at the low speeds desirable during landing and takeoff. Since lift is proportional to speed squared, it is common to size the wings of an aircraft to be much larger than needed while flying at cruise speeds so that the angles of attack needed at low speed can be accommodated. On simple UAVs that lack high lift systems of slats and multipart flaps, this can be a particularly important part of the wing design. Again, a compromise is needed – if the wing has to be oversized to provided adequate lift at low speed, it is important that the wing drag be low at the shallow angles of attack used during cruise – implying the use of thin sections. However, to maximize lift at low speed, a high angle of attack will be needed, where a thicker section is less likely to stall. The location of where stall begins along the wing is also an important aspect for handling – if the wing stalls at the root before the tip, the aircraft is likely to have more benign flight

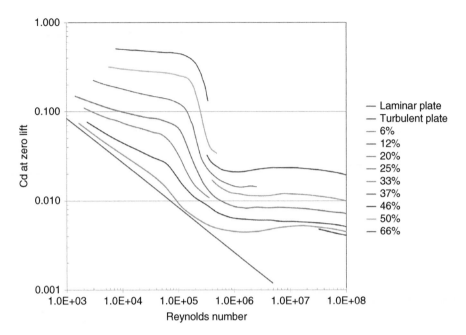

Figure 3.1 Variation of airfoil section drag at zero lift with section Reynolds number and thickness-to-chord ratio. After Hoerner [9].

characteristics – thus wings are commonly twisted (have washout) or have reduced camber toward the tip.

A brief consultation of any of the widely available airfoil section libraries on the Web will reveal that there are literally thousands of potential sections that can be adopted. The section lift, drag, and moment coefficients for the more popular sections are widely available – most will lift perfectly adequately up to 10° angle of attack where, as predicted by the classical wing theory noted earlier, they typically have lift coefficients of 1.1. Their zero-lift drags basically follow the data presented in Figure 3.1, varying quadratically away from this as angle of attack increases. Where significant differences do reveal themselves is in the maximum lift coefficients achieved at stall and the way the lift and drag vary once stall has begun.

This leads to the final key consideration in wing design – the behavior of the wing as lift starts to break down at high angles of attack. Unless the UAV under consideration is likely to fly at very high speeds where drag minimization is crucial, adopting very thin wing sections is likely to be counterproductive, especially if complicated and expensive leading-edge high-lift devices cannot be used. We generally see little point in adopting thickness to chord ratios below 15%. In our experience, the precise choice of the section geometry and camber matters rather less in slow to moderate speed UAV design than accurate manufacture of the wing so that the sections chosen are realized in practice – there is little point adopting a highly optimized airfoil section if the build process is unable to follow the prescribed section data over its entire surface.

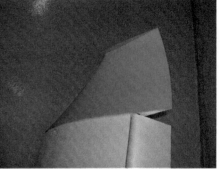

Figure 3.2   A UAV with significant FDM ABS winglets (this aircraft also has Custer ducted fans).

When building small UAVs, it is also very convenient to have a straight main spar and simple two-dimensional curvature of the lifting surfaces, so we have generally opted for tapered wings of fixed camber but often with suitable winglets mounted at the tips, see, for example, Figure 3.2. To ensure that the wings have the correct aerodynamic shape, we now rely on precision-cut closed-cell foam cores for all our wings (and tail surfaces). We have access to several digitally controlled hot-wire foam cutters of various sizes and capabilities, which is why we prefer to use straight-line generators to loft the wing surfaces where possible. These foam cores are typically hollow but also have a circular spar hole or load transfer region cut into them to aid transfer of aerodynamic loads to the spars, see Figures 3.3 and 3.4. We will return to more detailed considerations of aerodynamics in subsequent chapters.

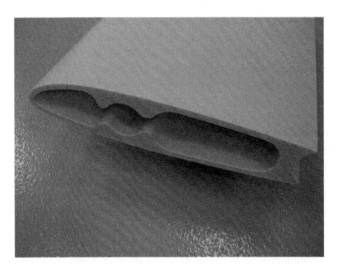

Figure 3.3   Wing foam core prior to covering or rib insertion – note strengthened section in way of main wing spar.

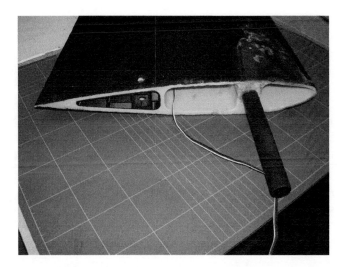

Figure 3.4 Covered wing with spar and rib – in this case, the rib just acts to transfer the wing twisting moment while the spar is bonded directly to the foam without additional strengthening.

3.2 Spars

The high strength to weight and stiffness of carbon-fiber-reinforced tubes have made them the obvious choice for spars in small UAVs. They are cheap, and available in a wide variety of diameters, thicknesses, and ply orientations. Perhaps the only issue we find when using such tubes for spars is that they are not simple to drill or machine since the material tends to splinter rather than cut cleanly – therefore we try and avoid making holes in them or doing anything other than cutting them to the required length. Instead, we typically use selective laser-sintered (SLS) nylon clamps and bushes to join them to the rest of the aircraft – since nylon is very much less stiff than typical carbon-reinforced plastic, this provides suitable cushioning at any joints and reduces stress concentrations in the spars. The loads that can be carried by such spars when correctly engineered into designs is prodigious – Figure 3.5 shows the main spar and wing assembly of the SPOTTER UAV under sandbag static loading – a single 35 mm diameter spar is here carrying a total mass of nearly 100 kg on each wing, equivalent to a 7 g gust loading. Ultimately, the foam parts of the wing failed before the carbon spar in this test.

3.3 Covers

In conventional aircraft, the outer surface of the wing, the so-called wing cover, is often added to the completed internal structure of the wing, typically by riveting. On model aircraft, this covering was originally provided by doped tissue paper stretched over a balsa wood structure. After experimenting with many types of wing structure, we have finally settled on using closed-cell foam to form the aerodynamic shape to which we add a covering of either a thin Mylar film (held in place with spray adhesive) or fine glass fiber tissue (adhered with a water based resin). The covering adds to torsional and bending strength and at the same time gives

Figure 3.5 A SPOTTER UAV wing spar under static sandbag test.

resilience to ground handling. In either case, great care is taken to use the minimum amount of adhesive or resin so as to control weight build-up. All of the wings illustrated in this chapter have been built following this approach.

3.4 Ribs

To help join the foam cores to the main spar and to provide hard points for features such as servos or hinges, we typically add a few SLS nylon ribs to the wing build-up, see Figure 3.6 for an example. The rib has large attachment areas for bonding to the foam wing core, a hard point for a trailing edge flap, and a hole that is a push-fit to the main wing spar. A wing might have four or five such ribs depending on the number of individual control surfaces in use and the number of pieces the foam core has been made from.

3.5 Fuselage Attachments

Most conventional layout UAVs need to have removable wings for storage and transportation. If carbon-fiber spars have been adopted for the main structural elements of the wings, the most natural way to incorporate them into the fuselage (or engine nacelles) is to provide a suitable tubular opening in the fuselage into which an extension of the wing main spar slides and can be clamped. Alternatively, a continuous wing spar that spans both wings can be used,

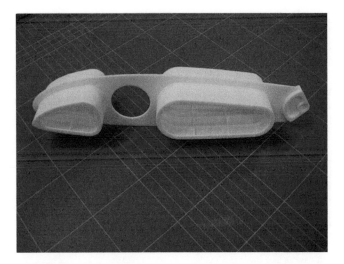

Figure 3.6 SLS nylon wing rib with spar hole – note the extended load transfer elements that are bonded to the main foam parts of the wing and also flap hinge point.

running through a tube in the fuselage and being clamped into the wings at either end. To carry wing moments into the main structure, some kind of torque peg near the rear of the wing will also be required – since this acts in shear, it can be quite small provided the wing moments are adequately controlled along the span of the wing. We have used both a secondary spar with matching fuselage tube and a simple SLS nylon peg formed onto the innermost wing rib with acceptable results in both cases. Figure 3.7 shows the carbon spars protruding from

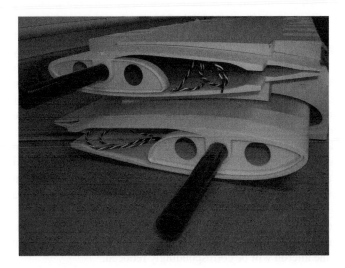

Figure 3.7 Two wing foam cores with end rib and spar inserted – note in this case the rib does not extend to the rear of the section, as a separate wing morphing mechanism will be fitted to the rear of the wing.

Figure 3.8 SPOTTER UAV wing under construction showing the two-part aileron plus flap, all hinged off a common rear wing spar – note also the nylon torque peg on the rib nearest the camera.

a pair of morphing wings, while Figure 3.8 shows the use of a nylon peg printed into the innermost rib.

3.6 Ailerons/Roll Control

Most conventional winged aircraft have aileron control surfaces toward the outer ends of the main wings. The ailerons require some kind of hinge plus an actuator mechanism. If extra redundancy is required, each one can be multipart with its own actuator. Since they are in continuous motion during flight, it is important that they be low-drag devices, yet they need to be large enough to provide sufficient roll authority, especially at the low speeds encountered during takeoff and landing. Figure 3.8 shows a typical aileron arrangement, in this case with a two-part control surface but sharing a common hinge spar that supports both aileron parts and the inner wing flap. Notice that planform taper has been achieved by reducing the depth of each of the three control surfaces when moving from root to tip.

We typically design the ailerons to extend over the final 30% of the chord and the outer 50% of the wing. However, if very low landing speeds are required, this may necessitate large flaps extending over more than half of the wing span. In such cases, the available planform space for ailerons can be more limited – if smaller ailerons are to be fitted, care must be taken during the analysis stages to ensure sufficient roll authority will be available, perhaps by increasing the depth of the aileron surfaces to as much as 50% of chord. Note also that the outermost parts of the wing are most affected by the tip vortex, and so the amount or roll control contributed by these parts of the aileron will be correspondingly reduced.

Sometimes, to overcome the drag associated with a flapped aileron, roll control may be achieved by other means, such as via synthetic jets, tiperons, or wing warping – wing warping is how the original Wright brothers aircraft worked. We continue to research such schemes but

Figure 3.9 UAV that uses wing warping for roll control.

Figure 3.10 UAV that uses tiperons for roll control.

generally consider them to be too experimental for routine aircraft builds.[1] Figure 3.9 shows an aircraft that uses wing warping for roll control, and Figure 3.10 shows a tiperon system. In both cases, fences are used to help control tip vortex flows.

3.7 Flaps

Flaps are essentially similar to ailerons except that they are used to increase the wing lift coefficient and not to generate a roll moment (indeed, the roll caused by a jammed flap can be a serious control issue and must be considered when sizing ailerons). Since flaps are generally only fully deployed during landing, a low drag result is not that important in flap design so that the so-called split flaps can sometimes be fitted to good effect. Since we aim to simplify the wing design where possible, we normally just use a series of movable surfaces along the

---

[1] See, for example, the FLAVIIR project that investigated the use of synthetic jets to control roll, https://www .theengineer.co.uk/issues/september-2010-online/first-flight-for-flapless-uav/.

Figure 3.11   Fowler flap – note the complex mechanism required to deploy the flap.

full length of the wing trailing edge with the innermost one taking the role of the flap and the outer ones that of aileron, that is, having sized the ailerons, we generally use the remaining real estate on the wing to fit as large a flap as possible – this is what is seen in Figure 3.8. Note also that there is little to gain from extending any flap further forward than the trailing 30% of the wing chord (see, e.g., Hoerner [10]). Although we commonly used plane flaps, we have tried more complex systems such as Fowler flaps, see Figure 3.11. While traditional flaps will generate section lift coefficients up to 2.0, single-part Fowler flaps will give coefficients as high as 3.0 and two-part systems even more (as long as the flow remains attached to the flap).

3.8   Wing Tips

Because of the pressure difference at the ends of all finite-span wings, wing tip vortices are generated in flight, leading to downwash and so-called induced drag. To reduce the impact of this phenomenon, many types of wing planforms and tip shapes have been flown over the years. It is now common practice in civil airliners to have complex wing tip devices to help the aerodynamic performance of the aircraft, albeit with some consequences for structural loading. Since, generally, we use full span spars and need to terminate them in some structurally efficient way, we routinely place a wing rib at the outermost edge of each wing. It is then a simple matter using SLS or fused deposition modeling (FDM) printing to create quite complex additional tip devices to improve the aerodynamics of the wing. We generally use computational fluid dynamics (CFD) models and wind tunnel testing to refine these shapes for best performance, see, again, Figure 3.2 and also Figure 3.12. It is often possible to reduce the overall induced drag of a wing by 20% with careful tip and planform taper design.

3.9   Wing-housed Retractable Undercarriage

Although it is common practice in large aircraft to use a main undercarriage the retracts into the wing body, this is rarely seen on small UAVs because of weight, cost, and complexity issues – the most common exceptions being where aero-modelers are building accurate scale copies of large aircraft. It is, instead, more conventional to fix the undercarriage to the main fuselage structure (with tail-wheels on the empennage in some configurations). If, however, the drag from the main undercarriage is a significant issue, then electrical or pneumatic retract

Figure 3.12 Simple FDM-printed wing tip incorporated into the outermost wing rib.

Figure 3.13 UAV with pneumatically retractable undercarriage – the main wheels retract into the wings while the nose wheel tucks up under the fuselage (wing cut-out shown prior to undercarriage installation).

systems can be fitted. Figure 3.13 shows one such aircraft that was fitted with a pneumatic retract system – in flight, the sudden acceleration caused by the reduction in drag on retracting the undercarriage was very noticeable. Suitable pockets in the underside of the wing are required to house the whole system along with structural elements to transfer the landing loads to the main spar and onward to the rest of the airframe. We make such components from SLS nylon and also increase spar diameters to allow for the loads seen – we have routinely measured accelerations of 20g in undercarriage elements during landing and even seen low-cycle fatigue failure in undercarriage axles.

3.10   Integral Fuel Tanks

While it is quite normal in full-sized aircraft to place fuel tanks within the main wing structure, this is rarely seen on small UAVs. The issue is that the materials used to construct small UAV wings are rarely compatible with storing fuels. We have, however, built integral fuel tanks into the main lifting surfaces by using SLS nylon. This is only really justifiable where there is some other structural requirement being met that calls for the extra weight and strength of the nylon over our conventional foam wings.[2] On the SPOTTER aircraft, the central wing between the two nacelles is used to support the payload, and thus it has to be able to support considerable masses. Given this requirement, we made the part from SLS nylon, fitted a rear flap to it, and made the entire hollow structure into a large fuel tank, see Figure 3.14 and also Figures 2.3 and 2.6.

Figure 3.14   Integral fuel tank in central wing section for SPOTTER UAV.

---

[2] The inertia relief gained by placing fuel directly into wing tanks is not sufficient, on its own, to overcome the weight penalty incurred from using SLS nylon elements to form UAV wings in our experience.

4

Fuselages and Tails (Empennage)

After the wings, the next most fundamental parts of the aircraft are the fuselage and tail/empennage. The fuselage generally houses the avionics and payload, supports the engines or motors (sometimes in the form of nacelles), and provides attachment points for all the flying surfaces and undercarriage elements. The tail surfaces provide flight control and stability. All must be engineered for low drag and light weight while being robust and readily maintainable.

4.1 Main Fuselage/Nacelle Structure

In most aircraft, there is a well-defined main fuselage or nacelle structure that is elongated in the direction of flight, suitably streamlined and with some means of access to the interior. These provide the space to house the myriad components needed for flight control and payload operation. They can be made in a wide range of materials, always bearing in mind the desire for light weight, high strength, and good rigidity. The most common construction materials are plywood, carbon-fiber-reinforced plastic (CFRP), rigid foam, and polymers. We have tried all types, including space-frame and monocoque methods and now prefer to use selective laser-sintered (SLS) printed nylon monocoques. Although SLS nylon has an inferior strength-to-weight ratio when compared to CFRP, the ability to adopt complex geometrical reinforcing allows competitive structures to be produced, see Figure 4.1. Also such 3D-printed structures do not require molds or other tooling for their construction and permit the inclusion of fittings for bayonets, hatches, switch gear, and sensors with comparative ease, see Figure 4.2.

When very light structures are needed, the use of rigid foams can be attractive, though special care then has to be taken to provide hard points for fitting highly loaded components such as engines or undercarriage – typically by gluing load-spreading plates to the foam, usually made from thin sheets of plywood or plastic (see Figure 4.3) – this can also be necessary when using nylon for the main fuselage material, and even some CFRP structures include metal hard points within the molding process. Foam can, however, be used for the entire structure of lightweight designs, see, for example, Figure 4.4. One way of combining the attributes of CFRP, foam, and SLS nylon is in the form of a foam-covered space frame that is made of CFRP tubes joined together with SLS nylon clamps, see Figure 4.5.

Small Unmanned Fixed-wing Aircraft Design: A Practical Approach, First Edition.
Andrew J. Keane, András Sóbester and James P. Scanlan.
© 2017 John Wiley & Sons Ltd. Published 2017 by John Wiley & Sons Ltd.

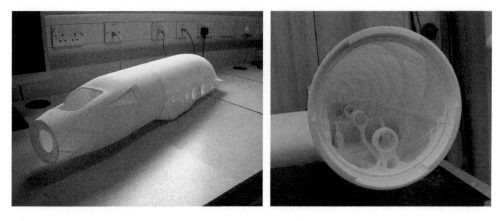

Figure 4.1   SPOTTER SLS nylon engine nacelle/fuselage and interior structure.

Figure 4.2   Bayonet system for access to internal avionics (a) and fuselage-mounted switch and voltage indicators (b).

Figure 4.3   Load spreader plate on Mylar-clad foam core aileron.

Figure 4.4 Commercially produced model aircraft with foam fuselage (and wings).

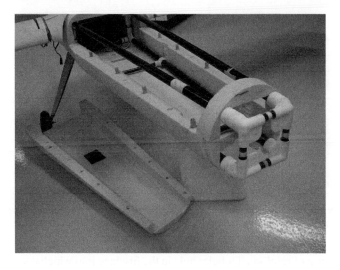

Figure 4.5 Space frame structure made of CFRP tubes with SLS nylon joints and foam cladding.

In cases where experimental aircraft are being developed, we sometimes opt for modular fuselage sections that allow the length of the fuselage to change. This allows the longitudinal position of the payload and avionics to be widely varied during development. In our experience, good control of the longitudinal center of gravity (CoG) can be difficult in a rapidly changing research airframe, so this capability can be very useful – note the repeated modules used in Figure 4.6, here held together by longitudinal tension rods that can easily be varied in length.

4.2 Wing Attachment

As noted in the previous chapter, we typically attach wings to the fuselage using carbon fiber spars to carry the principal loads, with pegs of some form to resist moments. This means

Figure 4.6   DECODE aircraft with modular fuselage elements.

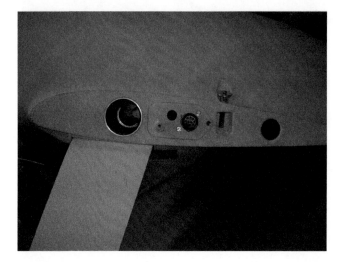

Figure 4.7   Wing attachment on SPOTTER fuselage. Note the recess for square torque peg with locking pin between main and rear spar holes.

providing some form of reinforced tubular hole in the fuselage into which the wing spar slides. It is important that the sliding fit be a good one to ensure a wide load transfer surface that reduces stress concentrations. It is also important to diffuse the local stresses stemming from flight loads and running through the spars into the wider fuselage structure, particularly toward the main undercarriage pick-up points, since the connection between wing spars and undercarriage is probably the most highly stressed part of the fuselage, see Figure 4.7.

4.3   Engine and Motor Mountings

Engines and motors, of course, provide the thrust for powered flight, and this thrust must be carried into the airframe via the engine mount. Engines and motors are also, however, among the highest density items in the aircraft, and so due allowance must be made for the inertia forces they generate during flight such as during gust events. Finally, when dealing with internal combustion engines, allowance must be made for the heat and vibrations caused by engine operation. High temperatures can adversely affect polymer structures, while vibrations can lead to fatigue failures in any mountings made of material such as aluminum that do not have lower stress limits on fatigue (we try and avoid using aluminum components in tension near engines – steel is much preferred for such loads). We have found that, provided suitable care is taken to prevent heat transfer, engines and motors can be bolted directly to sufficiently thick SLS nylon structures without problems. We do, however, often include antivibration mountings between the engine and the main fuselage to insulate avionics and payloads from such vibrations. Figure 4.8 shows some typical engine and motor mounts.

Figure 4.8 Typical engine and motor mounts for SLS nylon fuselages and nacelles. Note the steel engine bearer in first view, engine hours meter in second image, and vibration isolation in third setup.

4.4 Avionics Trays

All UAVs carry an avionics fit. This can be as simple as a basic model aircraft receiver, battery, and wiring to control servos. On a long-range, multiply redundant UAV, such as the SPOTTER, quite complex avionics is employed with multiple receivers, autopilots, batteries, generators, and appropriate change over systems to cope with failures. All this avionics equipment must be held firmly inside the UAV, permit access for maintenance and assembly while being lightweight and offering suitable support. Sometimes it is important to use antivibration mountings to insulate devices such as autopilots from engine vibrations. Our preferred approach to housing such systems is via laser-cut plywood base boards, which can be quickly and cheaply designed and cut while being highly customized to the equipment fit being used. Typically, such boards slide into groves in the SLS fuselage or are carried on dedicated antivibration mounts. We find it expedient to enable each board to be readily removable from the aircraft for maintenance – it can be extremely frustrating and time consuming if maintenance in the field has to be carried out through hatches that are inevitably always smaller than one would wish and quite often prevent access with the desired tools (we have found having team members who have very small hands can be most welcome when dealing with some UAVs we have operated, see Figure 4.9). Figure 4.10 shows a typical range of avionics boards designed following this approach.

Figure 4.9   Frustratingly small fuselage access hatch.

Figure 4.10   Typical plywood avionics boards with equipment mounted. Note dual layer system with antivibration mounts in last image.

4.5   Payloads – Camera Mountings

In our work, payloads have generally been sensors or dumb cargoes – we have not been involved with munitions and military UAVs so far. The most common payloads we have operated have been cameras, either still or video, both with and without steerable gimbals. Sometimes, these have simply been used to record images to on-board data cards, and sometimes these have been used to echo to radio downlinks. Housing static cameras is in most respects entirely similar to dealing with other avionics items, except for the need to provide a suitable window for lenses to see through – such windows can themselves become complex if they are to deal with poor weather or to be proof against dirt accumulated during flight. Static cameras typically require little more than a power supply and a relatively vibration-free location with a good field of view. We have two approaches to this. The simplest is to use a pusher configuration for the propulsion and to site the camera in the nose of the fuselage (in front of the forward undercarriage leg if fitted). Here it has an unobstructed field of view and is well away from likely sources of dirt that might otherwise foul the lens (as in the SULSA aircraft, see Figure 4.11). Alternatively, we fit an under-slung and removable payload pod placed between twin tractor engines (as in the SPOTTER aircraft, see Figure 4.12).

Figure 4.11   SULSA forward-looking video camera.

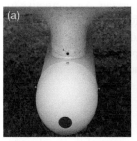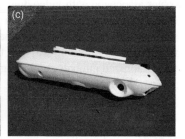

Figure 4.12   SPOTTER payload pods with fixed aperture for video camera (a) and downward and sideways cameras (b and c).

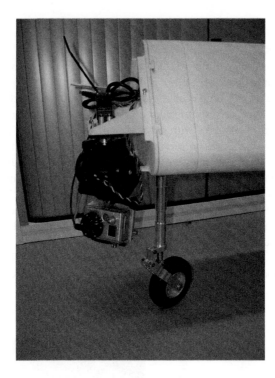

Figure 4.13   Simple two axis gimbal system and Hero2 video camera mounted in front of nose wheel.

If a steerable camera (or other sensor) is required, some form of gimbal may have to be included. We have designed and built these ourselves and also fitted those made by others, see Figures 4.13 and 4.14. High-quality systems are readily available commercially but they can be easily as expensive as the rest of the aircraft in total (it is, of course, not at all uncommon for the payload to be as expensive as the rest of the system in all classes of aircraft). Currently, stepper-motor-based systems linked to some means of assessing aircraft position and orientation are generally required if accurate pointing at targets is to be maintained. If the target itself is moving, then some form of image recognition capability will be required to maintain lock on the target and this will generally have to be on the aircraft to avoid the latency and bandwidth issues associated with video downlinks for ground-based image processing. An alternative to mechanically steering the camera is to use wide-angle high-resolution cameras and then use software to isolate the required part of the image as the aircraft and target maneuver. If the target is small or a long way off, then such systems are rarely competitive with high-quality well-stabilized gimbals supporting a powerful zoom-lens-based camera.

4.6   Integral Fuel Tanks

If SLS-printed fuselages are in use, this opens the possibility of using part of the fuselage to form an integral fuel tank, see Figure 4.15. Nylon is quite tolerant of gasoline and other fuels, but when formed using selective laser sintering, it is slightly porous. We thus first prepare

Figure 4.14   Three-axis gimbal system and Sony video camera mounted in front of the nose wheel. Note the video receiver system on the bench that links to the camera via a dedicated radio channel.

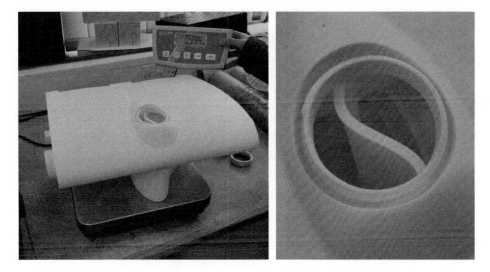

Figure 4.15   SPOTTER integral fuel tank. Note internal baffle and very small breather port (top left) in the close-up view of the filler neck.

any structures for fuel tank usage by a thorough depowdering process with mechanical agitation to loosen any material remaining from the manufacturing process in the parts delivered to us. This is followed by a sealing process where we fill the tank with a fuel-proof sealer (we use products supplied by Kreem[1]). Once the tank has been emptied and the sealer cured,

---

[1] See http://www.kreem.com.

the tank is then ready for use. Any tank fittings such as sensors, gauges, or fuel lines then need to have fuel-proof sealing rings applied where they penetrate the tank. We also routinely include significant internal baffling to control sloshing because integral tanks tend to be rather large. We also take care to ensure that fuel consumption does not induce significant adverse center-of-gravity shifts.

4.7 Assembly Mechanisms and Access Hatches

When working with a fully molded CFRP-type fuselage, it is quite normal for the bulk of the fuselage to be formed in a single piece. The use of SLS nylon as a structural material tends to work against this approach: first, we typically want structures that are larger than can be printed in the available laser sintering machines; and second, we commonly wish to break the fuselage down into parts for access and portability reasons. We have found two straightforward means of dealing with these issues: either we use bayonet-type joints designed directly into the SLS nylon parts or we thread a series of parts onto longitudinal tension rods that pull the components into a single hull. Sometimes we adopt both approaches in a single fuselage. Figure 4.16 shows an aircraft built in this way (see also Figure 4.6, where again tension rods and bayonets are used). When using bayonet systems in fuselage parts, it is wise to print several

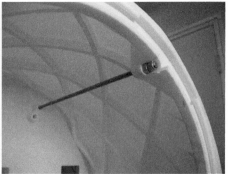

Figure 4.16 Aircraft with SLS nylon fuselage formed in three parts: front camera section attached by bayonet to rear two sections joined by tension rods. Note the steel tension rod inside the hull just behind bayonet in the right-hand image.

 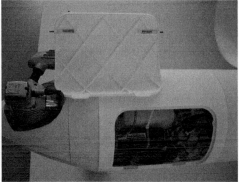

Figure 4.17 Example hatches in SLS nylon fuselages. Note the locking pins and location tabs on the right-hand hatch.

trial parts to get these to work as desired before committing to large and expensive prints. In particular, we find that with SLS nylon, the first few operations of a bayonet will slightly polish the mating surfaces so they need to be quite tight when first made if they are not to become too loose when used repeatedly.

In most cases, the fuselages and nacelles used in UAVs should not be open to the elements since they contain sensitive avionics components, wring looms, batteries, and so on. The only exceptions tend to be for cooling requirements around engines, motors, and speed controllers. In consequence, some means of internal access on the airfield is then almost always necessary. Bayonets form one means of achieving this, and we often do this for payload items where we may have multiple payloads capable of fitting onto a single fuselage or other bayonet. The alternative is some form of hatch and cover. Hatches can be readily SLS-printed with closely conforming covers and duplicate locking mechanisms, see the examples in Figure 4.17. When designing hatches, we ensure structural continuity by forming a suitable flange around the opening, and we also sometimes add neoprene O-rings to these to provide watertight closure.

4.8 Undercarriage Attachment

Most of the UAVs we build have conventional wheeled undercarriages either with nose- or rear-wheel steering. Obviously, the size and strength of the undercarriage is set depending on the maximum take-off weight (MTOW) of the aircraft. Wheel diameters, however, must also reflect the likely runway conditions – even very light aircraft must have sufficiently large wheels so that they do not have problems rolling over any unevenness in the runway. This is particularly true when dealing with grass strip runways.

The design loads used in sizing the undercarriage and its mounting points stem from two cases: first, one must allow for impact loads on landing. Although the pilot or automated control system will, of course, attempt a smooth landing, significant forces will still arise and care must also be taken to be able to cope with less than perfect approaches. Second, allowance must be made for faults in the runway surface. Even on well-maintained grass strips, it is not uncommon to encounter small mounds or divots that can generate significant and sudden impact loads while taxiing at speed or during landing and takeoff. Suitable suspension systems can help mitigate these loads, but nonetheless the structural mounting points for undercarriages usually need to be among the strongest points in the airframe. When working with SLS nylon structures, we sometimes find it necessary to use metallic load spreaders to cope with the loads being generated. Alternatively, rather large and well-reinforced areas of nylon will be required. Figure 4.18 shows a nose wheel mounting with metallic reinforcing, while Figure 4.19 shows a plain nylon attachment point using two nylon bearings to permit steering.

Figure 4.18 Metal-reinforced nose wheel attachment with steering and retract hinge in aluminum frame attached to SLS nylon fuselage. Note the nose wheel leg sized to protect the antenna.

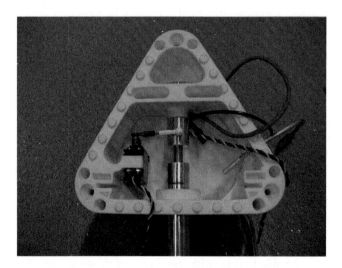

Figure 4.19 Nylon nose wheel attachment. Note the significant reinforcement around the lower and upper strut bearings.

Figure 4.20 Tails attached directly to the fuselage. The right-hand aircraft is a heavily modified commercial kit used for piggy-back launches of gliders.

4.9 Tails (Empennage)

All the UAVs we build have conventional rear-mounted tail surfaces. We use both normal rudders and elevators and also combined V or inverted-V elevons, in most cases made of closed-cell foam, supported by small CFRP spars and covered with Mylar or glass-fiber cloth. These can either be directly attached to the rear of an elongated fuselage or are attached to long CFRP tubular booms. Booms generally allow for smaller tail surfaces because of the increased moments that can be generated by shifting the tail further aft, but this can limit rotation angles during takeoff. Figures 4.20 and 4.21 show a selection of the arrangements we have used. In most designs, we use conventional flaps on the tail surfaces, but we have made use

Figure 4.21 Tails attached using CFRP booms, both circular and square in cross-section.

Figure 4.22 All-moving horizontal stabilizer with port/starboard split to augment roll control and provide redundancy.

of all-moving horizontal stabilizers and even split horizontal stabilizers that can augment roll control, see Figure 4.22.

When sizing tail surfaces, we follow the idea of vertical and horizontal tail volume coefficients – these coefficients are given by the projected vertical or horizontal areas multiplied by the distance between the quarter chord points of the tail surface and the main wing, all divided by the main wing area and either the main wing span (for vertical coefficient) or the main wing mean chord (for horizontal coefficient). For aircraft with conventional operating characteristics, we generally choose a horizontal coefficient of 0.5 and a vertical coefficient of 0.04 following the advice offered in Raymer [11]. If we wish to have a particularly stable aircraft or one that has significant shifts of the CoG in flight or has to operate well at very low flight speeds, we increase these, sometimes to as much as twice the normal values.

Whether a tail boom is in use or not, suitable attachment points for the tail items must be provided, which are sufficiently rigid to avoid flexibility and flutter issues in the tail. When attaching tail surfaces directly to the fuselage, this is rarely a problem, but equally it is then almost inevitable that one uses V or T configurations of some sort. Booms offer much greater flexibility in tail arrangements, although we now rarely use a T or inverted-T tail with a single tail boom because it is then difficult to generate sufficient torsional rigidity without very large boom diameters. For directly attached surfaces, we generally place the servo actuators inside the fuselage and then use small cantilevered CFRP tubes to provide rigid hinges for the flaps. When using tail booms, one has to provide both a stiff location inside the main fuselage to mount the boom (see again, for example, Figure 4.1) and a small self-contained structure to support the tail and attach it to the end of the boom, see Figure 4.23. Sometimes, these small tail structures are also used to house steerable tail wheels.

Figure 4.23   SLS nylon part to attach tail surfaces to a CFRP tail boom.

5

Propulsion

All (non-glider) aircraft require powerful, lightweight, efficient propulsion. For medium and small unmanned air vehicles (UAVs), this almost always means some form of propeller-based system. As it is difficult to engineer variable-pitch propellers on the scales required, fixed pitch systems are generally used. Combined with the high costs of developing bespoke prime movers, this means that the designer of small and medium UAVs is generally faced with rather difficult compromises in selecting the appropriate propulsion system from components already available on the market. We have found no concrete solutions: for long-endurance aircraft, we tend toward gasoline-fueled internal combustion (IC) engines with multiple cylinders, via multicylinder engines, multiple engines, or both. For short-endurance systems, especially if low vibration is important, battery-powered electric motors are often preferable, and as battery technology improves, electric propulsion will become more useful for longer endurance systems. Here we consider both these forms of propulsion in various configurations and also hybrid systems where both IC engines and electric motors are fitted to a single airframe.

In sizing and assessing propulsion systems, we rely on various sets of results we have obtained from a dedicated engine/propeller test facility we have built and also the first-rate data produced at the University of Illinois at Urbana-Champaign (UIUC) by Brandt and Selig [12]. It has been our experience that it is wisest to actually measure static and dynamic thrust, fuel, and battery consumption, and so on, for ourselves whenever possible. Dedicated test cells also provide an ideal place for setting up and running in new engines in controlled conditions, see Figures 5.1 and 5.2. We deal with the range of engines and motors we have direct experience with in the following sections. See also Section 11.6 for various expressions we use in estimating likely propeller and engine sizes.

5.1 Liquid-Fueled IC Engines

Liquid-fueled engines available for small and medium-sized UAVs are essentially two- or four-stroke gasoline, methanol, or diesel units. Major considerations are the availability, weight, cost, power, and efficiency. Few UAV programs can afford to develop their own prime movers, so generally one is forced to select from those already on the market. For

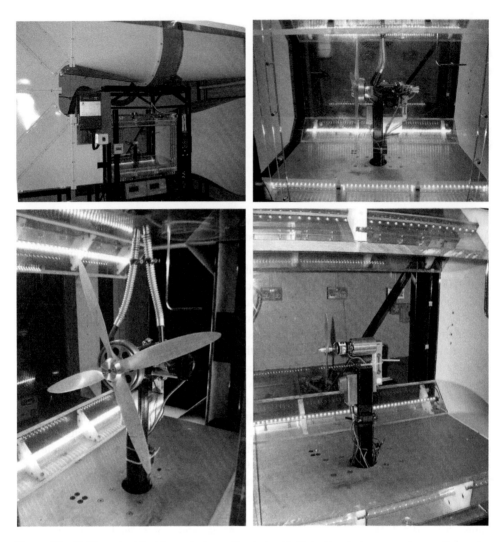

Figure 5.1 UAV engine/electric motor/propeller test cell. Note the starter generator on the engine behind the four-bladed propeller.

example, for airframes in the 10–50 kg maximum take-off weight (MTOW) range, there are probably fewer than 20 suitable spark ignition engines and almost no widely available diesel engines (though developments of new UAV engines are under way in many countries). For UAVs of less than 10 kg MTOW, the widest selection of IC engines comprise glow-plug ignition running on methanol-based fuels. These are cheap and powerful but tend to have poor fuel consumption. In our experience, the simplest way to get reasonable economy is via four-stroke gasoline systems using conventional carburetors or, at higher cost, two-stroke gasoline engines with manifold fuel injection. The latter are more involved but offer the prospect of improved fuel management, though a well set up four-stroke carburetor engine tuned to operate at a fixed cruise speed/rpm can be reasonably competitive. The conventional poppet valve gear on most

Figure 5.2   UAV engine dynamometer.

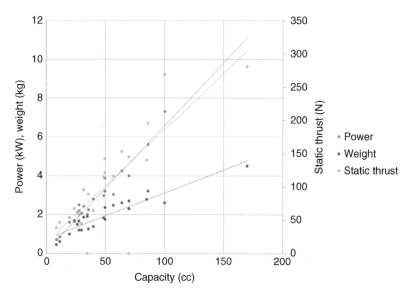

Figure 5.3   Typical maximum powers, weights, and estimated peak static thrusts of engines for UAVs in the 2–150 kg MTOW range.

four-stroke engines can, however, be a weakness in such systems, generally requiring adjustment after any extended flight. Figure 5.3 shows typical powers, weights, and estimated static thrusts of IC engine for UAVs in the 2–150 kg MTOW range. Note the relative paucity of data for the larger engine sizes.

Figure 5.4   OS Gemini FT-160 glow-plug engine in pusher configuration. Note the permanent wiring for glow-plugs.

5.1.1   Glow-plug IC Engines

The IC engines most commonly used in small model aircraft are methanol-fueled glow-plug engines. A vast range of these are cheaply available running two- and four-stroke cycles with capacities ranging from a few cubic centimeters (cc) up to over 100 cc in configurations spanning single cylinder, flat twin, flat four, inline four, and radial combinations with as many as nine cylinders. Even the largest and most complex versions from the best makers typically cost no more than $3500. We have made extensive use of the products manufactured by OS in Japan.[1] We find these to be well made and reliable at acceptable cost. In particular, we have flown many hours with their four-stroke flat-twin FT-160 Gemini engines, see Figure 5.4. Typical fuel consumption can, however, be over 1 l/h with this 26.5 cc engine, even though it is a four-stroke design (two-stroke glow-plug engines are even more thirsty). While this is not an issue for the aero-modelers for whom they are designed, it means that for long-endurance UAVs, a significant part of the aircraft take-off weight is fuel when powered by these types of engines.

5.1.2   Spark Ignition Gasoline IC Engines

The primary reason why glow-plug engines are so thirsty for fuel is that there is no control over when ignition takes place in the engine cycle. To overcome this limitation, one must switch

---

[1] See http://www.osengines.com/ – the airplane engines tab.

Figure 5.5 OS 30 cc GF30 four-stroke engine installed in a hybrid powered UAV. Note the significant size of the exhaust system.

to spark ignition systems (or direct injection). The advent of compact, Hall-effect-triggered ignition modules and tiny spark plugs has made spark ignition engines readily available at modest cost, though the need to incorporate a functioning spark plug in the cylinder head limits how small these engines can be. Currently, the smallest OS spark ignition engine is their single-cylinder, 15 cc, two-stroke GT15 engine. If a four-stroke engine is desired, the need to accommodate valve gear as well as the spark plug means that their smallest engine of this type is the 30 cc GF30 (see Figure 5.5). For larger aircraft, we have used either larger single-cylinder engines or flat twin- and flat four-cylinder variants. For added operational redundancy, we currently opt for twin single-cylinder engines for aircraft up to 40 kg MTOW. Above that, multicylinder engines become more appropriate. Figures 5.6, 5.7, and 5.8 show three of our larger UAVs, one with a single twin-cylinder engine and the others with two singles, in the latter cases running two- and four-stroke cycles. We have not tried rotary Wankel engines because of their generally poorer fuel efficiency, though they are available in a range of sizes and may offer lower vibration levels.

The choice between two-stroke and four-stroke spark ignition engines is not an entirely straightforward one. In theory, the two-stroke cycle gives more power at a given rpm because there are twice as many firing pulses at any engine speed. However, the volumetric efficiency of two-stroke cycles is not as well controlled as that of four-stroke cycles, which permit accurate control of valve timings. This tends to make two-stroke engines more thirsty than their four-stroke counterparts and not as powerful as theory would predict, though size for size they are generally lighter as they do not need cam shafts and valve gear. They also have the great benefit of simplicity. Maintaining accurate valve seating in such small engines is a real difficulty with conventional four-stroke systems that use poppet valves: if not regularly maintained, it is very easy to burn exhaust valves and lose compression. To take advantage of two-strokes mechanical simplicity while retaining good fuel efficiency, it is possible to use manifold injection on these engines. The UAV factory UAV28-EFI Turnkey Fuel Injected Engine does this to good effect, although this inevitably sacrifices some of the simplicity of the overall

Figure 5.6   Saito 57 cc twin four-stroke engine in pusher configuration. Note the pancake starter generator fitted to this engine.

Figure 5.7   Twin 3W-28i CS single-cylinder two-stroke engines fitted to 2Seas UAV. Note again the significant size of the exhaust systems.

Figure 5.8   Twin OS 40 cc GF40 four-stroke engines installed in SPOTTER UAV, with and without engine cowlings.

system.[2] Alternatively, more sophisticated valve arrangements may be used on four-stroke engines such as those fitted to the RCV DF 35 and DF 70 UAV systems.[3] Their rotary valves give high reliability and very good fuel consumption, particularly when combined with manifold fuel injection, although this level of sophistication does not come cheap. At the most extreme end of this spectrum lie engines such as the AE1 developed by Cosworth, which is a direct-injection two-stroke system using compression ignition rather a spark plug, permitting operation of the Diesel as opposed to Otto cycle, with the resulting higher compression ratio and better thermodynamic efficiency. Currently, such systems are not widely available and are extremely expensive when compared to a simple OS two-stroke system, but thermodynamic efficiencies as high as 0.335 kg/kWh have been reported.

5.1.3   IC Engine Testing

As has already been noted, we prefer to test each IC engine we are considering in our own test cells rather than simply relying on manufacturers' data. In the dynamometer cell, seen in Figure 5.1, the basic measurements we take are fuel consumption and torque at varying speeds and throttle settings, often with both gravity-fed and fully pumped fuel lines (a number of small UAV engines will give greater power when used with pressurized fuel lines, while some have built-in fuel pumps). Figure 5.9 shows a typical plot of raw data taken from our dynamometer. We average across the raw data aiming to complete tables such as the one shown in Table 5.1, from which we can then derive the engine power, brake-specific fuel consumption (BSFC), and brake mean effective pressure (BMEP) plots. When combined with the engine mass (including standard exhaust system and ignition system), one can then make like-for-like comparisons between alternative engines. Prior to running tests, we carefully calibrate the torque sensor with a weighted balance bar and also the throttle servo to ensure that 0% and 100% throttle openings are correctly achieved when set by the control system of the dynamometer.

BMEP, which is given from torque T by BMEP= $2\pi nT/displacement$ ($n = 1$ for two-stroke and = 2 for four-stroke engines), allows a comparison of the quality of the combustion system between engines of differing sizes and between different cycle types. Table 5.2 shows typical values of BMEP for a range of engine types, higher values of BMEP being better: the values achieved by the supercharged Rolls-Royce Merlin were clearly outstanding. In the end, however, it is usually BSFC and power-to-weight ratios that drive design calculations and engine selection.

Having completed a dynamometer test, we then run engine and propeller combinations in our dedicated wind tunnel system, also shown in Figure 5.1. This allows static thrust and thrust at varying air speeds to be recorded along with the reaction torque experienced at the engine mountings. The test cell also allows fuel flow rates to be measured. This allows an assessment of the overall efficiency of the propulsive system. Careful matching of propellers to engines is vital in achieving best performance from such systems. Moreover, the propulsive efficiency of small UAV propellers varies quite widely: the extensive test data collected by UIUC demonstrates this most clearly [12].

---

[2] See http://www.uavfactory.com/ – the Fuel Injected Engines tab.
[3] See http://www.rcvengines.com/ – the UAV Engines tab.

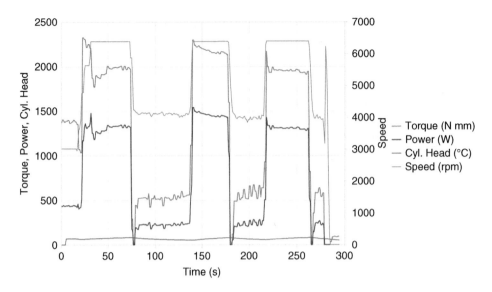

Figure 5.9   Raw performance data taken from an engine under test in our dynamometer.

Table 5.1   Typical liquid-fueled IC engine test recording table (maximum rpms are of course engine-dependent).

Fueling type	40% Throttle		60% Throttle		80% Throttle		100% Throttle	
Speed (rpm)	Fuel consumption (l/min)	Torque (Nm)	Fuel consumption (l/min)	Torque (Nm)	Fuel consumption (l/min)	Torque (Nm)	Fuel consumption (l/min)	Torque (Nm)
1500								
…								
4500	0.44	0.45	0.51	0.170	0.57	0.93	0.69	1.22
…								
8000								
8500								

Note that all engine test results are subject to variability depending on the engine set up and wear, test cell operating conditions, and any losses due to induction, exhaust, or coupling factors. If precise details of the tests carried out are not available, such results should be treated with caution during the design phase.

5.2   Rare-earth Brushless Electric Motors

While we most often use IC engines for our UAVs, small, powerful brushless electric motors based on rare-earth magnets are now available in a wide range of sizes so there is a much greater choice of units than for IC engines. Obviously, such motors must be supported by

Table 5.2   Typical IC engine BMEP values taken from various sources.

Engine type	BMEP at peak power (MPa)
Naturally aspirated spark-ignition engines	0.85–1.05
Boosted spark ignition engines	1.25–1.7
Naturally aspirated four-stroke diesels	0.7–0.9
Boosted automotive four-stroke diesels	1.4–1.8
Very large low speed diesels	1.9
Napier Sabre 7 (3055 HP at 3850 rpm)	1.94
Rolls-Royce Merlin 130/131 (2030 HP at 2900 rpm)	2.31
Pratt& Whitney R-4360 Wasp Major (3600 HP at 2700 rpm)	1.72
Typical small aero-modeler-based four-stroke UAV gasoline engines	0.85–1.15
Typical small aero-modeler-based two-stroke UAV gasoline engines	0.6–0.9

Note that BMEP values are typically 10–15% higher at the maximum torque speed as compared to maximum power speed and also that by assuming a BMEP of 1.0 MPa, it is possible to estimate the likely achievable peak power of a four-stroke UAV engine at a range of engine speeds.

Figure 5.10   Hacker brushless electric motor.

battery packs, fuel cells, or on-board generators (hybrid systems). In the world of multi-rotor UAVs, they are extremely widely used, and many commercial camera-carrying systems adopt these technologies. They are smooth, powerful, and highly reliable devices since they involve so few moving parts. Their Achilles heel remains the short endurances that are possible with conventional LiPo battery packs, though these continue to improve. We tend to use the motors made by Hacker[4] along with their Master Spin controllers. Brushless motors readily provide powers from a few tens of watts up to 15 kW, see Figure 5.10.

---

[4] http://www.hacker–motor–shop.com/ – the Motors tab.

5.3 Propellers

Perhaps the single biggest variable affecting the efficiency of the propulsion system is the choice of propeller (it is worth recalling that the Wright brothers cited good propeller design as one of the key developments that enabled them to fly their first aircraft successfully). There are very many manufacturers offering propellers in the more popular sizes, and while it might be assumed that all modern aero-modeler propellers would be essentially similar, this is far from the case. Their design philosophy, geometry, materials, and manufacturing accuracy vary considerably – all of which impact on the efficiency. Brandt and Selig [12][5] show that peak efficiencies vary for such propellers from 43% to 65% for two-bladed devices of similar diameters, even when operating at the correct advance ratio for the chosen pitch. If at all possible, it is best to use two bladed propellers since these have best efficiency, but, particularly for pusher configurations, installation difficulties may force the choice of three-, four-, or even five-bladed propellers. At large sizes and in pusher configuration, the choice of material and manufacturer becomes more limited. Where possible, we adopt APC Thin Electric propellers because of their very good efficiencies.[6] See also the spreadsheet available from RC advisor.[7]

Of course, propeller thrust also varies very significantly with variations in advance ratio (i.e., airspeed compared to propeller speed). As the airspeed seen by the propeller changes, if the rotational speed does not, the various aerodynamic sections of the propeller see different angles of attack, and just like wings, these sections stall if the angles are too high or low. Thus a fixed-pitch propeller will have to compromise between thrust during takeoff and while at cruise speed and at maximum airspeed. On large aircraft, this difficulty is overcome by using variable-pitch propellers, but as already noted, these are not generally available for small UAVs (we have had mixed success in trying to build such devices). The freely available code JavaProp[8] can be used to rapidly assess how propeller performance is likely to change with operating conditions.

To use JavaProp, one specifies at least the propeller outer and hub diameters, number of blades, rotational speed, forward velocity, and then one of torque, thrust, or power. It is also possible to insert sectional properties for the blades, but we generally leave these as Clark Y airfoils operating at a Reynold's number of 100 000. A design is then produced for the given forward velocity along with details of the performance at the input operating point. By next shifting to the "Multi Analysis" tab and selecting one of rpm, power, thrust, torque, or forward velocity, it is possible to generate tables of thrust (and other properties) for a range of conditions for the just designed propeller. Since most engines tend to be fixed-torque devices, we mostly choose that option both to design the propeller and to analyze it at other operating points (if the torque is not known for a given IC engine, this can be estimated from appropriate BMEP values; 1.0 MPa for four-stroke engines and 0.85 MPa for two-stroke engines). Figure 5.11 shows a typical set of outputs from JavaProp.

---

[5] See also http://m–selig.ae.illinois.edu/props/propDB.html.
[6] https://www.apcprop.com/.
[7] http://rcadvisor.com/understanding–propellers.
[8] http://www.mh–aerotools.de/airfoils/jp_applet.htm/.

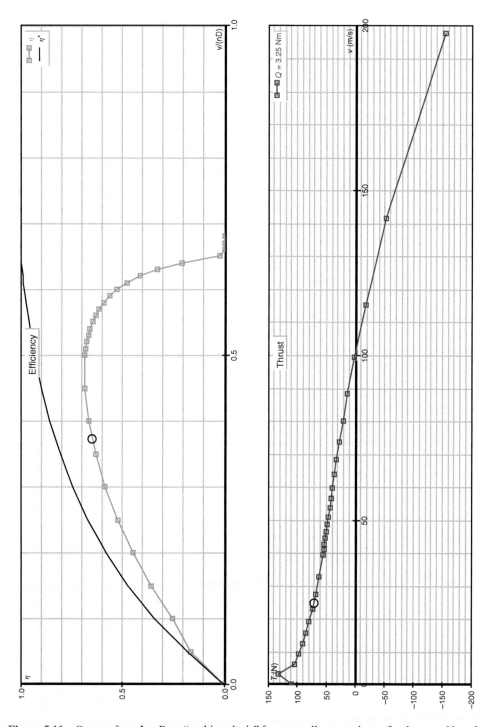

Figure 5.11 Outputs from JavaProp "multi analysis" for a propeller operating at fixed torque. Note the differing horizontal scales. Note the design point: here a typical cruise speed of 25 m/s is shown by the small circle and is slightly below the peak efficiency for the design. Peak thrust occurs at about 4 m/s, ensuring a good ability to start the aircraft rolling on a grass field.

5.4   Engine/Motor Control

All engines or motors require some form of control. At its simplest, this is a radio-controlled servo linked to the throttle on the carburetor (preferably by a ball-jointed, backlash-free link-age). Even such a basic system requires some form of fail-safe capability that shuts the engine down if control of the aircraft is lost, allowing controlled descent glide-down to the ground. For spark ignition systems, some form of spark generation and distribution is required; for multi-cylinder engines, it is common to fire multiple plugs simultaneously even though not needed and/or combined with multiple ignition systems timed for different cylinders. For fuel-injected systems, some form of engine control unit (ECU), which monitors at least engine speed and perhaps other parameters, will be necessary, again duplicated if multiple cylinders are in use. For electric motors, a digitally controlled three-phase output speed controller is the most common solution. These can be as expensive as the motors they control, since typically they must convert high-current DC sources to high-current, variable-frequency three-phase supplies.

5.5   Fuel Systems

If a wet fuel engine is used, some form of fuel tank becomes necessary. Commonly, these are separate items with some form of closure where supply, filler, and air lines are installed (with the filler readily accessible for refueling). They are available in a range of sizes and can be obtained for methanol- or gasoline-based systems (generally not interchangeable - note also that fuel pipes must be chosen to suit the fuel in use; gasoline rapidly degrades many forms of plastic pipe), see Figure 5.12. If SLS nylon structures are being used on the aircraft as we commonly do, the fuel tank can be formed as an integral part of the structure, as already noted in Chapter 4, allowing for very large capacities with minimal tank weight. In all cases, some form of fuel pick-up must be included that allows for the aircraft to roll and pitch without lead-ing to engine fuel starvation. This is commonly in the form of a "clunk" – a heavy-weighted end that fixes to the pipe inside the tank so that it always lies at the bottom with the remaining fuel.

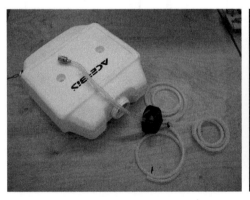 

Figure 5.12   Large UAV fuel tanks. Note clunks and fuel level sensor fitting at rear left-hand corner of one tank.

The fuel tank should be sited either close to the engine (minimizing pipe runs) or near the center of lift (minimizing the impact of fuel usage on trim). The choice largely depends on the total fuel load: if only modest endurance is needed, then placing the tank close to the engine is good practice; for long-endurance aircraft where the weight of fuel may be many times that of the engine, a location near the main wing center becomes important. If a pusher propeller configuration is adopted, sometimes it is possible to achieve both aims at once. The aircraft of Figure 5.7 has a separate fuel tank placed just behind the main spar and is thus only a few centimeters forward of the engine. In the case of the SPOTTER aircraft of Figure 5.8, a large central monocoque tank is used. Since the OS GF40 engines used on this aircraft incorporate diaphragm pumps, the lengths of fuel pipe are not an issue. For engines without such systems, it may be necessary to install a dedicated fuel pump to pressurize the fuel lines, though this can lead to dangers of engine flooding at idle settings. It is important to test all such installations at a range of throttle settings and pitch and roll angles before flying, to ensure that fuel starvation or flooding is not a problem.

The final part of the fuel system is some means of checking on the amount of fuel remaining during flight. At the most basic level, the time in the air can be recorded and compared against known fuel consumption rates from earlier flights along with visual inspection of the amount of fuel on board before commencing flight. While this may suffice for short tests and smaller aircraft, it is not workable for long-endurance systems where some kind of in-flight fuel monitor is necessary. While float systems can be used for these purposes, we prefer to use optical sensors that detect low fuel levels and trigger alarms, typically at 25% and 10% remaining levels, see Figure 5.13. These sit on metal inserts glued into the nylon and are sealed with O-rings.

5.6   Batteries and Generators

All UAVs need on-board electrical power for avionics and control; those with electric motors also need high-power supplies for propulsion. Most commonly, these needs are met by battery

Figure 5.13 SPOTTER fuel tank level sensors. One sensor lies behind the central flap in the upper wider part of the tank (just visible in the right-hand image), while the second one lies at the bottom just above the payload interface.

Figure 5.14   Engine-powered brushless generators driven directly or by toothed belt.

supplies, optionally augmented by engine-powered generators for in-flight recharging. For most purposes, we prefer the ruggedness and resilience of NiMH or LiFe batteries wherever possible. For main propulsion batteries, LiPo systems become necessary to achieve the required power densities. Great care is called for when using LiPo batteries, as they are a major fire risk if they are damaged or misused. A wide range of capacities and voltages are readily available, though the best quality high-density LiPo batteries are expensive.

With all batteries, maintenance is very important, and we keep dedicated battery use and charge logs for all flight-critical batteries. Batteries should not become discharged too far, be overcharged, or be stored in inappropriate charge states if they are to give their best. It is also important to note that batteries represent a real fire risk, and in many countries transport of battery packs can constitute a hazardous load and be subject to regulation. Ideally, charging should be carried out in dedicated environments with suitable fire control measures in place.

For long-endurance UAVs, it is important to segregate batteries to ensure that critical systems never lose power. This can also simplify wiring and reduce unwanted noise on important control lines. For example, using separate batteries for receivers/servos and ignition systems is often desirable. If a single charge of a battery will not support an entire mission, then on-board recharging becomes necessary: we use brushless motors driven directly or by toothed belts from Wet-fueled engines to provide the required power, along with suitable battery conditioning circuits to guard against overcharging, see Figure 5.14.

6

Airframe Avionics and Systems

Even without considering payloads, the avionics in all but the simplest unmanned air vehicle (UAV) can become fearsomly complex. Figures 6.1 and 6.2 show the outline avionics diagram for the SPOTTER aircraft and the level of complexity potentially present in a UAV with a maximum take-off weight (MTOW) of ≈40 kg. In this chapter, we step through the main building blocks of such systems and explain the rationale behind our designs.

6.1 Primary Control Transmitter and Receivers

Although the whole point of UAVs is that they are *not* flown directly by a pilot operating the control surfaces with manual inputs, we always equip our systems with a primary receiver that can be operated by a pilot with a standard aero-modeler hand-held transmitter. We now almost always use Futaba products for this purpose since we find them to be highly robust, well built, and with good short-range communications. Such systems are typically effective only out to ≈1 km from the pilot since their range is limited: anyway, at any greater ranges, the ground-based pilot is unable to see what the aircraft is doing sufficiently well to be able to fly it safely. This primary control system is used for initial test flights of new aircraft and for overriding autopilot control if the flight team thinks this best (e.g., in difficult cross-wind landings or to force-ditch the aircraft).

The systems we use typically make use of between 6 and 14 channels. Table 6.1 sets out a typical set of assignments, in this case for a system with twin engines and dual receivers. We commonly fit dual receivers on our more complex aircraft, both bound to the same transmitter, with a switch-over system that allows the second receiver to take control if the first loses stable connection to the transmitter. Figure 6.3 shows a typical avionics fit with dual receivers and a switch-over system. Note that because each channel has its own independent wiring using neutral, power, and signal lines, this requires a very significant quantity of cables to connect the two receivers, switch-over unit, and autopilot together. The incoming signals pass from the receiver, via the switch-over unit, through the autopilot before heading on to the relevant servos connected to the control surfaces, throttle, and so on.

Small Unmanned Fixed-wing Aircraft Design: A Practical Approach, First Edition.
Andrew J. Keane, András Sóbester and James P. Scanlan.
© 2017 John Wiley & Sons Ltd. Published 2017 by John Wiley & Sons Ltd.

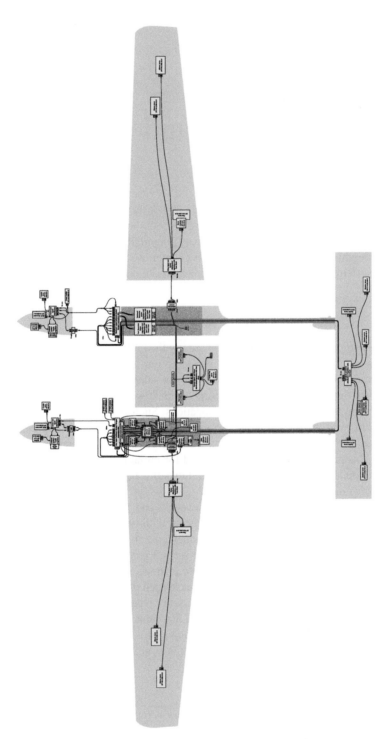

Figure 6.1 Outline avionics diagram for SPOTTER UAV.

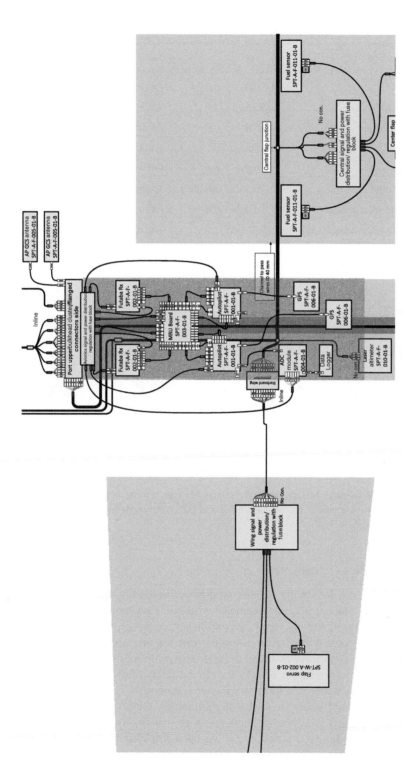

Figure 6.2 Outline avionics diagram for SPOTTER UAV (detail) – note switch-over unit linking dual receivers and dual autopilots.

Table 6.1 Typical primary transmitter/receiver channel assignments.

Channel	Function	Type	Normal position	Fail-safe position
1	Roll	Stick	As demanded	2°–3° port
2	Pitch	Stick	As demanded	Neutral
3	Throttle 1	Stick	As demanded	Idle
4	Yaw	Stick	As demanded	Neutral
5	Flaps	Switch (flaps)	Off	Off
6	Throttle 2	Mixed to throttle 1	As demanded	Idle
7	Auto-pilot engage	Switch (AUX1)	Off	Mission dependent
8	Redundant receiver system	Fail-safe	Signal < 1.5 ms	Signal > 1.5 ms

 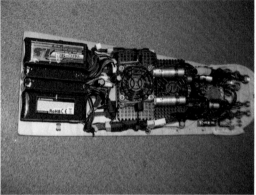

Figure 6.3 Typical avionics boards. Note the use of MilSpec connectors (the Futaba receivers are marked 1, the switch-over unit 2, the SC2 autopilot and GPS antenna 3, and the avionics and ignition batteries 4 and 5, respectively).

6.2 Avionics Power Supplies

At the most basic level, we supply the electrical power needed by the avionics from one or more battery packs. Since it is quite common for the same power supply that feeds the primary receiver and autopilot to also feed the control surface servos, the batteries need to have sufficient capacity to deal with any servo torques that may be needed in flight. The forces on large flaps can be significant and lead to large current draws by their servo systems, for example. Since the avionics supply is so critical to the safety of the aircraft, it is common to have multiple batteries with some form of fail-over system to switch between them. In addition, the operators should assess over what flight time any battery goes from 100% to 50% capacity in typical flying conditions so that they can estimate likely battery condition directly by timing the flight. Although it is possible to use telemetry systems to monitor battery voltages from the ground, this should never be the sole mechanism for assessing battery health.

We also commonly fit LED strips to our airframes that are visible from the outside so that there is a clear and unambiguous voltage status visible before takeoff, see, for example, Figure 6.4.

If batteries do not offer sufficient endurance for the UAV under consideration, some form of engine-powered generation system will be needed. This can either be in the form of a generator fixed to the main propulsion engine(s) or from a dedicated power generation system. If a sufficiently large power generation capacity is provided, it can even be possible to use electric motors for the main form of propulsion, supplied from the power generator to yield a hybrid system where batteries then only augment the operations of the airframe. When such generators are large enough, they can also be configured to allow for in-flight (re)starting of the prime mover given a suitably sized battery. Gasoline-powered engine systems with attached generators are commercially available, see, for example, those sold by the UAV Factory, Figure 6.5. We have also built our own systems based on brushless motors, see, for example, Figure 6.6.

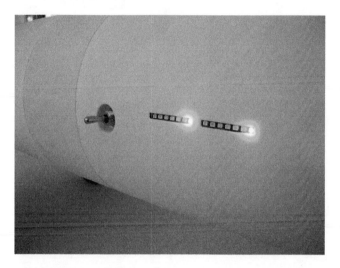

Figure 6.4 Fuselage with externally visible LED voltage monitor strips. Here, one is for the avionics system and the second for the ignition system.

Figure 6.5 Aircraft with twin on-board, belt-driven generators as supplied by the UAV Factory and a close-up of UAV Factory system.

Figure 6.6   On-board, belt-driven brushless motor used as generators.

Figure 6.7   Aircraft with a Sullivan pancake starter–generator system.

As already noted, if sufficiently large systems are used, they can generate enough torque to provide on-board engine starting, as seen in Figure 6.7, although the losses in such large systems
often create significant reductions in the propulsive power available, even when not generating
any useful electrical power.

6.3   Servos

Unless very radical systems are being used, almost all UAVs rely on flaps or articulating surfaces for pitch, roll, and yaw control. These are activated via servos that are readily available
in a bewildering range of shapes, sizes, and torques. Prices vary from a few dollars to hundreds

Figure 6.8 A selection of aircraft servos from three different manufacturers.

per servo: typical examples of the sort we use are shown in Figure 6.8, see also Table 6.2. We tend to always specify digital as opposed to analog servos, even though these are more expensive because of the more precise control they offer. http://www.servodatabase.com/ servos/all provides a comprehensive database of servos with information on sizes, types, torques, typical costs, and so on, currently listing nearly 2500 different options! Figure 6.9 shows how servo torque typically varies with servo weight. Regression of this data suggests that weight may be deduced from the required torque using $w(\text{oz.}) = 0.0489T(\text{oz.in.})^{0.7562}$.

If compact, high-torque and power, metal-geared devices are used, large surfaces can be reliably controlled by each one. If digital position control feedback servos are used, they can be relied on to very accurately hold control surfaces at demanded angles provided sufficient

Table 6.2 Typical servo properties.

Type	Make	Depth (mm)	Width (mm)	Height (mm)	Weight incl. horn and fixings (g)
Metal	Hitec	30	46	60	100
Metal SHT	MKS	22	40	40	79
Metal HT	Savox	22	40	40	67
Metal	Hitec	22	40	40	58
Metal	MG	22	40	40	57
Plastic	Futaba	20	40	35	45
Metal	MKS	10	30	35	34
Metal	Blue Bird	10	30	35	33

HT, high torque; SHT, super-high torque.

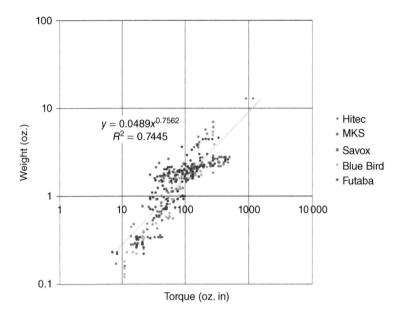

Figure 6.9   Variation of servo torque with weight for various manuafcturers' servos.

electric power is available. If, however, two servos are connected to a single surface, issues can arise when they are not completely aligned to each other, leading to the servos fighting and very heavy current flows. To avoid this, it is best to ensure that some form of compliance is designed into the system, either by making the flap slightly flexible in twist or by using the flexible rubber servo mounts typically supplied with the units when purchased. In practice, we strive to avoid ever having multiple servos on a single control since this does not in any case provide redundancy. A jammed servo will prevent a surface moving even if its mate is still working reliably. Instead, if redundancy is important, we typically divide up the control surfaces, with each part having its own dedicated servo. Thus, on the SPOTTER aircraft we use four ailerons and four elevators, each with a dedicated servo, see Figure 6.10.

The servos used to control moving aerodynamic surfaces need to be solidly attached to the airframe, as very significant loads can be experienced during flight. This means bolting down the servo body itself to a suitable set of hard points, see Figure 6.11, and also attaching the (sufficiently stiff) actuating linkage securely at the other end. We typically use two-part selective laser sintering (SLS) nylon servo mounting boxes to house our servos. The base of the box is glued into the wing with epoxy, and after the servo is fitted, a cover plate is screwed over the servo just leaving the servo arm exposed. The actuating linkage joined to the servo arm is attached to what is commonly termed a servo horn – an element that sticks out from the surface of the part being moved, see Figure 6.12. The servo horn also needs to be attached to a suitable hard point. Since our control surfaces are typically made of foam, we fit load-spreader plates between the foam and the horn, also seen in Figure 6.12. It is also possible to place the servo in the moving element and attach the horn to the main airframe: we sometimes do this when we are using all moving flying surfaces, for example.

Kinematically, servos are normally configured in what are known as four-bar chains: the servo body and the structure it is bolted to forms the first bar (often called the ground link), the

Figure 6.10 SPOTTER aircraft showing multiple redundant ailerons and elevators.

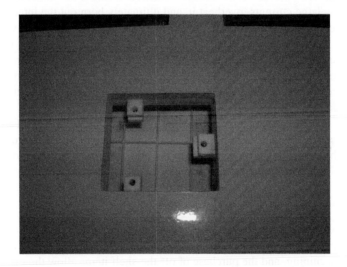

Figure 6.11 Servo cut-out in wing with SLS nylon reinforcement box.

rotating servo arm the second, the (adjustable-length metal) linkage the third, and the servo horn and moving surface the fourth. For such mechanisms to work well, the linkage needs ideally to form a rectangle with 90° corners in the neutral position. Moreover, it is good practice for the horn not to be mounted too far from the hinge line to avoid unnecessary compliance in the linkage, which can lead to control instabilities. The link between the two arms must also be sufficiently stiff to prevent buckling: on 20–40 kg aircraft, we use 3 mm diameter steel links, for example. Care must be taken to ensure that there are no clashes between moving elements, such as where rudder flaps and elevators are close to each other. It is also good practice to

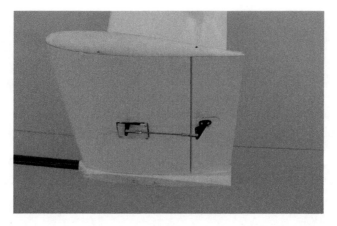

Figure 6.12   Typical servo linkage. Note the servo arm, linkage, and servo horn (with reinforcing pad).

avoid forcing servo arms into position by twisting them (and the internal servo mechanism) manually: rather, it is better to position them electrically. It is also critically important that before being powered up, care is taken to ensure that the servo will not try and adopt a position that cannot be reached because of the linkages attached. If this is not done, it is very easy to stall a servo, draw large currents, overheat it, and ultimately burn out the internal wiring. On high-power servos, this can happen in just a few seconds, resulting in expensive damage. We recommend that the servo arm or linkage should be fitted after the servo has been powered, placed in its neutral position, and then shut down again. Then, once the linkage has been attached, attention can be paid to trimming the servo end positions to match the mechanism design and control system settings (Ideally, the initial mechanical design of the linkage will allow the servo to move through the bulk of its range of movement while the control surface does likewise. It is poor practice for a servo to only operate over a fraction of its operational range while the controlled surface moves through its entire sweep. Clearly, the reverse situation is even worse where the linkage design does not permit full control surface movement.)

6.4   Wiring, Buses, and Boards

In our experience, producing the wiring looms that connect all the avionics components in a UAV airframe remains the most labor-intensive task in UAV manufacture. While 3D print-ing, numerical machining, and the purchase of off-the-shelf items allow the majority of UAV components to be gathered together for assembly, largely without human intervention, the design and manufacture of wiring looms for low-volume production runs tends to work against computer-aided design (CAD)-based design and automated manufacture. Although CAD soft-ware for setting up wiring looms is readily available, we rarely use it: instead, we rely on the use of "iron birds" to prototype cable runs around the airframe. These are plywood full-scale mock-ups of the aircraft in plan view on which all the airframe avionics components are laid out. Wire runs can then be cut to suit the configuration, and functionality tests carried out before flight-ready harnesses are produced. We adopt this approach, as against CAD-based methods, because of the very large amount design effort needed to fully specify cable har-nesses in CAD systems, which even then rarely permit full functional testing and anyway take longer to complete than the making of a physical test harness. Figure 6.13 shows the "iron bird" for the SPOTTER aircraft, and Figure 6.14 shows one of the motor generator pairs used for functional testing. When the initial test harness is complete and functionally correct, we then use it as a template to specify the final flight-ready harnesses we need (which may

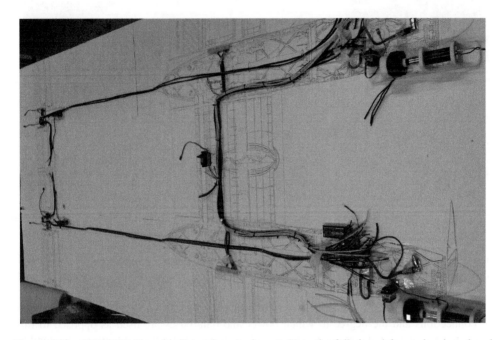

Figure 6.13 SPOTTER "iron bird" test harness layout. Note the full-size airframe drawing placed under the wiring.

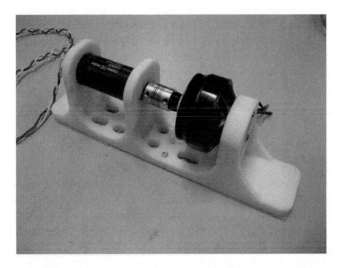

Figure 6.14 Generator and drive motor for "iron bird" testing.

Figure 6.15   SPOTTER "iron bird" with resulting professionally built harness in place.

then be drawn in a CAD environment). Figure 6.15 shows one of the resulting professionally built, flight-ready harnesses that results from following this approach, laid out on the original "iron bird."

Two key decisions in the design of wiring harnesses are the choice of plugs to be used for connections and the degree to which power and data buses are used as opposed to full wiring to each individual component. For simple airframes in lightweight aircraft, we tend to use individual three-part wires to connect each servo back to the control receiver and autopilot, with simple aero-modeler-style connections and safety clips, see Figure 6.16. For more complex and larger airfames, where a higher degree of integrity and redundancy is required, we use a power bus to supply all components and also adopt more sophisticated, self-locking plugs in which one-half can be bulkhead or baseboard mounted such as those that meet mil specs, see Figure 6.17. As yet, we have not routinely adopted one of the emerging proprietary standards for control data buses, largely to avoid committing to any particular supplier's range of components, although we have built aircraft using them to gather aircraft diagnostic data.

To locate the avionics components in an airframe, two approaches can be adopted: first, a separate avionics baseboard can be used that is populated and wired before being placed in the airframe, as already seen in Figures 6.16 and 6.17 and also in Figure 6.18. Second, when 3D-printed fuselage components are in use, sockets and fittings can be designed into the printed structure to accept the various components to be used directly, see Figure 6.19. Of course, a combination of these two approaches can be adopted: we typically locate the servos that operate aerodynamic surfaces via designed sockets in adjacent 3D-printed SLS nylon structure where this is possible, while the main set of receivers, autopilots, batteries, and so on, are attached to a dedicated baseboard and wired and tested before insertion into the aircraft. Low-volume manufacture of baseboards is readily achievable: we use digital laser cutting machines to make these trays.

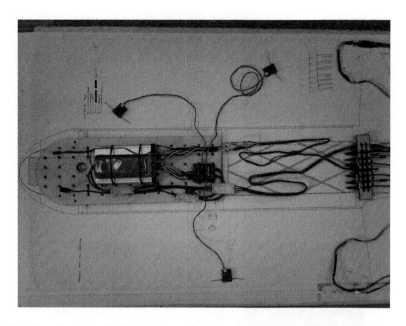

Figure 6.16 Decode-1 "iron bird" with harness that uses simple aero-modeler-based cable connections.

Figure 6.17 Baseboard with mil spec connections on left- and right-hand edges. Note SkyCircuits SC2 autopilot fitted top right with GPS antenna on top and switch-over unit in the center with very many wiring connections.

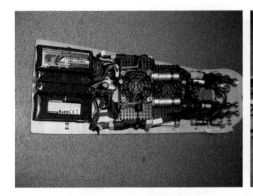 

Figure 6.18   Laser-cut plywood baseboards.

Figure 6.19   Components located directly into 3D SLS nylon printed structure. The servo is screwed to a clip-in SLS part, while the motor is bolted directly to the fuselage.

6.5 Autopilots

To allow flight beyond the line of site of the pilot, or even to automate flight within the pilot's field of vision, some form of autopilot must be fitted to the aircraft. Generally, this sits between the standard aircraft flight control receiver and the servos that operate the aircraft's aerodynamic control surfaces and throttle. In manual mode, the autopilot simply passes control signals through unchanged. When some form of autopilot control is required, the unit then substitutes its own servo commands for those coming from the receiver. The autopilot generally also has its own communications channel to a ground control station as well as connections for Pitot tubes and GPS aerials. A range of autopilots are commercially available, some with open-source codes such as those that use the Arduino family of hardware (Figure 6.20)[1] or the

---

[1] http://www.ardupilot.co.uk/.

Figure 6.20 Basic Arduino Uno autopilot components including GPS module on extension board, and accelerometer, barometer and three-axis gyro on daughter boards.

Figure 6.21 Pixhawk autopilot.

Pixhawk system (Figure 6.21)[2] and others that use proprietary approaches. The more sophisticated systems are typically subject to export control and can be as expensive as a small UAV in their own right. We mostly use the units developed by SkyCircuits[3] for our most complex aircraft, which we find to be a good compromise between capability, cost, and ease of use, see Figure 6.22. All allow the connection of a range of important sensors such as Pitot tubes, barometric altimeters, and GPS systems.

6.6 Payload Communications Systems

In addition to the primary radio link for the main control system and that for autopilot communications, it is often necessary to fit dedicated radio links for the payload system. Most

---

[2] https://pixhawk.org/start.

[3] http://www.skycircuits.com/.

(a)   (b)

Figure 6.22   The SkyCircuits SC2 autopilot (removed from its case (a), and with attached aerials and servo connection daughter board (b). See also Figure 6.3, where the SC2 is fitted with its case and a GPS aerial on top).

commonly, this is a downlink for a camera system, but on-board cameras may also need control inputs to point, zoom, and focus on objects of interest requiring some form of uplink as well. A number of commercial lightweight systems can be used for these roles, varying from low-cost analog items that will only cope with modest image resolution to high-bandwidth encrypted military-grade systems. The precise frequency band to be used and the maximum transmitted power will depend on the regulations applying in the area of operation. Most jurisdictions provide a few public frequency bands where low-powered transmissions can be made without a formal license: in the UK, transmissions can be made at 2.8 and 5.6 GHz for example, although transmission powers are strictly limited. Operators should always confirm the local regulations before transmitting on any frequency band.

Generally, all UAV radio links work on a line-of-sight basis, although this can be to a satellite if required (and affordable). Crucial to the performance of such links is the quality of the antennas used. Assuming that a steerable system cannot be fitted to the airframe, the on-board antenna will be a good-quality omnidirectional unit. On the ground, however, it is often possible to track the aircraft with a high-gain directional antenna so as to maximize reception quality, albeit at the cost of some added complexity. We have tried a range of systems, see for example Figures 6.23 and 6.24.

6.7   Ancillaries

Given suitable control systems, power supplies, wiring, and so on, the essential components of the airframe avionics systems are dealt with. On sophisticated platforms, many additional capabilities can be added, particularly if the autopilot or main control system supports a telemetry channel. It is very useful to be able to monitor fuel level, battery voltage, engine speed and temperature, and so on. This generally involves adding a suitable analog sensor, converting the output to digital form and transmitting the information back to the ground control station. Some of the more advanced aero-modeler radio control systems offer such capabilities ready-made,

Figure 6.23 A selection of professional-grade 5.8 GHz video radiolink equipment: (front) transmitter with omnidirectional antenna in ruggedized case, and (rear left to right) receiver, directional antenna, and combined receiver/high intensity screen unit.

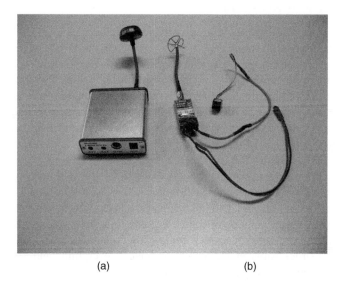

(a)   (b)

Figure 6.24 A hobby-grade 5.8 GHz video radiolink: (a) receiver with omni-directional antenna and (b) transmitter with similar unclad antenna and attached mini-camera.

with the user simply having to buy and fit appropriate sensors, see, for example, the Futaba s-bus units, Figure 6.25. Given such on-board sensors, deterioration in aircraft health can be detected before it becomes critical. This is particularly important when flying long-endurance or long-range missions where the pilot may otherwise have no knowledge of what is happening on board the aircraft.

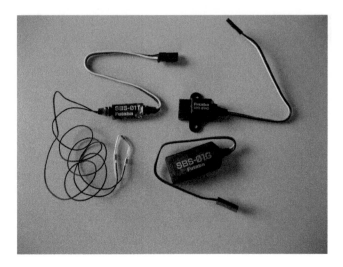

Figure 6.25   Futaba s-bus telemetry modules:(clockwise from left) temperature sensor, rpm sensor, and GPS receiver.

6.8   Resilience and Redundancy

The final issue to consider when designing aircraft systems is the degree of resilience required. Apart from using high-quality components and professionally produced and tested wiring harnesses, the best way to achieve good resilience is through the use of inbuilt redundancy so that single point-of-failure problems can be reduced as much as possible. Combined with aircraft health monitoring, this offers the best way of ensuring reliable aircraft. It should be noted, however, that the most commonly cited reason for loss of UAV systems is engine failure; so if a highly resilient aircraft is desired, it should always have two or more engines/motors and be capable of flying with the loss of one propulsion unit – something that is surprisingly rare in practice.

The main difficulty with trying to provide single point-of-failure tolerance is that ultimately the aircraft can only have one mission plan in play at any given time, and something has to decide which receiver or autopilot is in charge and connected to the aircraft controls, even if multiple receivers, autopilots, power supplies, control surfaces, and propulsion units are provided. It should also be noted that added complexity often leads to added vulnerability, so a balance has to be struck between attempting to duplicate everything and trying to maintain a simple overall avionics package. Our current practice on the SPOTTER series of aircraft, which are the most sophisticated we have built so far, is to have dual and independent engines, generators, batteries, control surfaces, receivers, and autopilots but to have a single switch-over system to decide which autopilot and receiver pair is connected to the airframe controls at any given time. This switching system thus becomes a weak spot in the system logic, for if it fails, control of the aircraft is lost.

Switch-over is accomplished in these units by plugging the control output from both (independent) control systems into the unit and then using a dedicated channel to control which system is connected to the outputs, which then feed on to the control surfaces, see again Figures 6.2 and 6.3. Generally, the switch-over units will themselves be fail-safe to a particular controller selection in the absence of a command to chose one or the other. It still means, however, that should a cable break occur between the output side of the switch-over unit and a control surface, that surface will become unusable. Worse, if the switch-over unit itself malfunctions, all control may be lost. Fortunately, switch-over units are typically simple solid-state boards with relatively few components. Even so, all electronic component boards can suffer from vibration-induced fatigue of component legs and solder joints or loosening of connections. Thus an essential part of ensuring resilience is a thorough inspection regime that examines all critical components before each set of missions and at specified numbers of flight hours.

7

Undercarriages

Undercarriage design can be something of a Cinderella subject when considering new aircraft. It may, however, dominate much of the inner wing and main fuselage layout, and for some types of UAVs become a significant design driver. Of course, not all UAVs have undercarriages, since for some applications catapult launch and belly or net-based landing can be an advantage. We have certainly built such aircraft, including SULSA (the Southampton University Laser Sintered Aircraft – the world's first fully SLS airplane), but generally it is more normal to operate from a conventional grass strip runway with a wheeled undercarriage. These are commonly of tricycle layout with soft wheels and some form of suspension. Since there is a large drag penalty associated with leaving wheels in the airflow, it can be desirable to retract the undercarriage in flight, but this brings much added complexity and additional failure modes, weight, and cost. Again, we have used retract systems, but mostly choose not to do so since our flight cruise speeds are rarely above 40 m/s (for vehicles that must operate in adverse weather conditions where significant headwinds are probable, then retract systems are all but inevitable). In addition, it is normal to have some form of steerable system, generally linked to the rudder, so that ground maneuvering is easier. Using a nose wheel arrangement protects the nose of the aircraft (and the propeller if a tractor arrangement is in use), while tail wheels generally require shorter struts and are typically lighter and simpler to design and do not obstruct any forward-looking under-hung sensors (but can restrict rotation angles on take-off and also need to allow for any overturning moments caused by the combination of propeller thrust and main wheel drag); the final choice is often dependent on the particular preferences of the design team. Figure 7.1 shows a range of typical small UAV undercarriage systems.

7.1 Wheels

Small UAV wheels are commonly made from injection-molded plastic or spun or turned aluminum and fitted with rubber tires (either solid foam or pneumatic). They can be purchased in a range of sizes, though the choice gets rapidly limited above 150 mm diameter. In the larger sizes, roller or ball bearings may be fitted as may wheel brakes. We tend to avoid both unless the aircraft is substantially over 30 kg in weight. Assuming a conventional undercarriage is to

Small Unmanned Fixed-wing Aircraft Design: A Practical Approach, First Edition.
Andrew J. Keane, András Sóbester and James P. Scanlan.
© 2017 John Wiley & Sons Ltd. Published 2017 by John Wiley & Sons Ltd.

Figure 7.1   Some typical small UAV undercarriages.

be fitted the main wheel, diameter represents a significant design variable; large wheels give better ground handling, lower rolling resistance (thus aiding takeoff), and a greater ability to deal with less than perfect runways. The penalties of large wheels are twofold: most obviously, they weigh more, and as with all aircraft components, weight is a key factor; perhaps more important, however, is the added drag large wheels generate if they are not retracted out of the airflow. It is a well-known fact of aerodynamics that circular objects have startlingly more drag than those of an airfoil shape but similar frontal area. According to Hoerner [9], typical drag coefficients, based on wheel outer diameter and width, vary from 0.12 to 0.35 as compared to streamlined shapes where values less than 0.01 are normal. Thus the drag of the

Figure 7.2   An aircraft with spats fitted to its main wheels to reduce drag.

whole fuselage is typically less than that of the main wheels when unretracted. This drag can be reduced if the wheels avoid having hollow centers (as most do) and fairings (commonly termed spats) are placed over the upper part of the wheel, see Figure 7.2.

7.2 Suspension

Although the tires fitted to the wheels will provide some form of cushioning during ground roll, it is also normal to add some form of suspension between the wheels and the main aircraft body. At its simplest, this will take the form of some compliance in the wheel struts, either in the form of a deliberately wound spring element or by relying on bending of the main strut in the form of a compliant beam subject to a tip load. The problem with such a simple approach is that very little energy dissipation can be achieved by using only the elastic behavior of the strut. To get better dynamics, it is necessary to add some form of shock absorber; this may take the form of a rubber compression element or be as complex as coil over gas strut of the form now commonly seen on off-road bicycles, although the more complex devices are again seldom warranted on lighter aircraft. Figure 7.3 shows a nose wheel and strut with integral suspension elements. Figure 7.4 shows a simple spring-based tail wheel suspension.

7.3 Steering

With tricycle undercarriage systems, the small nose or tail wheel is generally steerable and also provided with modest self-centering by using an arrangement with wheel caster, also seen in

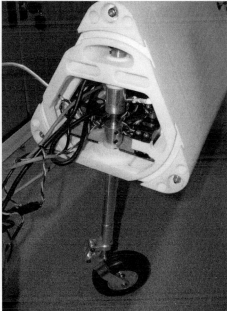

Figure 7.3 Nose wheel and strut showing suspension elements, main bearings, control servo, and caster.

Figure 7.4   Tail wheel showing suspension spring.

Figure 7.5   Nose wheel mechanism with combined spring-coupled steering and vertical suspension spring.

Figure 7.3. This is typically connected to the steering servo via stiff coupling springs rather than the solid linkages used elsewhere, because wheels are subject to many impact loads during ground roll and it is desirable to protect the servo mechanism from such loads. Figure 7.5 shows the internal workings of a simple spring-coupled nose wheel steering system which also has a vertical suspension spring. When dealing with nose wheels, it is not uncommon to find the wheel strut to be quite long so as to provide the correct attitude during takeoff; this inevitably means the strut is subject to quite large bending loads and so it and its bearings in the fuselage must be sized accordingly. On the other hand, it is wise to avoid overly stiff nose wheel struts because during heavy landings it can be desirable for the strut to bend or even collapse if this reduces damage to the main fuselage or limits deceleration in any sensitive on-board systems such as cameras or other sensors.

7.4 Retractable Systems

As already mentioned earlier and in the previous chapter on wings, it is sometimes desirable to reduce drag by retracting the undercarriage, particularly the main wheels. We rarely do this

Figure 7.6 UAV with a pneumatic, fully retractable undercarriage system. Note also the nose camera that has been added to the aircraft shown in the image with undercarriage retracted.

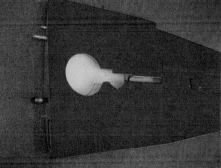

Figure 7.7 Details of fully retractable undercarriage system.

because of concerns of cost, weight, and particularly reliability. If part of the undercarriage fails to deploy for landing, it is almost inevitable that serious airframe damage will result. However, Figure 7.6 shows one UAV that we built with fully retractable undercarriage, while Figure 7.7 shows the details of its wing-mounted gear. Significant reduction in drag can be achieved by such means, which can more than compensate for the increased aircraft weight. Such systems are typically pneumatically powered or use electric motors.

Part III

Designing UAVs

Part III

Designing UAVs

8

The Process of Design

As has already been pointed out, the whole purpose of design is to achieve a set of goals while meeting various constraints, usually in an iterative process proceeding through the various stages of design described earlier. The goals in unmanned air vehicle (UAV) design are related to mission and payload while the constraints are those associated with the various ways aircraft can go wrong. In this part of the book, we set out in more detail how designers can use analysis to achieve good designs, that is, those that flexibly achieve the mission goals and avoid the common forms of failure that all aircraft are subject to.

A typical aircraft design project will go through the stages illustrated in Figure 8.1. For a very low cost, or student, fixed-wing UAV project, the aircraft is generally defined by perhaps 10 parameters at the conceptual stage, several tens of parameters at the preliminary design stage, and hundreds of parameters at the detailed design stage. There is an exponential growth in the information content. This is one of the reasons why design has to be carried out in a systematic manner to ensure that reliable decisions are made before moving onto the next stage in the design process. A detailed description of the overall design-stage logic is given in Appendix A.

Having decided on the goals and constraints and selected the toolsets to be used, we start with concept design based around constraint analysis, spreadsheets, and simple sketches before moving on to physics-based analysis applied to computer-aided design (CAD) geometries using computational fluid dynamics (CFD) and structural finite element analysis (FEA). We then describe lab-based experimental work before concluding this part with a chapter on detailed design.

8.1 Goals and Constraints

At the most basic level, a UAV aims to carry a payload over a specified distance or for a specified length of time, often at some desired cruise speed and height. The payload may be a simple dumb mass or may be active in some way, requiring power supplies or the ability, in the widest sense, to "see" its environment and perhaps communicate with the ground or other aircraft. It is also commonly the case that the precise payload and mission details will vary

Small Unmanned Fixed-wing Aircraft Design: A Practical Approach, First Edition.
Andrew J. Keane, András Sóbester and James P. Scanlan.
© 2017 John Wiley & Sons Ltd. Published 2017 by John Wiley & Sons Ltd.

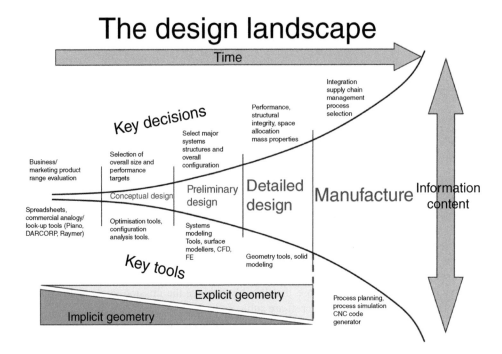

Figure 8.1 Explosion of information content as design progresses.

over the life of the design and even the life of an individual airframe. Thus designers are also keenly interested in the impact of payload and mission changes on aircraft performance. It is often useful to carry out concept evaluation using a range of missions and payloads so as to ensure that a suitably flexible design emerges. These can then be considered using a raft of multiobjective decision-making aids so as to identify the concepts worthy of more detailed and time-consuming assessment. We have, for example, used event-based simulation tools to assess maritime search aircraft in realistic scenarios. It must be remembered though, that all vehicle design leads to multiple goals and that whatever aids are used, the eventual trade-off choices between performance across these goals will have to be taken by the designer, usually in collaboration with the eventual end users. It is also important to note that many goals are expressed in the form of constraints: for example, the maximum speed must be at least 40 m/s, or the service interval must be at least 100 h of flight. The most commonly encountered user goals in UAV design (in no particular order) are as follows:

- maximum speed
- maximum operational ceiling
- maximum range
- maximum endurance
- maximum service interval
- maximum payload mass
- minimum purchase and service costs.

To optimize one or more of these goals, the designer will probably work with more technical quantities, such as wing and overall airframe lift to drag ratios, stall speeds, wing aspect ratio and loading, and so on.

8.2   Airworthiness

While it is natural for the users to focus on their mission goals, it will go without saying that they also expect a serviceable and reliable aircraft to be built. This will involve satisfying a whole raft of constraints that the designer must be aware of even if they do not at first occur to the eventual user. For example, it is no good producing a really low-drag design that on landing needs a very long runway because no brakes have been fitted or a very smooth surface because the landing speed is high, yet the user only has access to much less ideal facilities – the user simply assumes the aircraft must be able to land safely and without damage. The general requirement that an aircraft be fit for flight is normally termed "airworthiness." There are various guides to establishing the airworthiness of new aircraft; perhaps the most relevant to the aircraft considered in this book is the NATO document STANAG 4703.[1] This identifies a series of essential requirements (ERs) broken down into seven main headings:

- *ER.1 System integrity.* System integrity must be assured for all anticipated flight conditions and ground operations for the operational life of the unmanned air system (UAS). Compliance with all requirements must be shown by assessment or analysis, supported, where necessary, by tests;
- *ER.1.1 Structures and materials.* The integrity of the structure must be ensured throughout, and by a defined margin beyond, the operational envelope for the UA, including its propulsion system, and maintained for the operational life of the UA;
- *ER.1.2 Propulsion.* The integrity of the propulsion system (i.e., engine and, where appropriate, propeller) must be demonstrated throughout, and by a defined margin beyond, the operational envelope of the propulsion system and must be maintained for the operational life of the propulsion system;
- ER.1.3 Systems and equipment;
- ER.1.4 Continued airworthiness of the UAS;
- ER.2 Airworthiness aspects of system operation;
- ER.3 Organizations.

Those familiar with the Federal Aviation Authority requirements[2] or the European Aviation Safety Authority certification specifications[3] will recognize significant chunks of this document. The basic aim is to set out an "acceptable means of compliance" by which a new aircraft may be deemed airworthy. This will be made up of "detailed arguments" and "means of evidence." STANAG 4703 gives considerable detail in a series of tables for each of the seven areas it covers.

---

[1] NATO Standard AEP-83, Light Unmanned Aircraft Systems Airworthiness Requirements: https://www.gov.uk/government/uploads/system/uploads/attachment_data/file/391827/20140916-STANAG-4703_AEP-83_A__1_.pdf.
[2] Such as FAR-23 Small Airplanes Regulations, Policies & Guidance, Part 23, Airworthiness Standards: Normal, Utility, Acrobatic, and Commuter Category Airplanes.
[3] Such as CS-23 for Normal, Utility, Aerobatic and Commuter Aeroplanes.

For example, Essential Requirement ER.1.1.2 says: "The UA must be free from any aeroservo-elastic instability and excessive vibration." The detailed arguments required for this are set out as

> Aeroservoelastic effects – A rational compelling set of arguments must be provided to the satisfaction of the Certifying Authority, in order to show that the UA is free from flutter, control reversal, and divergence in all configurations. A margin ≥ 1.22 VD should be applied. Simplified analytical or computational conservative methods may be used. Though specific flutter flight tests with appropriate excitation are not mandatory, flight tests survey should not reveal excessive airframe vibrations, flutter, or control divergence at any speed within the design usage spectrum as per UL.0.

while the acceptable means of evidence are listed as "A combination of assumptions, tests and analyses."

Typical constraints that must be checked in the earliest stages of design include the following:

1. Sufficient flight stability and control authority to carry out desired maneuvers (static margin in pitch is a key factor here);
2. Landing speed low enough to enable repeated damage-free landings on the available airstrips;
3. Suitably reliable propulsion system with sufficient installed power and thrust to be able to take off and maneuver in likely adverse conditions;
4. Suitable fuel tanks can be fitted that will safely contain the required amount of fuel and supply it in a controlled and reliable manner to the engine (while not upsetting center-of-gravity (CoG) requirements);
5. Structurally sound airframe able to withstand likely gust and maneuver loads and to avoid aeroelastic problems such as flutter, over the life of the aircraft;
6. Avionics capable of maintaining safe command and control during missions (sufficient radio range, autopilot capability, etc.);
7. Sufficient on-board battery life or generator capacity to meet mission endurance achievable from fully fueled aircraft;
8. Repeatable manufacturing processes making use of suitable materials that make due allowance for fatigue, corrosion, contact with fuels, maintenance, repair, and so on;
9. Construction and assembly standards robust enough to ensure reliable operation given user capabilities;
10. The possibility of the airframe to be sufficiently broken down for ground transport and storage.

These constraints arise mainly from a failure mode analysis of the aircraft, essentially a list of the things that might go wrong if sufficient design care is not taken.

8.3 Likely Failure Modes

It is a fundamental (and often legal) duty of any engineer to ensure that products are safe and fit for purpose (in that order). Given that there are no crew or passengers to be concerned with, here safety means that of other aircraft and people and structures on the ground. This directly leads to two areas of failure that the UAV designer must consider from the outset: first, how to sense and avoid other aircraft, and, second, how to avoid uncontrolled descent and ditching (it being assumed that normal ground maneuvers take place in strictly controlled areas where the safety of operators and spectators can be ensured). In what follows, we consider how things might go wrong and compromise the safety of operators or others who find themselves in the vicinity of the UAV.

8.3.1 Aerodynamic and Stability Failure

Aircraft flight is now so commonplace that it is worth recalling that it is only a little over a 100 years ago that powered flight was first achieved. One of the most important aspects of early flights was learning to deal with the subjects of aerodynamics and control. Just because an aircraft has wings and control surfaces, it will not necessarily fly safely, or indeed at all. The wings must be large enough to generate sufficient lift to carry the UAV, while drag must not be so great as to prevent flight given the installed power. Aerodynamic stability must be assured during flight, most obviously in pitch and roll, though there are numerous ways that aircraft can exhibit unstable flight behavior, including those linked to structural flexibility. This must allow for any changes in fuel or payload weight or their effects on the longitudinal CoG.

At the earliest stages of design, simply estimating likely wing area and installed power from similar aircraft for the desired landing and operational speeds will provide a start point. For conventional layouts, ensuring sufficient dihedral or a high wing design combined with a CoG at or in front of the main wing quarter chord point will go a long way toward ensuring stable flight. Refinements then include ensuring that the control surfaces are correctly placed and sufficiently large to allow typical maneuvers and that the main spars or equivalent structure will prevent divergence and flutter. Again, working from previous designs will help here, though it must be noted that UAV landing speeds are often much lower than for conventional manned aircraft, and this tends to mean that wings, empennage, and control surfaces may need to be on the large side – it is obviously better to have too much stability and control authority than too little.

If the aircraft is underpowered, it will not be able to take off (or if catapult-launched, maintain flight). Although this will be a very grave problem for the designer, it rarely places anyone in danger. This is a failure of function, not of safety. If the longitudinal pitch stability is insufficient, it is likely that the aircraft will stall and crash soon after takeoff. Given a suitably long test runway, again this will be unlikely to hurt anything but the designer's pride. The inability to provide adequate control once airborne in the face of gusts or any desired maneuver is a much more serious issue. It is also far from simple to guarantee adequate maneuver authority by direct calculation using the laws of physics. Instead, the designer must rely on following successful similar aircraft and the design guides established by others, wind tunnel tests (if a suitable tunnel is available), and flight trials. It is thus an almost mandatory requirement when producing any new aircraft that it be put through a controlled set of flight trials before it is declared safe for use. Such trials start with low speed and simple activities, proceeding through the entire flight envelope until the test pilot is sure that all is behaving as expected. Any failure of control during test flying may lead to an unplanned ditching of the aircraft. Tests should thus be conducted only over a suitably controlled test field where access is strictly limited to those involved in the flight program. Typical details of suitable test programs are detailed in Chapter 20.

8.3.2   Structural Failure

For small and medium-sized UAVs, ground handling damage to the airframe is as likely to occur as structural failure in flight. For larger aircraft ground damage becomes less of an issue since such aircraft are basically too large for manual lifting. In either case, a series of take-off, flight, and landing loads must be considered when designing the structure. Perhaps the most obvious structural calculation needed is to size the main spar or equivalent monocoque wing structure used to lift the UAV while airborne. Since the aerial environment is by its very nature a random one, designers typically design against "gust and maneuver loadings" expressed in the form of multiples of gravitational acceleration. Thus one might design a wing spar to deal with 5g loadings, that is, a simple weight load equal to 5 times that of the aircraft, assumed to be lifted by the pressure differential on the wings. Euler–Bernoulli beam theory will then allow one to assess likely the stresses and deflections in the wing. If the deflections are too great or the stresses too close to failure stresses, the design can be iterated to strengthen it. Similar estimates can be made for tail and fin loads linked to likely lift coefficients or inertia loads. Since carbon fiber tubing is such a structurally efficient and low-cost material, it is rarely a problem to provide sufficient strength in a design, even using simple booms and spars. More difficult is dealing with the way such elements are integrated into the rest of the aircraft structure – all highly stressed junctions will need careful detailing later on in the design process. One advantage of monocoque approaches is that local stress raisers are much less likely, though, conversely, it is then no longer simple to assess likely stresses at the outset of the design process.

After considering the main flight loads, the next most important loading that occurs during operation will be landing loads on the undercarriage. Because UAVs are typically smaller than manned aircraft and commonly use less well prepared runways, these loadings are typically much larger compared to flight loads than for manned aircraft. It is because of this that is usual to try and have rather low landing speeds for UAVs. In our experience, one of the main design choices made for any new small or medium UAV is the landing speed. If this can be held below 15 m/s, or even 12 m/s, then landing loads will be very much reduced. Even so, we have measured acceleration of as much as 200 m/s^2 on landing gear on a 25 kg aircraft. Thus the landing gear on such an aircraft may have to tolerate loads of half a ton, if only very briefly. These loads can be such a problem during heavy landings that on some designs we have included mechanical fuses that are deliberately sacrificial and protect the rest of the airframe structure by absorbing energy during the impact and which can then be easily replaced for subsequent flights. In any case, a key design aim will be to structurally connect the main landing gear to the wings and the heavier elements in the fuselage so as to effectively transmit deceleration loads. Again, much of this will be considered during detail design, but due allowance must be made to include sufficient space and weight budget in the initial concept design. Large wheels, suspension systems, and dampers all add mass and cost to a design, as well as possibly impacting on aerodynamic performance if not retracted.

The aircraft structure may also need to cope with fuel and oil spills, exhaust products, rain, and possibly salty environments. For example, we use foam to construct many aerodynamic surfaces but these are often not resistant to gasoline or exhaust products from internal combustion (IC) engines. We thus have to ensure that any exposed areas are clad in suitable protective layers of film of fiber-reinforced plastic. Aluminum parts will be vulnerable to salt damage if not washed off after flights over the sea; steel parts will be subject to rusting unless suitable stainless grades are used. Many aerospace materials are also vulnerable to fatigue failure stemming from vibrations or repeated loadings: engine mounts and wheel axles are two areas rather prone to such problems (particularly if of welded construction). Suitable care must be taken to guard against all these forms of structural failure.

Once the operational issues have been dealt with, consideration must be given to ground handling. While large UAVs will need treating just like manned aircraft and may need tow trucks and hangars, small and medium UAVs are typically packed up into shipping or storage crates after use. This may well involve operators undoing fixings and withdrawing wings, tail parts, and perhaps other components. Given that any structural junction will try and avoid play and slop, this may require noticeable effort, during which time a firm hold of the airframe may be necessary. To avoid the possibility of damage, suitable hard points will need designing into the UAV. Furthermore, any storage crates must be designed to hold delicate structures during transit, again by supporting at known hard points and/or by the liberal use of foam packaging. We typically have large flight cases custom-made for our more valuable or delicate aircraft. It is surprising how much damage can be accrued by trying to split down a UAV and transport it unprotected in an everyday automobile. Finally, it is worth noting that when unpacked and assembled and before flight, aircraft should be placed (and tethered) in locations where people will not accidentally trip over them or stumble onto them, or where they can be caught by unexpected gusts of wind.

8.3.3   Engine/Motor Failure

If the main propulsion system ceases to generate sufficient thrust, the aircraft will obviously be forced to land or ditch. Analysis of UAV losses shows that engine failure is a very common cause of accidents. Perhaps the simplest way to avoid such problems, and the one adopted by large civil airliners, is to have multiple engines such that if any one fails, the aircraft can still land safely. If this is not done, the best that can be hoped for if the propulsion fails is to glide down and ditch in a controlled manner. If the UAV is away from a suitable landing site, this will be a major safety hazard. In our experience, small IC engines are much less reliable than their bigger brothers found in light aircraft. They typically have much cruder fuel metering systems, smaller spark plugs, and often very small valve seat areas. Conversely, rare-earth brushless electric motors can be highly reliable, but then the issue becomes one of ensuring sufficient power supplies. Aside from simply fitting dual, independent engine/motor systems, one possible combination is to use an IC engine/generator to provide power to twin electric propulsion motors, backed up by a high-power density LiPo battery for emergency landings if the IC engine fails. Duplicated systems are always expensive and heavy but do offer much enhanced reliability. When flight beyond the line of sight is being planned, regulators may well insist on such approaches being followed.

8.3.4   Control System Failure

Clearly, safe flight operations will require the control system to maintain adequate authority over the UAV at all times. Apart from failures due to hardware faults in the avionics, the most likely cause of control failure will be due to communications failures with the operators. Another common cause of accidents is running out of electrical power: if batteries are being used to support the avionics, particular care must be taken to ensure their condition before and during each flight, both in terms of their state of charge and also their overall health and ability to maintain charge. This will include making choices about capacity and battery chemistry to be used. We tend to avoid using LiPo-type batteries for control systems, preferring less power dense but more reliable types such as LiFe systems. At the initial design stage, the designer will wish to plan the desired degree of redundancy and resilience to control failures, so as to correctly budget for weight and cost impacts on the design.

Primary Tx/Rx Failure

All the UAVs we build contain aero-modeler-based primary control systems to enable manual flight of the aircraft within the line of sight of the pilot. Since the systems used are mass produced and subject to very extensive user testing, these systems are typically very reliable provided they are carefully installed and they are used within the ranges specified by the manufacturers. In the UK, operators of aircraft using such systems are required to carry out range tests at reduced power settings before flight operations to ensure that radio communications are unlikely to be interrupted. The user must also set the fail-safe mode of the aircraft so that if communications fail, the aircraft enters a known and safe mode of flight, typically at low power with a shallow spiraling descent. This aims to prevent the aircraft flying away from the pilot and also to reduce flight speeds to minimize any damage caused in the event of uncontrolled ditching. To add to resilience, it is also possible to fit dual primary control receivers to the aircraft so that if one receiver fails, control is automatically handed to the second, though this, of course, makes the switching unit a single point of failure.

Control Surface Failure

Assuming that the primary control system is functioning, the next area of concern to be considered is operation of the control surfaces – these are usually servo-operated flaps on the wings and empennage. Failure of the control surfaces generally arises because of servo or linkages failing (assuming there is sufficient electric power available to the control system). If a servo jams in a significantly deflected position, control of the aircraft can be severely affected. If high levels of resilience are required, it may be necessary to fit multiple control surfaces to deal with such situations: the 2SEAS aircraft, for example, has four main wing ailerons and four elevators so that if any one is jammed in an adverse position, the aircraft can still fly safely, see Figure 8.2.

Autopilot Failure

Since the main point of UAVs is generally to fly under some sort of automated control, it is natural to fit an autopilot to the UAV. Such systems generally take over operation of all the control surfaces and navigate by using predefined global positioning system (GPS) way-points. The failure modes of these systems can be many and varied, ranging from simple mechanical problems to subtle software bugs. However, at the initial design stage the designer is generally seeking to just select an appropriate autopilot system so as to assess cost and communications requirements. Approvals to fly beyond the line of sight will require very significant discussions with the local aviation authority – lying outside the scope of this book – but they will heavily

Figure 8.2   2SEAS aircraft with redundant ailerons and elevators.

influence the final choice of system. Except for the very smallest UAVs, weight and power requirements are usually much less of a concern. Aside from selecting a high-quality system, delivering reliability requires ensuring that all aerials are well sited and that the gyros and accelerators are not exposed to significant levels of vibration. IC engines can prove to be quite troublesome in this regard, and we find great care has to be exercised in terms of engine and autopilot mountings to prevent vibration levels on the autopilot causing sensor difficulties. We typically fit multiple isolation mounts to our engines and suspend sensitive parts of the autopilot on dedicated and tuned antivibration systems, see Figures 8.3 and 8.4. Adding such treatments can take up significant amounts of space and add noticeably to the total weight.

Figure 8.3   Autopilot system on vibration test.

Figure 8.4   Treble isolated engine mounting.

8.4   Systems Engineering

In the aircraft industry, the collective phrase for the information pertaining to the aircraft is "product definition." As the product definition evolves, there are a number of challenges for the design team such as the following:

- how to manage the breakdown of tasks;
- how to maintain coherency of the product definition;
- how to manage the interfaces between parts of the aircraft.

Even a typical student group design project can thus become complex and need a disciplined approach.

The aerospace industry has, over the decades, developed a methodology called "systems engineering" to help manage complexity. A detailed explanation of systems engineering is beyond the scope of this book, but there are some core principles that should be adopted to help ensure that a design project progresses efficiently. The important principles include the following:

- creation of a clear work-breakdown structure (WBS);
- documentation of interface definitions;
- clear allocation of responsibilities;
- requirements flowdown;
- compliance testing.

8.4.1   Work-breakdown Structure

Essentially, the goal of creating a WBS is to break the overall task into manageable pieces. An example WBS for military UAV projects is given in Figure 8.5. These are often broken

AIRCRAFT SYSTEM

Air vehicle	Payload	Ground/host segment	UAV software release 1.....n	UAV system integration, assembly, test and checkout	System eng.	Program mang.	System test and evaluation	Training	Data	Peculiar support equipment	Common support equipment	Operational/ site activation	Industrial facilities	Initial spares and repair parts
Airframe	Payload integration, assembly, test and checkout	Ground segment integration, assembly, test and checkout					Develop. test and evaluation	Equipment	Technical publications	Test and measurement equipment	Test and measurement equipment	System assembly, installation and checkout on site	Construction/ conversion/ expansion	
Propulsion	Survivability payload 1.....n (specify)	Ground control systems					Operational test and evaluation	Services	Engineering data	Support and handling equipment	Support and handling equipment	Contractor technical support	Equipment acquisition or modernization	
Vehicle Subsystems	Reconnaissance payload 1.....n (specify)	Command and control subsystem					Mock-ups/ system integration labs (SILs)	Facilities	Management data			Site construction	Maintenance (industrial facilities)	
Avionics	Electronic warfare payload 1.....n (specify)	Launch and recovery equipment					Test and evaluation support		Support data			Site ship/ vehicle conversion		
Auxiliary equipment	Armament/ weapons delivery payload 1.....n (specify)	Transport vehicles					Test facilities		Data depository			Sustainment/ interim contractor support		
Air vehicle software release 1.....n	Payload software release 1.....n	Ground segment software release 1.....n												
Air vehicle integration, assembly, test and checkout	Other payload 1.....n (specify)	Other ground/host segment 1.....n (specify)												

Figure 8.5 Typical military UAV work-breakdown structure interface definitions, from MIL-HDBK-881C for UAVs.

down in one of three ways: by "object," or by "system," or by task or deliverable. An object breakdown, for example, might break the airframe into "front fuselage," "center fuselage," and "rear fuselage." In other words, it breaks a physical artifact into logical regional "chunks." A "systems" breakdown might include "electrical power" that might not be restricted to a particular region. A good WBS includes not only the artifact being designed but the contextual deliverables as shown in Figure 8.5. A WBS element that is very relevant to student UAV projects is "test and evaluation," which is also shown in Figure 8.5. Even a student UAV team needs to think about how the subsystems and overall system will be tested.

8.4.2   Interface Definitions

As responsibility for each WBS element is allocated to members of the team, it is important that the boundaries and connections between each element are unambiguously defined. The wing-to-fuselage joint, for example, is a typical interface where different team members might be responsible for the wing and the fuselage design. A good interface definition will ensure that both team members understand the geometry as well as the structural and functional interface requirements. In this example, there may also be a third team member who might be responsible for electrical systems, and the wing-to-fuselage interface definition will clearly need to include electrical connections. Examples of diagrams for the SPOTTER aircraft of the sort that can help define clear interface definitions have been given earlier in Figures 6.1 and 6.2.

It is often worth ensuring the each interface definition is owned by a specific team member. There is often a logical allocation of this ownership. For example, in the wing-to-fuselage interface mentioned above, it might be more logical for the wing designer to "own" the interface definition, as it is intimately connected with wing design decisions such as spar diameter and root chord.

8.4.3   Allocation of Responsibility

As the team progresses into the preliminary and detailed design phases, it is very important that there are clear allocations of responsibilities. This is essentially a project management function, but as the product definition becomes more detailed, this allocation becomes more important. Often, people use a "responsibility allocation matrix" (RAM), an example of which is given in Table 8.1. This example adds the sophistication of several roles as follows:

- responsible means actually undertaking the work;
- accountable means a third party who might check the work and be answerable for it;
- consulted means someone on the team who needs to be involved in decision making for that task;
- informed means a team member who needs to be alerted to progress/completion of a task.

8.4.4   Requirements Flowdown

All projects have a purpose, and even student projects will have hypothetical "customers." In general, projects have "stakeholders." In student projects, the participants themselves are

Table 8.1 Example responsibility allocation matrix for a maintenance team.

Tasks	Maint. supervisors	Maint. analyst	Maint. planner	Maint. technician	Maint. support	Rel. specialist	CMMS proj. engr.
Inputting failure data	A	C	I	R		C	C
Work order completion	R	C	C	C	A	I	I
Work order close out	C	R	C		I	I	A
QA of failure data input	C	R	I	C	I	C	A
Analysis failure reports	C	C	I	C	A	R	I
Maintenance strategy adjustments	C	I	I	C	A	R	R
Implementing new strategies	R	I	R	C	A	I	I

Responsible (R), the Doer; accountable (A), the buck stops here; consulted (C), in the loop; informed (I), kept in the picture.

stakeholders who probably want to get the best mark possible as well as gaining relevant knowledge and experience. The supervisor might want a deliverable as part of a wider academic research roadmap.

In order to ensure that the needs of the stakeholders are addressed, it is worth documenting the goals, requirements, and deliverables at an early stage, that is, the design brief. In the discipline of systems engineering, these requirements are then broken down logically and functionally such that each of the items in the WBS has a clear set of auditable requirements that "flow down" from the top level requirements, as shown in Figure 8.6.

8.4.5 Compliance Testing

In order to demonstrate that the overall system meets the requirements, it needs to be tested. Waiting until the overall system has been built is a very high risk way of carrying out compliance testing, and hence systems engineers tend to follow the "V" model of systems engineering[4] illustrated in Figure 8.7.

The "V" model ensures that engineers (a) know what the requirements flow down for their subsystem or part is and (b) think about how they can demonstrate compliance (i.e., test their contribution). This should result in a lower risk overall system. When the overall systems test takes place, it should involve a set of already tested subsystems.

---

[4] https://en.wikipedia.org/wiki/V-ModelFederalHighwayAdministration.

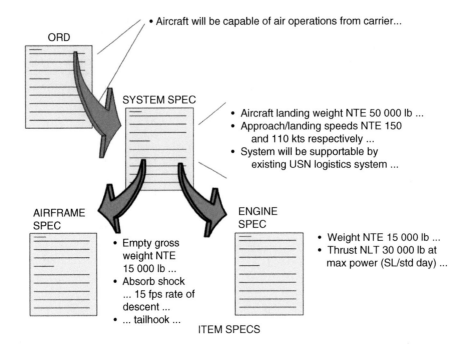

Figure 8.6   Example military system requirements flowdown [13]. Defence Acquisition University.

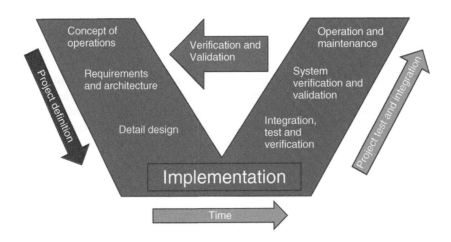

Figure 8.7   Systems engineering "V" model.

8.4.6   *Cost and Weight Management*

For an aerospace team-design activity, it is extremely important to manage holistic parameters such as weight and cost very closely. It is often useful to allocate management and monitoring of weight and cost to a single individual.

Weight and cost are aggregating characteristics and can easily exceed aircraft-level targets unless a very disciplined approach is taken. They have to be managed continuously throughout the design process by using different techniques at each design stage depending on the level of detail available. During the concept design stage, for example, parametric or empirical methods are necessary. This involves the use of actual historical data for products that are sufficiently similar to the concept being worked on. During the preliminary design phase, the product definition becomes more refined and allows some of the parts and subsystems to be based on "bottom-up" calculations. This might involve the use of supplier cost and weight figures for "bought-out" parts and subsystems as well as estimates of structural parts based on emerging geometry and material selection.

Finally, in the detail design phase a full product definition starts to emerge and accurate "bottom-up" estimates can be developed. Even at this stage it is hard to make calculations which are 100% accurate, particularly for weight estimates. For example, Figure 8.8 shows a plot of the predicted weight of the SPOTTER UAV calculated during the detail design phase against the actual measured weight. This shows that most of the time engineers underpredict weight mainly because of missing detail. Hence it is vital to use a mix of "bottom-up" estimates which are validated against relevant "top-down" historical actual data.

Figure 8.9 shows the same data plotted in a pie chart form. For anyone developing a similar class of UAV to SPOTTER, the actual data is very useful to help set weight targets during the concept design phase of a new design.

Figure 8.10 shows an extract from the SPOTTER weight breakdown table. This illustrates a useful approach whereby "traffic-light" color coding is used to indicate the fidelity of the estimate. Green indicates that the figure used is very accurate/reliable (ideally the actual mass of a part/subsystem validated using calibrated weighing scales). Amber is an estimate that is reasonably accurate (ideally within 10%). Finally, red is used to signify rough estimates which

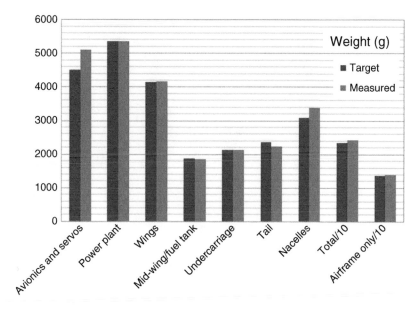

Figure 8.8   Weight prediction of SPOTTER UAV.

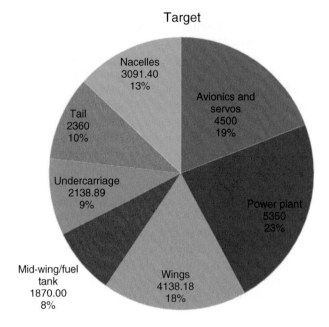

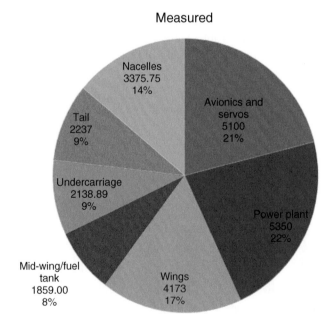

Figure 8.9 Pie chart plots of SPOTTER weight.

	Note	Mass (g)	QTY	Total (g)	Unit cost	Total
Power plant						
Engines + generators	Supplier data	2314	2	4628	£ 2800.00	£ 5600.00
Propeller	Measured	150	2	300	£ 40.00	£ 80.00
Exhaust		78	2	156	20	40
Ignition unit		133	2	266	50	100
			Total	5350		£ 5820.00
AIRFRAME						
Wings						
Main spar	Measured	447	2	894	£ 51.00	£ 102.00
Foam main	Rough estimate	331	2	662	£ 80.00	£ 160.00
Aileron foam	CAD model	30	2	60	£ 25.00	£ 50.00
Flap foam	CAD model	61	2	122	£ 30.00	£ 60.00
Flap spar	CAD model	30	2	60	£ 20.00	£ 40.00
Nylon parts	Rough estimate	-	14		£ 26.12	£ 365.66
Other	Rough estimate		2			
			Total	2798		£ 827.66
Mid-wing/fuel tank						
Main spar	Measured	228	1	228	£ 58.60	£ 58.60
Rear spar	Measured	133	1	133	£ 39.20	£ 39.20
Fuel tank	Rough estimate		1	1000	£ 985.00	£ 985.00
Fuel tank rear	CAD model	30	1	30	£ 13.09	£ 13.09
Foam flap	CAD model	100	1	100	£ 20.00	£ 20.00
Other	CAD model	150	1	150	£ 50.00	£ 50.00
			Total	1641		£ 1165.89

Figure 8.10   Example weight and cost breakdown.

are a source of significant uncertainty (and therefore risk). An important ambition during the preliminary design stage should be to eliminate all significant mass red entries. Similarly, during the detail design phase all the significant amber entries should be eliminated.

8.4.7   Design "Checklist"

Before committing to the geometry creation (sometimes known as embodiment) stage of design, it is worth going through a checklist to ensure that the detail design result is well founded. At this stage of design, the overall system should have been broken down into logical subsystems for which subteams or individuals have been tasked. Broadly speaking, in the context of a fixed-wing UAV, each subsystem will be composed of the following elements or combinations of them:

- primary structure (in manned aircraft this is defined as "The structure that carries flight, or ground loads, and whose failure would reduce the structural integrity of the airplane or may result in injury or death"[5]);

---

[5] http://www.discovery-aviation.com/wp-content/uploads/135A-970-100-51ir_Stanard_Practices-Structure.pdf.

- secondary structure (again in manned aviation the same source defines this as "… not primary load carrying members and failure would not reduce the structural integrity of the airframe. Such components do not form an integral part of the airframe, e.g., access panels.");
- mechanisms such as hinged control surfaces and actuators;
- functional electrical, pneumatic, hydraulic, or fuel subsystems or parts such as fuel storage and distribution or electrical power distribution subsystems.

9

Tool Selection

When using computational methods to underpin the design of any reasonably complex system, a bewildering array of potential software tools presents itself. These span everything from the computer aided design (CAD) tools that are used to define the geometry to be manufactured through a range of analysis codes to support computational fluid dynamics (CFD) or computational structural mechanics (CSM) calculations to simple spreadsheets where overall sizing and performance calculations can be made. If a modern approach is also taken to the management of knowledge and decision making, a further range of somewhat less familiar tools must be considered, such as databases and optimizers. Almost inevitably, it will not be clear which are the best ones to choose for the project at hand, and, moreover, the range of tools available will grow and change throughout the design process; the work at Southampton has been no different in this respect from any other design project. We can, however, set out a few main principles that have guided our choices:

1. Where possible, we have opted for tools that are either freely available on the Web or are low cost or are almost ubiquitous in their availability. This has not always been possible, but if a relatively expensive commercial tool has been adopted, it has had to provide a clear and convincing benefit.
2. We have aimed to build systems that are not locked too tightly into any given tool choice, since one never knows when there will be compelling reasons for a change of tool.
3. We have opted to use the Microsoft Windows operating system as the background to our work since it is almost the *de facto* standard in most technical organizations these days; this extends even to the large numerical clusters we have used to support CFD calculations (note, however, that many of the tools we have used are available for other operating systems).
4. Since we are fundamentally academics interested in the design process, consideration of the design system itself and the value any tool brings to it is of just as great an importance to us as the resulting designs being produced, the *raison d'etre* of the DECODE project, for instance, being to investigate the interplay between design tools, designers, and the resulting designs.

Small Unmanned Fixed-wing Aircraft Design: A Practical Approach, First Edition.
Andrew J. Keane, András Sóbester and James P. Scanlan.
© 2017 John Wiley & Sons Ltd. Published 2017 by John Wiley & Sons Ltd.

It is useful when considering tool choice to sketch out what we term the "design workflow": that is, the sequence of steps and tasks that designers must undertake when moving from initial design goals to a product or series of products to meet those goals. This is rather different from the design spiral introduced in Chapter 1: here we are interested in data and knowledge flows between tools rather than the iterative consideration of disciplines as a design becomes more refined. Figure 9.1 sets out this workflow in an abstract sense without reference to particular methods; it essentially deals with the questions "why?," "how?," "what?," and "what if?". These map onto a series of tasks that must be tackled and justified so that a balanced design can be achieved. Note that "why?" relates to issues of design rationale and supports "how?," which deals with the process of turning rationale into possible geometry. "What?" describes the current designs being compared, and "what if?" is supported by making changes to these designs so that consequent performance variations can be assessed. The overall aim is, of course, to gradually evolve and detail a design description suitable for manufacture that will deliver high-value products, with value being taken in a very broad sense to encompass monetary and mission performance issues through the life of the products being designed and the mission under consideration. This is the role of "design search" in Figure 9.1. Fleet mix and characteristics are a mixture of information relating to the tasks at hand and the way possible assets are used to address them. They are thus not really concerned with "what" the air vehicles are, as opposed to the use they are put to. Here we treat these aspects as part of the mission planning and value assessment tools. These various functions and the possible tool choices to support each are next considered in turn. Figure 9.2 shows how this workflow can be mapped onto a range of analysis domains and tasks along with possible tool choices.

9.1 Geometry/CAD Codes

The heart of any aerospace engineering design system are the databases that store the geometric details of the products being considered. We begin with this consideration rather than the tools used for concept design since we believe that an understanding of the geometry environment to be used is the most important first step in tool selection. At the point where manufacture begins, this is now almost universally one of the mainstream CAD packages such as Dassault-Systeme's Catia™ or Siemens' NX™. Here we use Rhinoceros™ for initial concept definition, which is a very low cost tool and ideally suited for parametric programming, and Solidworks™ for detail design, since this is a modest-cost detailing option but one that still offers very significant functionality. While there are some public-domain CAD codes capable of production-level drawing, the CAD description is so central to detail design that the use of a commercial code always seems to us to be justified. Commercial codes also interface well with the range of modern rapid manufacturing technologies being used to source components for the Southampton air vehicles.

Solidworks™ is, of course, capable of parametric modeling of the components being designed but, in common with all today's generation of CAD tools, is not sufficiently intelligent to allow a complete parameterization of every dimension and choice in the design of a UAV: there are simply too many interactions between parts, surfaces, and functions for this to be possible. As soon as any significant change is made to several parameters at once, the models typically fail to rebuild. This causes a fundamental difficulty in realizing a decision support vision where the aim is for rapid and radical changes to be made possible with strictly limited human intervention. Accordingly, we have restricted the use of a fully parametric

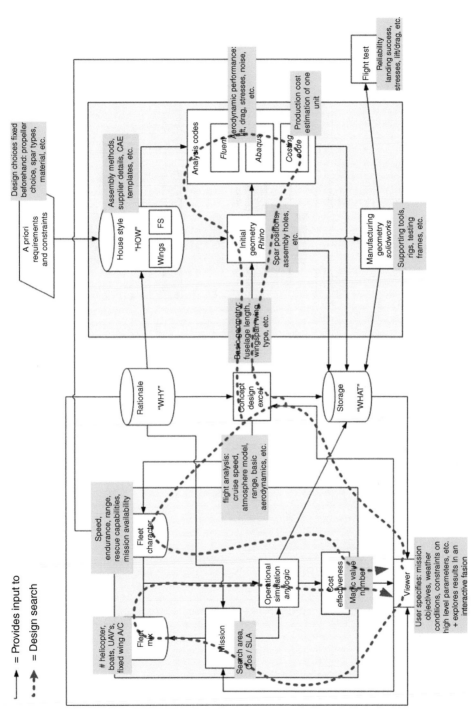

= Provides input to

= Design search

Figure 9.1 Outline design workflow.

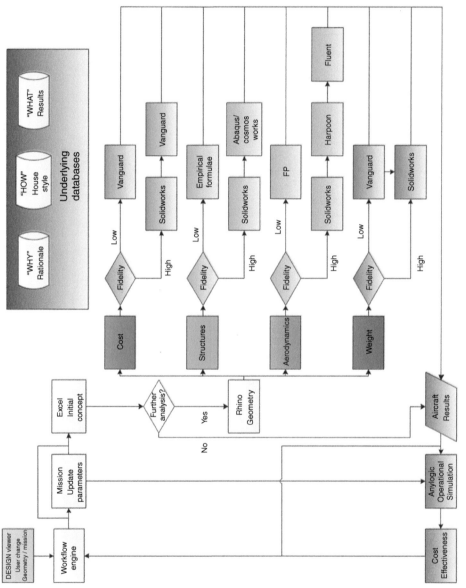

Figure 9.2 Analysis tool logic.

approach to those places where it is convenient/most important; there we use Rhinoceros™ and have had to accept that this implies a subsequent and more manual detailing phase for almost all aspects of the design after concept decision making and prior to manufacture, and where only limited parametric approaches can be taken. This means that analysis codes have to be able to work with both a fully detailed and a nondetailed set of geometries – the former to allow final design checks on stressing and aerodynamic performance, and the latter when carrying out design searches to achieve the best design balance. Although research is currently under way in several universities and companies to address such shortfalls, it seems likely that some manual, CAD-based, detailing, based on engineering experience and judgment, will still be required before production, for many years to come. The assumption made here is that such work does not impact significantly on the overall design balance being struck by the main decision support system which uses a simpler fully parametric geometry.

9.2   Concept Design

At the outset of any design process, a series of initial choices and calculations are necessary to bootstrap the whole process. These typically span the design brief and the selection of overall design topology as well as a sequence of preliminary sizing decisions. A number of specialist aircraft concept tools are available to support this process, both commercially as well as the many empirical processes detailed in various text books that can be set up in simple programming environments such as spreadsheets, see, for example, ACSYNT [14], ADS,[1] Gudmundsson [15], PaceLab APD,[2] OpenVSP,[3] Raymer [11], Stinton [5], Torenbeek [1], and so on. There are also some specialist software development environments that can be used to build concept design tools that combine some form of limited geometrical modeling along with the ability to build a linked series of aircraft analysis modules (PaceLab Suite[4]). Initial concept calculations do not generally need *detailed* geometrical information and so can be considered outside of any CAD-based model. They do, however, often consider geometrical quantities, so some form of geometric sketching to support concept development remains important. In most of our work, we proceed from basic principles, since small and medium-sized UAVs are not so common that the available tools are well suited to this class of aircraft, although some empirical relationships were drawn from the aforementioned texts at the outset, before our first series of aircraft had been built and flown and from which more detailed data could be taken. It is also possible to make very crude aircraft sketches using the limited drawing capabilities of some spreadsheet tools; here we do just that using Microsoft Excel.

To enable rapid prototyping of the concept model, our initial work is carried out using a simple constraint analysis followed by the use of an Excel spreadsheet that contains several likely topologies, which are then balanced using the internal "solver" optimization tool within Excel to satisfy a series of design requirements and constraints. The basic design calculation that we chose to tackle at this stage in the design process is the maximization of range given a fixed total maximum take-off weight (MTOW) at a given cruise speed. The fixed weight is achieved by simply adding fuel to the base airframe. The principal design variables are chosen

---

[1] http://www.pca2000.com/en/pca2000/main.htm.
[2] http://www.pace.de/en – PaceLab Aircraft Preliminary Design (APD).
[3] http://openvsp.org/.
[4] http://www.pace.de/en – PaceLab Suite.

as the main wing area and the cruise lift coefficient along with the longitudinal positions of the front bulkhead and tailplane (to control trim). To this are added a range of constraints that ensure a feasible overall concept emerges. By way of example, this basic design task is illustrated in Table 9.1 for UAVs of 7–20 kg MTOW. Note that inside Excel, the solver optimizer will be used in global mode to balance the design by finding a wing area and coefficient of lift that match the take-off weight. Moreover a table of available engine types is provided that the design process selects from as the overall size and performance requirements change. As set up at Southampton, our concept process gives a reliable initial estimate of leading dimensions for UAVs in the 2–150 kg category, and assumes conventional propeller-based propulsion and elevator/rudder/aileron control surfaces. Airframe weights are based on a collection of data built up by the designers at Southampton over a number of years and assume carbon-fiber main spars, foam-cored wings, and selective laser-sintered (SLS) nylon or foam and carbon-fiber space-frame fuselages.

Table 9.1 Concept design requirements.

Item	Lower limit	Typical value	Upper limit	Units	Type	Comment
Range		100		km	Goal	Goal to be maximized
Wing area	0.5	1.2	3	m^2	Design variables	Sets size of main lifting surface
CL cruise	0.1	0.25	0.5	—		Defines style of wing section
M fuel	0	0.5		kg		Mass of fuel (ex reserve)
X forward bulkhead	200	650	1000	mm		Position of forward bulkhead in front of main spar
X tail spar	500	900	1500	mm		Position of tail spar behind main spar
MTOW		20	20	kg	Constraints	Fundamental design requirement
CL/CD cruise		15	15	—		Prevent unrealistically optimistic wing performance
L/D cruise		4	8	—		
CL landing		1.5	1.5	—		
Thrust takeoff		60	Installed static thrust	N		Allow for limits of selected power plant
Power max.		1.5	Installed power	kW		
Available take-off rotation	15	18		°		To allow take-off
LCoG		25	25	mm		Pitch stability
Static margin	0.25	0.3		—		

9.3   Operational Simulation and Mission Planning

At the most basic level, the mission plan for our UAVs is to cruise at a fixed speed and altitude for a given time or over a given distance (fixed endurance or range). This requirement can be easily fed into the concept process using the Breguet range equation for propeller-driven aircraft as detailed in any of the major aircraft design texts already cited. If a more realistic mission is required, then ground maneuver, takeoff, climb, descent, and landing phases can be added to the calculation, either using simple empirical equations or Euler integration of the flight equations that allows for fuel consumption. The integration can be made more realistic by using actual engine fuel consumption data at various throttle settings and measurements of ground rolling performance, though this is seldom warranted during concept work.

At the next level or realism, account can be taken for variations in weather, either from mission to mission or even during a single mission. Also way-point-based navigation can be introduced and the routes chosen based on stochastic-event-driven scenarios such as air–sea surveillance and rescue missions. If fleet mixes and costs are of interest, stochastic maintenance models can be introduced based on service intervals and assumed variations of wear and tear on airframes flying multiple missions. We use the AnyLogic[5] suite for such work, see Figure 9.3.

9.4   Aerodynamic and Structural Analysis Codes

Physics-based analysis plays a key underpinning role in all engineering design; without analysis, design moves into the realm of art and guesswork. Since air vehicle design requires a great deal of analysis, such calculations are now universally carried out using computers, ranging from simple spreadsheets to full Navier–Stokes codes for CFD. At Southampton we have identified four main analysis domains that need to be addressed:

1. *Aerodynamics.* Essentially the calculation of lift, drag, and pitching moment coefficients, the capabilities of control surfaces and assessment of basic flight stability;
2. *Structures.* The assessment of stresses and deflections in all parts of the UAV for comparison with acceptable limits so as to ensure structural integrity;
3. *Weights.* The prediction of all-up weight and center of mass at takeoff, which is critical to assessing aircraft performance;
4. *Costs.* Full costing of all designs including through-life costs to allow meaningful comparisons of design choices.

9.5   Design and Decision Viewing

Key aspects of all design work are the selection between competing concepts and the auditing of analysis and design work carried out to date. To do this, a number of plots showing how design choices affect performance, often illustrated with sketch geometries, will be required. It is very helpful if their production is automated and the results are provided in standardized form. These can then be considered in design review meetings where key choices are made. Typically, this means the production of sets of Microsoft PowerPoint slides, which should be archived as the design proceeds.

---

[5] http://www.anylogic.com/.

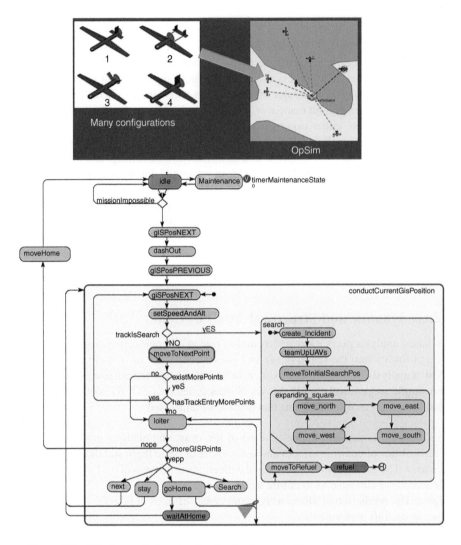

Figure 9.3   Mission analysis using the AnyLogic event-driven simulation environment.

9.6   Supporting Databases

The various codes and functions described thus far all require quantities of data to be stored and manipulated. In most cases, these tools read data files from some source and operate on these stores; that is, they implicitly have internal databases. Even so, it useful to structure the data being accessed by these tools in some well-defined and logical way so that the design team has good control over what is available and also so that the designers can browse through this data, both from function to function and also back over previous iterations of the design process. We use TortoiseSVN,[6] which is an Apache™ Subversion (SVN) client, implemented as a Windows shell extension, to manage and version control our repository of design work.

---

[6] http://tortoisesvn.net/.

10

Concept Design: Initial Constraint Analysis

Having set out all the components we are interested in and the basic design process and tool choice options, we now begin to consider how to design an unmanned air vehicle (UAV) in earnest. As will already be obvious, the boundaries between the various phases of the design process – concept, preliminary, detail – are not particularly clear-cut, nor is a very precise terminology essential here. For the purposes of this chapter, we shall loosely define concept design as the phase of the design process leading up to (and ending with) the first detailed geometry model of the aircraft. The subsequent phases may still alter almost any feature of the design (in addition to adding more detail to its definition), but two major elements should be nailed down in the concept design process: the topology (or basic layout) of the airframe (including, in general, its fundamental structural philosophy, e.g., foam wings and carbon spars) and a first-order estimate of its scale. These are the two aspects we deal with in this chapter. First, however, we discuss the starting point of the unmanned aircraft design process – indeed, of any design process – the design brief.

10.1 The Design Brief

10.1.1 Drawing up a Good Design Brief

No rational design process can start without a specification of the desired final product. This is a list of requirements put together by the customer (or by someone playing the role of a potential customer), though the engineer's input may already be required here too. The fundamentals of the design brief may be relatively easy to formulate for some classes of aircraft: for example, an airline may wish to express a requirement for a new aircraft capable of transporting 300 passengers and 10 t of freight at Mach 0.8 over 3000 nautical miles when departing from an airport at sea level experiencing temperatures of 25 °C above the International Standard Atmosphere value and admitting aircraft with a wingspan under 65 m. However, with unmanned

Small Unmanned Fixed-wing Aircraft Design: A Practical Approach, First Edition.
Andrew J. Keane, András Sóbester and James P. Scanlan.
© 2017 John Wiley & Sons Ltd. Published 2017 by John Wiley & Sons Ltd.

systems, formulating a good original specification may be thornier (somewhat paradoxically, considering that the result may be orders of magnitude simpler than in the example above).

In practice, the greatest stumbling block might be that the customer is often prejudiced by having half a solution in mind, which they are liable to weave into the formulation of the design brief. A typical *bad* design brief begins with something like "we need a multicopter to conduct surveillance of· · ·." The role of the engineer here is to guide the customer toward specifying the problem and the problem alone ("I need an unmanned system capable of pointing a camera at..."). After all, if offered a helium-balloon-based platform or a fixed-wing aircraft with a slow loiter capable of the same surveillance task, but costing less than the multicopter, they would no doubt accept it – they just may not have thought of that solution when picturing the result of the unmanned aircraft development process, perhaps biased by their existing system or by whatever they may have seen a competitor operate. Even if, in this notional example, the best solution was rotorcraft-based, what if this solution involved a swarm of 10 very small rotorcraft? A system-level view does not come naturally to many customers, but it should do to an engineer.

This also underlines the importance of a dialogue between the customer and the engineer, where the initial concept is the result of an iterative process. Such an exchange may also refine the design brief itself, down to some of its numerical elements; for example, the engineer may be able to provide feedback on which component of the brief is the strongest cost driver (we shall return to this later in this chapter).

10.1.2   Environment and Mission

The environment in which the aircraft is expected to operate plays a very important role in the concept design process. For example, an aircraft designed to carry a certain payload over a certain range may look significantly different (in terms of its wing and powerplant, which are the main quantitative considerations in this early phase of the process) if it is to fly out of "mile high" Denver in August from what it would look like if it was to shuttle between two sea-level research stations in Antarctica.

Incidentally, playing with these numbers as part of the concept phase of the design could reveal the importance (or otherwise) of bespoke airframes. In other words, it may help answer questions like: 'do we need to provide different wings for different applications or do the added development and life-cycle costs outweigh the performance gains that can be achieved through customization?'

The physics of the atmospheric environment in which aircraft operate is complicated and, in most cases, significantly variable with season, weather, and climate. However, in keeping with the general fidelity level of conceptual studies, it is generally considered acceptable to use the simple model of the International Standard Atmosphere (ISA) as an acceptable approximation. The ISA is generally very close to the true "mean" atmosphere at mid-latitudes and anywhere on the globe at low altitudes.

ISA conditions generally constitute a good baseline, but once a feasible design is found, it may be good practice to revisit the design calculations discussed in the latter half of this chapter and try a range of, say, take-off altitudes and temperatures representative of the operating conditions of the aircraft and check that the chosen design still meets the performance requirements.

10.1.3   Constraints

Once in possession of a design brief, the engineer can proceed to the next step, which is to convert the design specification into algebra, and in the case of aircraft design, this typically means drawing up a clear, constrained optimization problem.

Such clarity is relatively easily achieved in the case of some of the fundamental performance constraints, such as not allowing the design to exceed a given landing run; we shall deal with finding the chunk of the design space left feasible by these constraints in Section 10.3. Other constraints require more subtlety. Their formulation may be hindered by the fact that, at this stage, we do not have a layout.

A typical example may be that the aircraft should fit into a case no larger than $l \times w \times d$ for transportation purposes. There may be a number of ways of solving this. The trivial solution is to limit the bounding box of the design to the size of the case. We may, however, wish to consider an aircraft that can easily be disassembled into a number of components that can be accommodated by the specified transportation case. We may also end up breaking the mission down into subtasks executed by a half a dozen small aircraft, which would all have to fit into the case (there may well be scope here for interesting 3D packing layout optimization – essentially a game of *Tetris* integrated into the concept design process!).

Because of the enormous range of possible missions unmanned aircraft could be designed for, an all-encompassing design specification template will always end up being· · · well, not all-encompassing. The following list of questions is simply meant to be a series of prompts to assist in drawing up the brief, the overarching document that will guide the design process.

1. *Regulatory framework.* Which document governs the operation of the aircraft? Which certification standard(s) will have to be met?
2. *Take-off constraints.* How will the aircraft be launched? Is a strip available for normal rolling takeoff? If so, will it always be a hard surface or can it also be a grass strip? What is the maximum expected cross-wind at takeoff?
3. *Landing constraints.* Is a runway available? (Ship-based operations are typical of this not being the case.) What is the maximum allowable landing roll? What is the maximum safe touch-down speed that will ensure damage-free landings? Will the aircraft always land into a headwind? (This is quite common when the landing roll is very short compared to the size of the airfield.)
4. *Climb constraints.* What should the rate of climb versus payload/fuel mass trade space look like? (Typically, the most stringent number is specified here.) What is the minimum service ceiling/maximum operating altitude?
5. *Turn performance.* What is the minimum acceptable rate of turn (constant velocity and instantaneous)?
6. *Mission profile.* This is a catch-all term for the fundamental performance figures of almost any aircraft: still air range, endurance in the hold/loiter, contingencies/diversion, cruise speed/Mach number.
7. *Gliding and "unpremeditated descent".*[1] What is the minimum rate of descent the aircraft would have to maintain without propulsion? (Note that this may vary with density and hence altitude.)

---

[1] Regulatory bodies can always be trusted to supply engaging euphemisms for "crash."

8. *Stall performance.* What should the maximum allowable clean/dirty stall speed be constrained at? Is the aircraft expected to have "benign" stall characteristics? (The latter is a relatively rare requirement in the world of autonomous aircraft.)

9. *Stability and handling.* Constraints on static stability, control forces, dynamic stability, and general handling characteristics may be considered, though in a system designed for autonomous operations these are generally of less importance than in the case of manned aircraft.

10. *Engine out constraints.* Is the aircraft expected to safely become airborne in case of an engine failure at takeoff? Is the aircraft expected to climb following the failure of one or more of its powerplants?

11. *Ditching.* Is the aircraft expected to float on water for a significant amount of time? Does it have to be operational immediately following recovery?

12. *Constraints imposed by the payload.* What are the environmental extremes the payload may be subjected to? What are the dynamic extremes (e.g., maximum acceleration)?

13. *Ground transportation/handling constraints.* What is the bounding box the aircraft will have to fit into? What ground handling overloads can be expected? This is to differentiate between a surveillance drone carried in a soldier's backpack to be deployed in the heat of battle and a science platform that is carefully wheeled out of the lab on a calm day by operators wearing white gloves. Is there a maximum mass constraint dictated by transportation considerations? Will one person have to be able to pick it up?

14. *Aerosols, pollutants, and operation in harsh environments.* Will the aircraft be expected to operate in polluted environments, for example, in airborne sand, volcanic ash? Will the aircraft have to operate in extreme heat/cold or be exposed to nuclear fallout? Is there any risk of airframe/engine icing?

15. *Environmental impact.* What are the constraints on the environmental impact of the aircraft in terms of noise (very important, e.g., in covert surveillance applications or wildlife monitoring) and emissions? This may end up setting the agenda in terms of propulsion system design. Is there a significant chance that the aircraft will land in a remote area and not be recoverable – if so, which materials, battery types, and so on, should not be used?

16. *Ground personnel.* What is the maximum number of personnel available to operate the system? How many people will be available to unpack/assemble the aircraft and conduct preflight checks? Will the aircraft be operated by qualified, experienced personnel? The answer to the latter is generally "yes," but there may be exceptions, for example, a delivery/cargo drone that may have to be unloaded by the recipient.

Once the design brief is complete, the stage is set for the first "real" design task: the definition of the topology (or layout) of the airframe. This is what we turn our attention to next.

10.2   Airframe Topology

10.2.1   Unmanned versus Manned – Rethinking Topology

Commercial transport aircraft – in particular large jet airliners – are phenomenal demonstrations of technological progress. Indeed, over the last half century they have seen steady improvements in performance (most importantly in terms of fuel burn), environmental impact (they are getting quieter and their emissions have fallen significantly), safety, dispatch reliability, and so on. This progress is the result of the advent of digital avionics, leaps in material science, careful refinement of aerodynamic surfaces (both external and internal), and so on, and advances in the design tools that enable the effective and efficient integration of all of these aspects of aeronautical engineering.

There is one aspect of transport aircraft design, however, that has *not* changed over half a century: *airframe topology*. In other words, the airframe is made up of the same major components and these components connect to each other in the same way as they did in the 1950s. Modern commercial transport aircraft trace their lineage back to the Boeing 367-80, sometimes referred to simply as the "Dash-80," which rolled out of The Boeing Company's Seattle assembly plant in 1954, see Figure 10.1. If the Dash-80 were sprayed in a contemporary livery and pulled up at a gate at Heathrow, few passengers would notice anything out of the ordinary.

Of course, aeronautical engineers and camera-wielding plane spotters would immediately notice the relatively low aspect ratio wings and the slender nacelles housing now obsolete turbojet gas turbines but, ultimately, the Dash-80 is a "tube-and-wing" design,

Figure 10.1 On May 14, 1954, Boeing officially rolled out the Dash-80, the prototype of the company's 707 jet transport. Source: This photo, by John M. 'Hack' Miller, was taken during the rollout (Image courtesy of the Museum of History & Industry, Seattle https://creativecommons.org/licenses/by-sa/2.0/ – no copyright is asserted by the inclusion of this image).

with pylon-mounted nacelles slung under the leading edges of the swept wings and fuselage-mounted tailplanes and fin – much like a twenty-first century transport jet.[2]

The multibillion dollar question is: why has airframe topology not changed in over 60 years when, for example, the avionics of a modern jet has about as much in common with that of the Dash-80 as a magnetic levitation train has with the Flying Scotsman?

There are three possible answers:

- The Dash-80 type configuration is the engineering equivalent of, to use an example from biological evolution, a shark. In other words, it is as good as a transport jet topology can be and will ever be.
- The complexity of a modern, clean-sheet design is such that the likes of Boeing and Airbus effectively bet the company on each new aircraft[3] and, in the face of such terminal commercial risks, deviating from the tried and tested template is seen as unacceptable by the respective boards of directors and shareholders.
- Public acceptability is often quoted as the reason for conservatism, though the only significant deviation – Concorde – was not noted for being shunned by prospective passengers due to its slender delta wings, rectangular nacelles, or elegant area-ruled body.

The truth is likely to contain elements of all of these. The "tube-and-wing" design may not be as good as a topology can be, but it *is* very good. It allows for easy stretching and shortening (i.e., it naturally forms the basis of a *family* of designs, whose members can be instantiated at relatively low costs – something a tapered fuselage would not easily permit, for example), it can accommodate a structurally efficient pressurized cabin, it has some manufacturing and maintenance advantages, it is easy to evacuate in an emergency, and so on.

At the same time, for example, there are few technical arguments against and many in favor of over-wing engines (a relatively minor deviation from the standard set by the Dash-80), and yet, with the exception of Honda's innovative HA-420, we are unlikely to see such aircraft in the skies before 2025 at least.

Wherever the exact truth might lie, how should we readjust our airframe topology design thinking when we remove the pilot (and any passengers) from the aircraft? Here are some of the relevant reasons why a design paradigm shift may be required:

- The development costs of unmanned aircraft tend to be orders of magnitude lower and so are the commercial risks of making the wrong topological decision.
- The fuselage rarely needs pressurization – this removes numerous structural engineering constraints.
- The ranking of design objectives and constraints is different, with cost often dominating ahead of, say, performance, in a way it rarely does in the case of manned transport aircraft.
- The range of propulsion system types worth considering is generally much broader than in the case of manned aircraft (e.g., distributed electrical propulsion, etc.).

---

[2] The reader wishing to get up close with this intellectual ancestor of the modern airliner can only do so at the Smithsonian Institute's Steven F. Udvar-Hazy Center at Dulles Airport, near Washington DC – only one Dash-80 was ever built.

[3] The order of magnitude of the costs of developing a clean sheet twin-aisle transport aircraft are currently estimated at tens of billions of dollars.

- The removal of the uncontained engine failure constraint adds significant topological freedom on unmanned aircraft (passengers on transport aircraft generally have to be shielded
  by the wing from detached turbine blades liberated by an uncontained failure of the turbomachinery).

10.2.2   Searching the Space of Topologies

One of the key thrusts of this book is the application of computational engineering tools to the
systematic search for the optimal solution to the problem of designing a fixed-wing unmanned
aircraft in response to a design brief. A neat storyline would demand, at this point, an immediate start along these lines – specifically, an algorithm that converts the design brief into a
topology, ideally without any intervention from the engineer who is there merely to oversee
the process, but allows a suite of physics-based analysis codes and optimizers to do the actual
decision making. For example, entering a service ceiling of 30 km and an endurance of 24 h
would result in an automated computational search yielding, eventually, the topology of the
Helios aircraft shown at the bottom of Figure 10.2 (or whatever the optimal topology may be).

Alas, we are unable to offer such an algorithm here. In fact, the aerospace industry is divided
even on the question of whether such computational machinery will ever exist. It is not incon-
ceivable that a layout like that of the Scaled Composites Proteus (top of Figure 10.2) will be
generated by an algorithm in response to a design brief – say, carry a one ton payload to a
15 km service ceiling – but for the algorithm to be truly a step forward from the status quo,
the expectation would be that it would also supply a guarantee that the Proteus-like layout will
be the *best* for an aircraft designed to carry a one ton payload to a 15 km service ceiling.[4] Of
course, this is an extremely hard multidisciplinary and multiobjective problem and it would be
unreasonable to hold the design tool to such a high standard; but then *what standard should
we hold it to?* What should the objectives be? What should the stopping criterion of the search
be?

The structural engineering community has made significant strides toward solving a
small subproblem of the vast optimal topology question, namely the single-discipline,
single-objective question of what is the topology of the minimum mass part designed to
withstand a given set of loading conditions. Consider the simple problem of holding a point
load at the end of a cantilever fixed at its root with two bolts. What network of nonlinear
members would carry this load in the most mass-efficient manner? How should these members
be connected to each other? As it happens, this problem even has an analytically verifiable
global optimum, which is shown in Figure 10.3.

This is an elegant solution, but the problem is a very simple one, even within the one load
case, one discipline, and one objective category. More complicated problems are solvable via
numerical methods (though generally not quite as neatly as in this example), typically based
on the iterative removal of underutilized material. Linear elastic structural design also receives
an enormous boost in the shape of a theorem which states that a framework whose members
all carry stresses equal to the "allowable stress" dictated by the material will be no heavier than
any other framework occupying the same region of space and subject to the same boundary
conditions. This is phenomenally useful, because if, by whatever means, we find the optimal

---

[4] Of course, these are just two headline numbers from the myriad targets and criteria the "real" aircraft would be
designed against – see Section 10.1.3.

Figure 10.2   Four semi-randomly chosen points in an immense space of unmanned aircraft topologies: (starting at the top) the Scaled Composites Proteus, the NASA Prandtl-D research aircraft, the AeroVironment RQ-11 Raven, and the NASA Helios (images courtesy of NASA and the USAF).

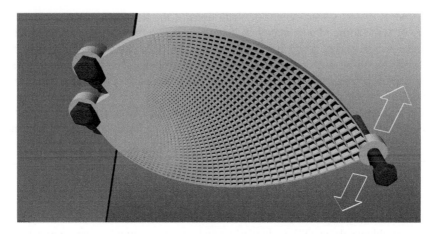

Figure 10.3   Minimum mass cantilever designed to carry a point load.

solution to the topology problem, *we will know that we have the optimum* – the familiar stress plot will all be a uniform color.[5] In other words, the route to the optimum may not be obvious, but the stopping criterion, at least, is (provided that the picture is not complicated by multiple load cases, buckling, etc.).

Stepping back to regard the big picture of the topology of the whole airplane, we have none of these luxuries: no analytical or numerical methods guiding us toward optima (we cannot start, say, with hundreds of wings and remove the unwanted ones iteratively), no neat theorem telling us that we have arrived at the solution, a plethora of disciplines and constraints shrouded in complex physics, mathematics, and economics, and, most pressingly, an objective function whose identity is not even clear in most cases (is it life-cycle cost? is it some performance metric?).

Looking to the future, all is not lost in terms of automating the concept design process. The use of genetic programming (GP) for *automated invention* has been advocated by Koza [16] and others since the 1980s and has seen some success in, for example, the automated discovery of electronic circuits (relatively simple physics compared to the flight of an aircraft!). In terms of applying this sort of evolutionary technology to aircraft concept design, there are two challenges:

- The physics-based analysis capability behind the computation of the GP fitness function has to be extremely fast (analytical models and look-up table methods are sensible candidates) – commensurate with the large number of putative designs offered by a typical evolutionary process.
- A parametric model has to be built that balances topological flexibility against the size of the design space and offers an encoding that is well suited to GP implementation.

As for the latter challenge, there is an obvious candidate that lends itself to GP: tree structures. It is also relatively easy to imagine a tree structure describing the connectivity of

---

[5] Of course, in "real" design practice, life is never quite as kind as this and the picture will always be a bit messier than that, but fundamentally the more uniform the stress field, the closer we are to the optimum.

the components of a UAV airframe. The definitions of the terminals of the tree might take advantage of some simple functional classification of the components that make up aircraft, for example, enclosure type components (whose function is to enclose a payload, fuel or some onboard system), lifting surfaces (including everything from the main wings to winglets, fins, etc.), and propulsion system components (see Ref. [17] for a proposed parameterization along these lines).

While the technology readiness level of such ideas remains largely between "speculation" and "trying a few things," we need to consider an alternative strategy that, while informed by engineering analysis, is steered directly by the engineer.

10.2.3 Systematic "invention" of UAV Concepts

Genrich Altshuller developed, in the course of the latter half of the twentieth century, a theory of inventive problem solving (TIPS – better known by the corresponding acronym in his native Russian, TRIZ). Perhaps the best known embodiment of his theory is what has evolved into "TRIZ 40," a series of basic principles of concept design, which may be viewed as a checklist of categories of possible design changes to be considered against an initial tentative topology or family of topologies.

Given a "back of a napkin" sketch of an early concept – perhaps an existing aircraft or a relatively conservative layout or perhaps one of several layouts produced by the concept design team – the following list of questions, inspired by a subset of the TIPS categories, may be used to guide the initial design review of the UAV concept design process. Some of these questions may stretch some way into what could be classed as preliminary design, but the dividing lines are often blurred and, in any case, the ideas listed below may be revisited in later stages of the design process.

Division

The principle of *divide et impera* (divide and conquer) is employed extensively in the solution *process* of larger scale design problems, and this may well be a good idea in many unmanned systems projects (e.g., allocating certain subsystems to subteams may not yield an optimal airplane however one may define optimality – but it may be the only way to expedite a design process to meet external deadlines). Here, however, we focus on the idea of division applied to the design itself. Typical questions include the following:

- Should the function of lift generation be divided between multiple surfaces, as on the Proteus aircraft shown in Figure 10.2 or (slightly more conventionally) on canard designs?
- What resources could be divided to increase redundancy? This is a good time to decide on the number of powerplants. In fact, how far can the division of power production be taken? A key advantage of electrical power is that powerplants can be highly distributed; for example, tens of small motors could be embedded in the leading edge of the wing, thus producing powered lift as a by-product of generating thrust.
- Should the fuel be divided between multiple tanks (either for redundancy or center of gravity considerations)?

- What topology would facilitate disassembly, compact storage, and transport?
- Could a modular construction facilitate operational flexibility? Should any of the lifting surfaces, payload pods, and so on, be designed to be unplugged and replaced with different size/shape ones to better suit a different mission?
- Should *the mission* be divided among several smaller aircraft, working together? For example, should one large volcanic ash cloud density sampling aircraft fly a complicated space-filling trajectory through a block of airspace or should a flock of smaller aircraft execute the same mission together by breaking the problem down in smaller, easier subproblems solvable with simpler trajectories (distributed autonomy)?

"Taking Out"

Unmanned aircraft are already the result of the ultimate "taking out" step in aircraft design: the shifting of the on-board human operator to the ground. Is there anything else that could be moved outside of its current location?

This is where a systems-level view may help. Which elements of the system have to be on board?

- Can the mass of the avionics be reduced by moving the heavier data processing tasks to the ground-based parts of the system or to the cloud? (What are the penalties of the additional communications bandwidth that may be required?)
- Can the power source be left on the ground? The most trivial case of this might be tethering the aircraft with a cable to a ground-based battery or generator. More boldly, could energy be transmitted wirelessly to the aircraft (e.g., via a laser beam)?
- Can a set of batteries be removed altogether and substituted with solar panels or some other energy harvesting system?
- Could the undercarriage be left behind to minimize weight (trolley take-off or U-2 style fall-away gears)?

Adapting the Design to Local Conditions

An obvious application of this idea, from a UAV concept design point of view, is to design systems or subsystems that can move the optimum operating point such that it matches the current conditions and mission. For example:

- A hybrid propulsion system that always operates at its "sweet spot" – such as an internal combustion engine that always operates at its most efficient throttle setting for a given altitude, ambient temperature, and so on, and which drives a generator, which, in turn, charges the battery that powers the electric motors that turn the propellers;
- Constant-speed, variable-pitch propellers;
- Morphing aircraft: could the wing shape change, for example, to enable the aircraft to operate efficiently at a range of flight conditions (landing flaps are a classic example).

Asymmetry

Subtle asymmetries abound in the vast majority of aircraft, including UAVs (at the very least, the positioning of the components of the avionics tends to be asymmetrical), but these are unlikely to be considered at the layout design phase. In these early stages, the engineer may wish to consider what advantages could be gained from *higher level asymmetries*. When sketching out early concepts, our pencil seems to instinctively default to a symmetrical layout and there is much to commend this (cost minimization through part commonality is easier to achieve, design effort is likely to be lower, etc.).

It is, nonetheless, worth considering whether anything in the design brief and the early sketches may warrant significant asymmetries. Only a handful of manned aircraft featured high-level asymmetries, and they were experimental designs (such as the Messerschmitt P.1109, a series of Blohm & Voss WWII designs, the NASA AD-1 and its precursor, the OWRA – shown in Figure 10.4, and two Rutan aircraft, the Boomerang and the ARES), but this may have much to do with public acceptability considerations. A few prompts the reader may wish to consider:

Figure 10.4 NASA oblique-wing research aircraft (images courtesy of NASA). Could your design benefit from asymmetry?

- On twin-engine designs, would an asymmetrical arrangement help mitigate one engine out thrust asymmetry concerns? (This was the main reasoning behind Rutan's Boomerang.)
- Could the fuselage be used to shield sensitive instruments from the heat and the exhaust of a single engine mounted on one side (in much the same way as the fuselage of the Rutan ARES shielded the intake of its single engine from the smoke of a machine gun mounted on its other side)?
- Would the desire for improved high-Mach-number performance warrant an oblique wing (like the NASA AD-1 and the OWRA shown in Figure 10.4)?

Incidentally, asymmetric aircraft are one example of the principle that topological concepts that have not found favor in the world of manned aircraft are worth revisiting in UAV design, as some of the reasons for their unpopularity may well have something to do with having passengers and/or the pilot on board.

Multiple Functions

The minimization of weight, complexity, and part count are almost universal concerns in aerospace engineering, and the design of unmanned aircraft is no exception. One way of achieving this – as early as the concept design phase – is by assigning multiple functions to as many components as possible. Here is a brief list of some ideas that may be considered at this point:

- the use of a V-tail instead of the more conventional stabilizer plus fin layout, replacing three surfaces with two and replacing two elevators and a rudder with two ruddervators;
- reducing the number of other control surfaces by merging their functions. This is where most aircraft-design-related portmanteaus (e.g., spoilerons, flaperons) live! Reciting them in the context of the concept layout design of the control system may be a good way of ensuring that the number of control surfaces (and thus ancillary equipment, like servos) is minimized. Of course, the trade-off is usually against redundancy, both in terms of merging the roles of control surfaces and sharing servos between multiple control surfaces;
- Should the fuselage generate some of the lift? This is not very common in manned transports for a variety of reasons (cylindrical cross-section fuselages tend to have to cruise at low angles of attack – meaning low C_L – to reduce the slope of the aisle, and blended wing bodies have a plethora of issues related to the "self-loading freight"[6]). Almost none of these reasons has any relevance to unmanned aircraft and all of the positive aspects of lifting fuselages do (chiefly, aerodynamic efficiency), so blended wing body or lifting body type configurations might well be worth considering. The typical trade-off is against manufacturing difficulties: blended wing bodies often require special tooling (though additive manufacturing can mitigate this issue at certain scales);
- Can large, external structural components (e.g., a wing brace) better earn their keep? For example, can they also generate lift or house sensors or fuel/batteries?
- Can any of the avionics components – for example, a printed circuit board (PCB) – have a structural role? This is a question worth asking especially when the size of the PCBs is the limiting factor in the miniaturization of a micro air vehicle.

---

[6] The passengers, to use a more conventional term.

Figure 10.5 Multifunctionality: the 3D-printed fuel tank (highlighted) of the SPOTTER unmanned aircraft does not only hold the fuel (with integral baffles) but also generates lift and it has a structural role too, see also Figures 2.3, 2.6, and 3.14.

- The cost of multifunctionality is often geometrical complexity. One way to solve this problem is the use of additive manufacturing techniques, the cost of which is independent of complexity. Consider, for example, the fuel tank of the SPOTTER aircraft, highlighted in Figure 10.5: it contains integral baffles, stiffeners, and a filler; it is a structural member connecting the two engine nacelles and the payload pod; and it generates a significant amount of lift. Such loading of a single part with a variety of functions may reduce weight, it reduces part count and facilitates assembly.

Nesting

Perhaps the most common form of nesting in aircraft design is the retractable undercarriage (seeing relatively limited use in small unmanned aircraft due to its relatively high cost and complexity) – but are there other forms of nesting worth considering at the layout stage?

- Telescopic components – would a telescopic boom or fuselage facilitate transportation of the aircraft?
- Telescopic wings could allow in-flight variation of the span. This could be useful either in allowing multiple roles the aircraft can switch between in flight (short, stubby wings for high-speed dash to station and extended high-aspect-ratio wings for low-speed, high-efficiency loiter on station) or simply to facilitate transportation and storage.

- Could the whole airframe be laid out such that it would facilitate the stacking of multiple airframes of the same layout? This may have the aim of optimizing the utilization of hangar floor space or, perhaps, in the case of small UAVs launched from a mother ship or a balloon, optimizing the volumetric utilization of the storage space on board the launch aircraft.
- Could a payload (e.g., an atmospheric science instrument) be stored inside the aircraft but attached to the end of a line that can be reeled in and out of the aircraft to ensure that it can take its measurements in the freestream, unaffected by the aircraft itself (like the trailing cone used for static pressure measurement in flight testing)?

Blessing in Disguise

There are some unavoidable ills, issues, and constraints in every aircraft design project. The early concept design of the vehicle may be a good time to ask: can we turn some of these to our advantage? The list below is by no means exhaustive (the possibilities are very design-dependent); they are merely examples of the type of question the designer may wish to ask.

- Can we use the heat of the exhaust gases of an internal combustion engine to keep the leading edges of the wings or the Pitot static system clear of ice? Can we use its waste heat to maintain a safe fuel temperature in a very cold (e.g., Arctic) operating environment?
- At high lift coefficients, most aircraft generate correspondingly large amounts of drag. Is it possible to design the approach and landing of the aircraft to take advantage of pushing this to the limit? For example, is a bird-like perched landing possible? Could the landing run be shortened by designing the aircraft to land in a steep nose-up attitude?
- The mass of many structural components is often driven up by severe landing loads. Could a concept be imagined where the landing loads are harnessed to disassemble the aircraft after flight? Unlike Elon Musk's concept of the RUD – or rapid unscheduled disassembly – this would be a scheduled, premeditated disassembly process, where, for example, some of the kinetic energy of the landing would be dissipated by the unclipping of wings, and so on.

Designing for Failure

Reliability modeling, as well as an analysis of the failure modes of a design and their effects, is an integral part of every unmanned aircraft design process and they form the backbone of the certification audit trail. However, most of this takes place later in the design process; what decisions can be made at the layout design stage?

- "Belts and braces": should the aircraft be equipped with a ballistic recovery system? Should an expensive payload be designed to be jettisoned with its own parachute in case of a malfunction leading to a likely loss of the aircraft?
- How to protect human operators? Can the layout be designed to prevent people on the ground from walking/reaching into rotating propellers? For example, could the propeller be ducted or surrounded by a "scaffolding" (booms, lifting surfaces, etc.) that forms a barrier? This may be a particularly important question in the case of UAVs designed to operate in close proximity of people, such as delivery drones or emergency response platforms.

- "Safe life" and "fail safe." The well-documented failure of the design of the de Havilland Comet airliner in the 1950s constituted a turning point in aircraft design philosophy. A crack originating in the dense pattern of rivets in the part of the skin of the aircraft near an antenna cover led to catastrophic decompression and explosion of the pressurized cabin. The misjudgment of the safe life of the structure was compounded by the criticality of the failure; modern engineering practice seeks to design for graceful (noncritical) failure to cover for potentially inadequate factors of safety. Are there any concept-level topology design decisions that may guard against "unknown unknowns" in the sense that they reduce the criticality of a failure or operational mishap? For example, can the undercarriage be designed to prevent loss of a wing in case of excessive roll on landing? Can the layout of the aircraft prevent a prop strike in case a belly landing becomes necessary?

The Other Way Around

Before proceeding to the preliminary design stage, it is worth pausing to consider whether a better layout is only an inversion or two away.

- Should the tail be inverted from a conventional layout to a "T" or from a "V" to an "A"?
- High wing or low wing?
- Inversion in operation. Could gravity be used to facilitate landing gear retraction by inverting the aircraft briefly after takeoff?
- Pitch control from the nose instead of from the tail (canard versus conventional layout)?
- Reverse pitch/reverse thrust for a shorter landing run?

Periodic Action

The propulsion systems of the overwhelming majority of small unmanned aircraft already rely on the periodic action of a propeller. But are there any other ways in which a concept could make use of periodic actions? Are there any applications where flapping wings may be warranted? We are still a long way away from replicating insect flight, but this may well be the future of nano air vehicles. The *fan wing* is another concept that makes use of periodic action to generate lift and thrust in conjunction with a fixed wing.

Rushing Through

The "rushing through" principle involves designing for spending the least possible amount of time in unfavorable design conditions – for example, a gas turbine engine spooling up quickly past a rotor speed that could cause resonance. Consider the following conceptual layout design example: an unmanned glider is released from a high altitude balloon and is designed to deliver a payload (e.g., a scientific instrument or an aid package) to a target location on the ground. Along the way, it has to fly through several layers of the atmosphere moving in different directions at different speeds, some of which are unfavorable. A possible solution may be to tuck the wings away upon passing the upper level of the unfavorable wind altitude range and fall through it at the highest possible rate of descent; then, the wings can be redeployed for an efficient glide upon emerging into a favorable layer.

Cheap/Disposable Components

This is a pair of concepts that used to be at home in aircraft design to the same extent as it is in the design of laser eye surgery equipment or that of nuclear reactor control rod actuators. The advent of unmanned aircraft has, however, changed that. This is largely because the risks to life and limb are often much lower than in the case of manned aircraft. Take, for example, the case of a 1 kg maximum take-off weight (MTOW) aircraft designed to collect aerial imagery of fields of crops to assist precision agriculture. Holding the servos of such an aircraft to the same standards as those of a 400-seat transport aircraft operating in and out of large urban sprawls clearly does not make sense and therefore savings can be made. Additionally, the operating life of a small UAV tends to be orders of magnitude lower than that of a manned aircraft – how can we best take advantage of this? What concept design choices can be made under this philosophy? Semidisposable landing skids? A single-use payload bay made out of a cheap foam?

Color Change

Color can have a significant functional role in unmanned aircraft design, mostly in terms of making the aircraft less conspicuous (camouflage) or more conspicuous for easier tracking from the ground (bright, contrasting colors, red/green wingtips to allow better judgment of attitude); but are there other concept-level choices we can make here? For example, could an externally placed camera (source of parasitic drag!) be replaced with one placed inside a fuselage or payload pod covered with transparent skin panels? Can a transparent "canopy" over the payload or the avionics bay expedite preflight visual system health checks?

Discarding (and Recovery?)

In the intensely weight-conscious mind of the aircraft engineer, the possibility of discarding components once they have fulfilled their function should always be present.

- Could external, podded fuel tanks be designed to be jettisoned when empty? Of course, this is feasible only when such an operation can be conducted in a safe environment where the falling object cannot cause harm and the tank can later be retrieved to avoid damage to the natural environment.
- Could the undercarriage be left behind after the take-off run (trolley take-off or Lockheed U-2 style fall-away gears)?
- Pushing this to the limit, could the whole aircraft be discarded after the mission? This is the natural mode of operation of many target drones used for artillery practice or missile testing, but is the mission being considered one where the aircraft could be made so cheaply that, upon completion of the mission, the aircraft could simply be discarded or recycled? A typical example might be humanitarian or disaster relief operations, where single-use gliders launched from balloons or "mother ship" aircraft could fly their payload to a crash landing in the disaster zone.[7]

---

[7] In 2015, the US Defense Advanced Research Projects Agency (DARPA) ran a call for an unmanned aircraft design with the brief that it would have to be balloon-launchable and, upon landing, it would not only be discarded, but *vanish*, leaving nothing but the payload behind (in practice, this would require all the airframe materials to sublimate/evaporate). In June 2016, PARC and DZYNE were selected for Phase I of the development project.

10.2.4   Managing the Concept Design Process

This list should not be regarded as a complete agenda for a concept design discussion – it is merely meant to guide it, to move this often dauntingly free-form process off its starting point. Nor is it necessary to limit the process to one tentative sketch. Indeed, every member of the design team might come up with his or her own suggestions, which may then be walked through these steps (and beyond) one by one. It is also useful to view this process as one akin to natural evolution, where parts of one design could be swapped with the equivalent part of another (crossover) and each design might branch off into multiple sub-trees.

   It is tempting to tackle all of the above armed with little more than pen and paper, but useful computational tools are available that can facilitate the process. On one hand, aircraft geometry sketches may be made in lightweight CAD software, or, perhaps even more effectively for aircraft design application, using a bespoke geometry tool, like *OpenVSP*[8] (the result of a collaboration between CalPoly and NASA). From the point of view of mapping and recording the thought processes that lead to the selection of the final concept, design rationale capture software tools may be used, which provide a searchable, clear, graphical picture of the decision tree. This can, of course, be continued into the preliminary design phase.

10.3   Airframe and Powerplant Scaling via Constraint Analysis

The roadmap (if that is not too ambitious a word) offered in the previous section will have, we hope, delivered the reader to a point where a clearly articulated design brief is in place and a conceptual layout has been derived from it. This is now a sketch that makes the major components (enclosures, main lifting surfaces, powerplants, and major structural members) and their connectivities clear, as well as perhaps containing a primitive shape definition of each. What this proto-aircraft lacks most conspicuously now is dimensions. This is the right time to return to the performance constraints introduced in Section 10.1.3 and convert them into numbers that scale the layout to provide a *geometry*.

10.3.1   The Role of Constraint Analysis

With a topology and rough shape (nondimensional design sketch) in hand, this must now be scaled to provide a first-order iteration toward the final geometry of the vehicle. The computation of this starting point of the preliminary design process is commonly achieved via a constraint analysis procedure, which is the subject of this section.

   The scale of the airframe has to be considered in conjunction with the scale (performance) of the propulsion system. To use a somewhat crude intuitive example, if a target climb rate can be achieved with a certain combination of wing area and engine thrust (or power), switching to a more powerful powerplant might allow us to reduce the wing area.[9]

   The goal is to determine the feasible region of the wing loading (W/S) versus thrust to weight ratio (T/W) space; the starting design point of the design process will have to live in this region. Two key design decisions can be made once we know the boundaries of this region:

---

[8] http://www.openvsp.org.
[9] Of course, both of these changes can have an impact on the overall weight, which may also affect the climb performance – aircraft design is riddled with such complicated webs of interactions.

the first-order selection of a wing area, and the choice of a powerplant (sometimes we will start with a given powerplant, in which case constraint analysis will simply tell us the feasible wing area range).

10.3.2 The Impact of Customer Requirements

Perhaps more useful still is the knowledge of the shape of the feasible region itself; this may give us a good sense of which of the design requirements is likely to have the strongest impact on subsequent design decisions. Is our wing area driven by the stall constraint and all other constraints are likely to be met comfortably? Or is the wing area driven by a strict rate-of-turn constraint and other constraints play no significant role? What type of high lift system is likely to be needed (a very weighty, as it were, concept design decision)?

At the beginning of this chapter we underlined the importance of a dialogue between the customer and the engineer as part of the process of drawing up a design brief. This conversation can be informed by the constraint diagram. At this stage the customer may not have a thorough understanding of the importance of certain numbers in the initial design brief they had drawn up. For example, they may have specified a "dirty configuration" approach speed based on little more than an educated guess, but they may be prepared to relax this requirement when told that, say, a 2 knot compromise may save them 10% on the wing area (as well as on the overall cost!) or even mean the difference between a plain flap and a sophisticated (and hugely expensive) multislotted flap system.

10.3.3 Concept Constraint Analysis – A Proposed Computational Implementation

An inevitable aspect of concept constraint analysis is that its fidelity is in keeping with its place in the design process: a stage where we often do not have a detailed geometry yet, hence nor do we have access to serious numerical analysis. Calculations at this point are based on empirical models and simple physics instead – first-order guesses for a first-order design iteration.

The reader may have a pet tool that may be wielded to code up these models and to perform the constraint analysis but, if not, there is a great variety to choose from (basically any sophisticated programmable platform with a ready plotting capability could be made to work – MATLAB, Python, etc.). Our own preference is for the *Jupyter*[10] web application, the power of which lies in its ability to create a *parametric notebook* – a "living" text document that is capable of updating its (code driven – specifically, in our case, Python code driven) calculations and graphs in response to changes in the inputs. From the point of view of the design process, this is a very neat way of performing concept design calculations, because the result is a document that, once complete, can simply be slotted into the design audit trail. The constraint analysis notebook is available with this book, and in Section 10.4 we have included a sample output document. The results of this model[11] will serve as the starting point of a more sophisticated (but narrower in scope) subsequent design iteration, based on spreadsheets and set out in the next chapter.

---

[10] http://www.jupyter.org.
[11] https://aircraftgeometrycodes.wordpress.com/airconics-uav/.

10.3.4 The Constraint Space

In Section 10.1.3 we provided a typical list of constraints the design brief of a fixed wing UAV may contain. Some of these may be a little hard to code up into a user-friendly, integrated mathematical model (we gave the example of the storage/transportation constraints), but the most important ones are, and here we focus on those. Namely, we shall discuss a unified treatment of the level, constant-velocity turn constraint, the rate of climb constraint, the ground run constraint, the cruise constraint, and the approach speed constraint (the importance of the latter, which is a good surrogate for the probability of damage on landing, is particularly great when unprepared landing strips are envisaged or the runway is in a confined area, such as on the deck of a vessel).

The challenge here is to find scale factors for the (at the moment) dimensionless aircraft sketch that will strike a suitable balance between the future aircraft meeting these constraints and its weight and cost remaining as low as possible (one way of satisfying all constraints might be, e.g., via an enormous wing area, but the wisdom of this is unlikely to be borne out by subsequent preliminary design calculations!). So why the plural in "scale factors"? Beyond the obvious geometrical scale – usually expressed in terms of the wing area, as this has relevance to all performance constraints – we also have to find the correct ballpark for the powerplant, most commonly in terms of the thrust it must be able to provide.

There is no unique "right" way of plugging these numbers into simple models of the constraints and solving the inequalities simultaneously, but there is a tried and tested convention that makes the solution process easy and facilitates the interpretation of the results. This involves the normalization of both numbers with the weight of the aircraft (let us say the MTOW for now). Specifically, in the interest of being able to plot all the constraints on the same diagram, we seek to express them in terms of weight-normalized thrust, that is, the *thrust to weight ratio*[12] T/W – and wing area-normalized weight – the *wing loading* W/S.

Defining the constraint space in terms of these two ratios allows the engineer to encapsulate the three most important numbers of the nascent concept, and it has the advantage that they are both relatively easy to sanity-check due to their intuitive and universal nature. The take-off thrust to weight ratio gives an immediate indication of the performance of the aircraft in the vertical dimension: typical numbers are close to 1 for air-superiority fighters, around 0.3–0.4 for transport aircraft, and under 0.3 for light aircraft and most unmanned aircraft. Typical wing loading values range from under $100 \, \text{kg/m}^2$ for light aircraft to $0.75 \, \text{t/m}^2$ for large transports. By comparison, unmanned aircraft tend to have far more lightly loaded wings – our UAVs typically have W/S of $16 \, \text{kg/m}^2$.

10.4 A Parametric Constraint Analysis Report

10.4.1 About This Document

This document captures the design algorithm used to place the first set of boundaries around the design domain of the aircraft. The algorithm is implemented in Python, embedded in a Jupyter notebook. You can open this notebook (`constraint_analysis.ipynb`) by starting the Jupyter server (type `jupyter notebook` into a terminal window opened in the

---

[12] Sometimes the related quantity of the power to weight ratio is used instead, using the relationship $P = TV/\eta_p$ – thrust times airspeed divided by propulsive efficiency – or, to get the result in horsepower $P_{\text{BHP}} = (1/746) \, TV/\eta_p$.

directory containing your notebook) – this should open Jupyter in the default browser, pointing at http://localhost:8888.[13]

The variable names have been chosen to be as intuitive as possible and additional explanations are provided in places to enhance clarity. The naming convention for all variables is to encapsulate the unit too, separated from the variable name by an underscore:

```
In [1]: VariableName_unit = 0
```

Variable names ending in an underscore are non-dimensional:

```
In [2]: VariableName_ = 0
```

This document captures one instance of a Jupyter notebook document. In other words, it is a parametric, self-building document, compiled from the output of an ipynb file. A similar document can be generated for a different design brief by editing and re-running the ipynb and compiling it with a tool called *nbconvert* by typing: `jupyter nbconvert --to latex constraint_analysis.ipynb` into the command window.

This document can then be included in the design audit trail of your aircraft and it can be re-instantiated later for a new design. It is a parametric, living, document.

We have annotated the document with explanatory notes like this one for convenience – of course, once the conceptual design process is complete, the reader may wish to delete these paragraphs to obtain a more concise document.

10.4.2   Design Brief

The conceptual design process described here starts from the design brief – essentially a broad brush mission profile – encapsulated in the following set of numbers.

Weight

At this stage in the design process it is not uncommon to only be able to provide an initial guess at the design gross weight. This may come from simply multiplying the desired payload weight with the payload fraction (payload weight divided by design gross weight) derived from looking at other aircraft designed for similar missions. In this case this "zeroth order" guess is:

```
In [3]: DesignGrossWeight_kg = 15
```

---

[13] To install *Python* load Anaconda version 2.7 onto your computer from https://www.continuum.io/downloads. Then download the *aerocalc* package from http://www.kilohotel.com/python/aerocalc/. To install *aerocalc* after you have set up Anaconda, first unpack the downloaded gz file into a directory, then start a command window in that directory and type "`python setup.py install`" there. If you also wish to locally convert files into *latex* documents you will need to load the *pandoc* package from http://pandoc.org/installing.html where installation instructions can be found. Once all the software is installed go to the directory containing the ipynb file and type "`jupyter notebook`" into the window there. Within Jupyter one can edit individual cells, typically by changing input values but also the formulae – having done this it is then sensible to use the "`Restart & Run All`" option on the cells from the "`Kernel`" tab in the Jupyter window. Once the analysis is finished either use the print options to generate hardcopy or close the Jupyter session and run *nbconvert* by typing `jupyter nbconvert --to latex constraint_analysis.ipynb` into the command window.

Take-off performance

```
In [4]: GroundRun_feet = 197
```

```
In [5]: TakeOffSpeed_KCAS = 31
```

```
In [6]: TakeOffElevation_feet = 0
```

Cruise

The cruising altitude may be viewed in two fundamental ways. First, it may be a constraint – for example, due to regulatory requirements the aircraft may have to cruise at, say, 350 feet. It can also be viewed as a design variable, in which case you may wish to return to this point in the document and revise it as part of an iterative process of optimization / refinement.

```
In [7]: CruisingAlt_feet = 400
```

```
In [8]: CruisingSpeed_KTAS = 58.3
```

Climb Performance

The climb performance of an aircraft and its variation with altitude is the result of a complex web of interactions between the aerodynamics of lift generation and the response of its powerplant to varying atmospheric conditions and airspeed. Typically a range of design points have to be considered, representing a variety of conditions, but at this early stage in the design process it is best to keep the number of these design points at a more manageable level. Here we use 80% of the cruise speed for the climb constraint.

```
In [9]: RateOfClimb_fpm = 591
```

```
In [10]: ClimbSpeed_KCAS = CruisingSpeed_KTAS * 0.8
```

The rate of climb constraint will be evaluated at this altitude:

```
In [11]: ROCAlt_feet = 0
```

Turn Performance

We define steady, level turn performance in terms of the load factor n (which represents the ratio of lift and weight). $n = 1/\cos\theta$, where θ is the bank angle (so $n = 1.41$ corresponds to $45°$, $n = 2$ corresponds to $60°$, etc.).

```
In [12]: n_cvt_ = 1.41
```

Service Ceiling

```
In [13]: ServiceCeiling_feet = 500
```

Approach and Landing

```
In [14]: ApproachSpeed_KTAS = 29.5
```

We define the margin by which the aircraft operates above its stall speed on final approach (e.g., a reserve factor of 1.2 – typical of manned military aircraft – means flying 20% above stall, a reserve factor of 1.3 – typical of civil aircraft, means 30% above stall; for small UAVs, lower values may be considered).

```
In [15]: StallReserveFactor = 1.1
```

```
In [16]: StallSpeedinApproachConf_KTAS = ApproachSpeed_KTAS\
              /StallReserveFactor
              print 'Stall speed in approach configuration: {:0.1f} KTAS'\
                      .format(StallSpeedinApproachConf_KTAS)
```

```
Stall speed in approach configuration: 26.8 KTAS
```

Maximum lift coefficient in landing configuration:

```
In [17]: CLmax_approach = 1.3
```

We also define the highest altitude AMSL where we would expect the aircraft to be established on a stable final approach in landing configuration:

```
In [18]: TopOfFinalApp_feet = 100
```

10.4.3   Unit Conversions

All constraint analysis calculations in this document are performed in SI units. However, it is more common to specify some elements of the design brief in the mix of SI and Imperial units traditionally used in aviation – here we perform the appropriate conversions.

```
In [19]: CruisingAlt_m = CruisingAlt_feet*0.3048
              print 'Cruising altitude: {:0.0f} m'.format(CruisingAlt_m)
```

```
Cruising altitude: 122 m
```

```
In [20]: TopOfFinalApp_m = TopOfFinalApp_feet*0.3048
              print 'Top of final approach: {:0.0f} m'\
                       .format(TopOfFinalApp_m)
```

```
Top of final approach: 30 m
```

```
In [21]: TakeOffElevation_m = TakeOffElevation_feet*0.3048
              print 'Take-off runway elevation: {:0.0f} m'\
                       .format(TakeOffElevation_m)
```

```
Take-off runway elevation: 0 m
```

```
In [22]: ServiceCeiling_m = ServiceCeiling_feet*0.3048
         print 'Service ceiling: {:0.0f} m'.format(ServiceCeiling_m)

Service ceiling: 152 m


In [23]: CruisingSpeed_mpsTAS = CruisingSpeed_KTAS*0.5144444444
         print 'Cruising speed: {:0.1f} m/s TAS'\
                     .format(CruisingSpeed_mpsTAS)

Cruising speed: 30.0 m/s TAS


In [24]: ClimbSpeed_mpsCAS = ClimbSpeed_KCAS*0.5144444444
         print 'Climb speed: {:0.1f} m/s CAS'.format(ClimbSpeed_mpsCAS)

Climb speed: 24.0 m/s CAS


In [25]: ApproachSpeed_mpsTAS = ApproachSpeed_KTAS*0.5144444444
         print 'Approach speed: {:0.1f} m/s TAS'\
                     .format(ApproachSpeed_mpsTAS)

Approach speed: 15.2 m/s TAS

In [26]: StallSpeedinApproachConf_mpsTAS = StallSpeedinApproachConf_KTAS\
                     *0.51444444444
         print 'Stall speed in approach configuration: {:0.1f} m/s TAS'\
                     .format(StallSpeedinApproachConf_mpsTAS)

Stall speed in approach configuration: 13.8 m/s TAS


In [27]: RateOfClimb_mps = RateOfClimb_fpm*0.00508
         print 'Rate of climb: {:0.1f} m/s'.format(RateOfClimb_mps)

Rate of climb: 3.0 m/s


In [28]: TakeOffSpeed_mpsCAS = TakeOffSpeed_KCAS*0.5144444444
         print 'Take-off speed: {:0.1f} m/s CAS'\
                     .format(TakeOffSpeed_mpsCAS)

Take-off speed: 15.9 m/s CAS


In [29]: GroundRun_m = GroundRun_feet*0.3048
         print 'Ground run: {:0.0f} m'.format(GroundRun_m)

Ground run: 60 m
```

10.4.4   Basic Geometry and Initial Guesses

Almost by definition, the early part of the conceptual design process is the only part of the product development where we do not yet have a geometry model to refer to. Thus, some of the all-important aerodynamic figures have to be guessed at this point, largely on the basis of high level geometrical parameters like the aspect ratio.

```
In [30]: AspectRatio_ = 9
```

```
In [31]: CDmin = 0.0418
```

```
In [32]: WSmax_kgm2 = 20
```

```
In [33]: TWmax = 0.6
```

```
In [34]: Pmax_kW = 4
```

Estimated take-off parameters

```
In [35]: CLTO = 0.97
         CDTO = 0.0898
         muTO = 0.17
```

10.4.5   Preamble

Some of the computations and visualizations performed in this document may require additional Python modules; these need to be loaded first as follows:

```
In [36]: %matplotlib inline
```

```
In [37]: from __future__ import division
         import math
         from aerocalc import std_atm as ISA
         import numpy as np
```

```
In [38]: import matplotlib
         import matplotlib.pylab as pylab
         import matplotlib.pyplot as plt
```

In the interest of conciseness and modularity, it is often useful to define repeated operations as functions. Let us first define a function for coloring in the unfeasible area underneath a constraint boundary:

```
In [39]: def ConstraintPoly(WSl,TWl,Col,al):
             WSl.append(WSl[-1])
             TWl.append(0)
             WSl.append(WSl[0])
             TWl.append(0)
             WSl.append(0)
             TWl.append(TWl[-2])
             zp = zip(WSl,TWl)
             pa = matplotlib.patches.Polygon(zp,closed = True\
                                    , color=Col, alpha = al)
             return pa
```

Next, we define a method for setting the appropriate bounds on each constraint diagram:

```
In [40]: def PlotSetUp(Xmin, Xmax, Ymin, Ymax, Xlabel, Ylabel):
             pylab.ylim([Ymin,Ymax])
             pylab.xlim([Xmin,Xmax])
             pylab.ylabel(Ylabel)
             pylab.xlabel(Xlabel)


In [41]: Resolution = 2000
         Start_Pa = 0.1
```

10.4.6   Preliminary Calculations

The Operating Environment

The environment in which the aircraft is expected to operate plays a very important role in many of the conceptual design calculations to follow. The conditions corresponding to the current design brief are computed as follows:

```
In [42]: SeaLevelDens_kgm3 = ISA.alt2density(0, alt_units='ft',
                                    density_units='kg/m**3')
         print 'ISA density at Sea level elevation: {:0.3f} kg/m^3'\
                       .format(SeaLevelDens_kgm3)

ISA density at Sea level elevation: 1.225 kg/m^3


In [43]: TakeOffDens_kgm3 = ISA.alt2density(TakeOffElevation_feet\
                       , alt_units='ft', density_units='kg/m**3')
         print 'ISA density at take-off elevation: {:0.3f} kg/m^3'\
                       .format(TakeOffDens_kgm3)

ISA density at take-off elevation: 1.225 kg/m^3


In [44]: ClimbAltDens_kgm3 = ISA.alt2density(ROCAlt_feet, alt_units='ft',
                                    density_units='kg/m**3')
         print 'ISA density at the climb constraint altitude: {:0.3f} kg/m^3'\
                       .format(ClimbAltDens_kgm3)

ISA density at the climb constraint altitude: 1.225 kg/m^3


In [45]: CruisingAltDens_kgm3 = ISA.alt2density(CruisingAlt_feet,\
                       alt_units='ft', density_units='kg/m**3')
         print 'ISA density at cruising altitude: {:0.3f} kg/m^3'\
                       .format(CruisingAltDens_kgm3)

ISA density at cruising altitude: 1.211 kg/m^3
```

```
In [46]: TopOfFinalAppDens_kgm3 = ISA.alt2density(TopOfFinalApp_feet,\
                        alt_units='ft', density_units='kg/m**3')
         print 'ISA density at the top of the final approach: {:0.3f}kg/m^3'\
                        .format(TopOfFinalAppDens_kgm3)
```

```
ISA density at the top of the final approach: 1.221 kg/m^3
```

Basic Aerodynamic Performance Calculations

In the absence of a geometry, at this stage any aerodynamic performance estimates will either be based on very basic physics or simple, empirical equations.

We begin with a very rough estimate of the Oswald span efficiency, only suitable for moderate aspect ratios and sweep angles below $30°$ (equation due to Raymer):

```
In [47]: e0 = 1.78*(1-0.045*AspectRatio_**0.68)-0.64
         print '{:0.3f} '.format(e0)
```

```
0.783
```

Lift induced drag factor k $(C_d = C_{d_0} + kC_l^2)$:

```
In [48]: k = 1.0/(math.pi*AspectRatio_*e0)
         print '{:0.3f}'.format(k)
```

```
0.045
```

Dynamic pressure at cruise

```
In [49]: q_cruise_Pa = 0.5*CruisingAltDens_kgm3*(CruisingSpeed_mpsTAS**2)
         print '{:0.1f} Pa'.format(q_cruise_Pa)
```

```
544.5 Pa
```

Dynamic pressure in the climb

```
In [50]: q_climb_Pa = 0.5*ClimbAltDens_kgm3*(ClimbSpeed_mpsCAS**2)
         print '{:0.1f} Pa'.format(q_climb_Pa)
```

```
352.6 Pa
```

Dynamic pressure at take-off conditions – for the purposes of this simple approximation we assume the acceleration during the take-off run to decrease linearly with v^2, so for the v^2 term we'll use half of the square of the liftoff velocity (i.e., $v = v_{TO}/\sqrt{2}$):

```
In [51]: q_TO_Pa = 0.5*TakeOffDens_kgm3*(TakeOffSpeed_mpsCAS\
                                    /math.sqrt(2))**2
         print '{:0.1f} Pa'.format(q_TO_Pa)
```

```
77.9 Pa
```

Dynamic pressure at the start of final approach, at stall speed:

```
In [52]: q_APP_Pa = 0.5*TopOfFinalAppDens_kgm3\
              *StallSpeedinApproachConf_mpsTAS**2
         print '{:0.1f} Pa'.format(q_APP_Pa)
```

116.2 Pa

10.4.7 Constraints

With the basic numbers of the current conceptual design iteration in place, we now draw up the boundaries of the wing loading W/S versus thrust to weight ratio T/W design domain. These boundaries are representations of the basic constraints that enforce the adherence of the design to the numbers specified in the design brief.

Constraint 1: Level, Constant Velocity Turn

First, we compute the thrust to weight ratio required to maintain a specific load factor n in a level turn at the cruise altitude:

$$\frac{T}{W} = q \left[\frac{C_{D_{min}}}{W/S} + k \left(\frac{n}{q} \right)^2 \left(\frac{W}{S} \right) \right]$$

... or, in Python:

```
In [53]: WSlistCVT_Pa = np.linspace(Start_Pa,8500,Resolution)
         TWlistCVT = []
         i = 0
         for WS in WSlistCVT_Pa:
             TW = q_cruise_Pa*(CDmin/WSlistCVT_Pa[i] + WSlistCVT_Pa[i]\
                          *k*(n_cvt_/q_cruise_Pa)**2)
             TWlistCVT.append(TW)
             i = i + 1
         WSlistCVT_kgm2 = [x*0.101971621 for x in WSlistCVT_Pa]
```

The load factor n is the inverse of the cosine of the bank angle (denoted here by θ) so the latter can be calculated as: $\theta = \cos^{-1} \left(\frac{1}{n} \right)$ so θ, in degrees, equals:

```
In [54]: theta_deg = math.acos(1/n_cvt_)*180/math.pi
         print '{:.0f}'u'\xb0'.format(theta_deg)
```

45°

We can now generate the constraint boundary over the wing loading range define above. The feasible range is above the curve, so we color in the region below this to begin carving into the feasible white space on the design domain chart.

```
In [55]: ConstVeloTurnPoly = ConstraintPoly(WSlistCVT_kgm2,TWlistCVT\
                              ,'magenta',0.1)
```

```
figCVT = plt.figure()
PlotSetUp(0, WSmax_kgm2, 0, TWmax, '$W/S\,[\,kg/m^2]$'\
         , '$T/W\,[\,\,]$')
axCVT = figCVT.add_subplot(111)
axCVT.add_patch(ConstVeloTurnPoly)
```

Out[55]: <matplotlib.patches.Polygon at 0x6cc9208>

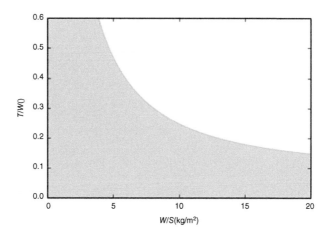

Constraint 2: Rate of Climb

Another constraint that can lead to interesting engine power versus wing area trade-offs is the rate of climb requirement. If q denotes the dynamic pressure in the environmental conditions specified earlier, V is the calibrated airspeed in the climb, and V_V is the rate of ascent, the required thrust to weight ratio T/W as a function of the wing loading W/S can be calculated as:

$$\frac{T}{W} = \frac{V_V}{V} + \frac{q}{W/S}C_{D_{min}} + k\frac{1}{q}\frac{W}{S}.$$

The Python implementation once again sweeps a sensible range of wing loading values to build the appropriate constraint diagram:

```
In [56]: WSlistROC_Pa = np.linspace(Start_Pa,8500,Resolution)
         TWlistROC = []
         i = 0
         for WS in WSlistROC_Pa:
             TW = RateOfClimb_mps/ClimbSpeed_mpsCAS + CDmin\
             *q_climb_Pa/WSlistROC_Pa[i] + k*WSlistROC_Pa[i]/q_climb_Pa
             TWlistROC.append(TW)
             i = i + 1
         WSlistROC_kgm2 = [x*0.101971621 for x in WSlistROC_Pa]

In [57]: RateOfClimbPoly = ConstraintPoly(WSlistROC_kgm2,TWlistROC\
                           ,'blue',0.1)
```

```
figROC = plt.figure()
PlotSetUp(0, WSmax_kgm2, 0, TWmax, '$W/S\,[\,kg/m^2]$'\
        , '$T/W\,[\,\,]$')
axROC = figROC.add_subplot(111)
axROC.add_patch(RateOfClimbPoly)
```

Out[57]: <matplotlib.patches.Polygon at 0x750f0b8>

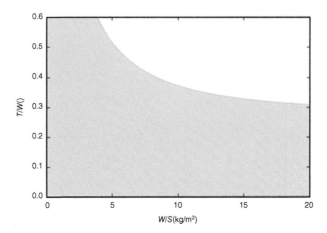

Constraint 3: Take-Off Ground Run Constraint

We next compute the thrust to weight ratio required for a target ground run distance on take-off. If C_L^{TO} and C_D^{TO} denote the take-off run lift and drag coefficients respectively, d_{GR} is the required ground run distance, V_L is the lift-off speed, μ_{TO} is the ground friction constant, the required thrust to weight ratio T/W as a function of the wing loading W/S can be calculated as:

$$\frac{T}{W} = \frac{V_L^2}{2g\, d_{GR}} + \frac{q\, C_D^{TO}}{W/S} + \mu_{TO}\left(1 - \frac{q\, C_L^{TO}}{W/S}\right)$$

Sweeping the range of wing loading values as before, in order to build the appropriate constraint diagram:

```
In [58]: WSlistGR_Pa = np.linspace(Start_Pa,8500,Resolution)
         TWlistGR = []
         i = 0
         for WS in WSlistGR_Pa:
             TW =(TakeOffSpeed_mpsCAS**2)/(2*9.81*GroundRun_m) + \
                 q_TO_Pa*CDTO/WSlistGR_Pa[i]\
                 + muTO*(1-q_TO_Pa*CLTO/WSlistGR_Pa[i])
             TWlistGR.append(TW)
             i = i + 1
         WSlistGR_kgm2 = [x*0.101971621 for x in WSlistGR_Pa]
```

```
In [59]: TORunPoly = ConstraintPoly(WSlistGR_kgm2,TWlistGR,'green',0.1)

         figTOR = plt.figure()
         PlotSetUp(0, WSmax_kgm2, 0, TWmax, '$W/S\,[\,kg/m^2]$'\
                  , '$T/W\,[\,\,]$')

         axTOR = figTOR.add_subplot(111)
         axTOR.add_patch(TORunPoly)

Out[59]: <matplotlib.patches.Polygon at 0x78f7908>
```

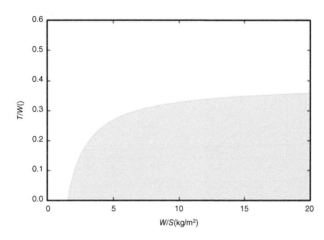

Constraint 4: Desired Cruise Airspeed

We next look at the cruise speed requirement. If q denotes the dynamic pressure at cruise conditions, the required thrust to weight ratio T/W as a function of the wing loading W/S can be calculated as:

$$\frac{T}{W} = qC_{D_{min}}\frac{1}{W/S} + k\frac{1}{q}\frac{W}{S}.$$

The Python implementation once again sweeps a sensible range of wing loading values to build the appropriate constraint diagram:

```
In [60]: WSlistCR_Pa = np.linspace(Start_Pa,8500,Resolution)
         TWlistCR = []
         i = 0
         for WS in WSlistCR_Pa:
             TW = q_cruise_Pa*CDmin*(1.0/WSlistCR_Pa[i])\
             + k*(1/q_cruise_Pa)*WSlistCR_Pa[i]
             TWlistCR.append(TW)
             i = i + 1
         WSlistCR_kgm2 = [x*0.101971621 for x in WSlistCR_Pa]

In [61]: CruisePoly = ConstraintPoly(WSlistCR_kgm2,TWlistCR,'red',0.1)
```

```
figCruise = plt.figure()
PlotSetUp(0, WSmax_kgm2, 0, TWmax, '$W/S\,[\,kg/m^2]$'\
          , '$T/W\,[\,\,]$')
axCruise = figCruise.add_subplot(111)
axCruise.add_patch(CruisePoly)
```

Out[61]: <matplotlib.patches.Polygon at 0x78fcc88>

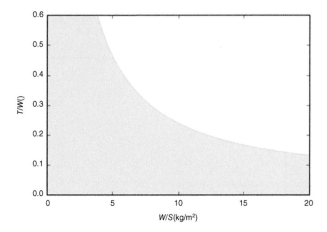

Constraint 5: Approach Speed

Assuming a given target approach speed (which, at the start of the typical final approach translates into a dynamic pressure q^{APP}) and a maximum lift coefficient C_L^{APP} achievable in the approach configuration (with the high lift system, if present, fully deployed), the wing loading constraint can be formulated as:

$$\frac{W}{S} \leq q^{APP}\, C_L^{APP}$$

The approach speed constraint will thus impose a right hand boundary in the thrust to weight versus wing loading space at:

```
In [62]: WS_APP_Pa = q_APP_Pa*CLmax_approach
         WS_APP_kgm2 = WS_APP_Pa*0.101971621
         print '{:03.2f} kg/m^2'.format(WS_APP_kgm2)
```

15.41 kg/m^2

```
In [63]: WSlistAPP_kgm2 = [WS_APP_kgm2, WSmax_kgm2, WSmax_kgm2\
                          , WS_APP_kgm2, WS_APP_kgm2 ]
         TWlistAPP = [0, 0, TWmax, TWmax, 0 ]
         AppStallPoly = ConstraintPoly(WSlistAPP_kgm2,TWlistAPP\
                                       ,'grey',0.1)

         figAPP = plt.figure()
         PlotSetUp(0, WSmax_kgm2, 0, TWmax, '$W/S\,[\,kg/m^2]$'\
```

```
                    ,  '$T/W\,[\,\,]$')

         axAPP = figAPP.add_subplot(111)
         axAPP.add_patch(AppStallPoly)
```

Out[63]: <matplotlib.patches.Polygon at 0x7e502e8>

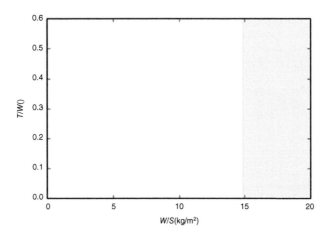

Combined Constraint Diagram

With all of the key constraints computed, we can now superimpose them to reveal what remains of the design space.

```
In [64]: figCOMP = plt.figure(figsize = (10,10))
         PlotSetUp(0, WSmax_kgm2, 0, TWmax, '$W/S\,[\,kg/m^2]$'\
                   , '$T/W\,[\,\,]$')
         axCOMP = figCOMP.add_subplot(111)

         ConstVeloTurnPoly = ConstraintPoly(WSlistCVT_kgm2\
                                          ,TWlistCVT,'magenta',0.1)
         axCOMP.add_patch(ConstVeloTurnPoly)

         RateOfClimbPoly = ConstraintPoly(WSlistROC_kgm2\
                                          ,TWlistROC,'blue',0.1)
         axCOMP.add_patch(RateOfClimbPoly)

         TORunPoly = ConstraintPoly(WSlistGR_kgm2,TWlistGR,'green',0.1)
         axCOMP.add_patch(TORunPoly)

         CruisePoly = ConstraintPoly(WSlistCR_kgm2,TWlistCR,'red',0.1)
         axCOMP.add_patch(CruisePoly)

         AppStallPoly = ConstraintPoly(WSlistAPP_kgm2,TWlistAPP\
```

```
                                       ,'grey',0.1)
    axCOMP.add_patch(AppStallPoly)

    axCOMP.legend(['Turn','Climb','T/O run','Cruise'\
                   , 'App Stall'])

    textstr = '\n                             The feasible aeroplane\
     lives\n                        in this white space'
    axCOMP.text(0.05, 0.95, textstr, transform=axCOMP.transAxes\
                , fontsize=14, verticalalignment='top')
```

Out[64]: <matplotlib.text.Text at 0x827c518>

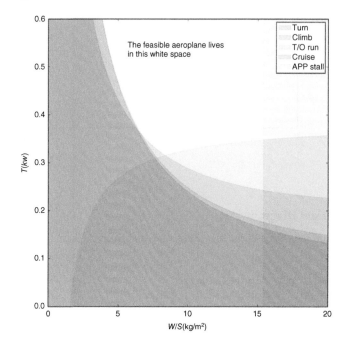

Since propeller and piston engine driven aircraft are normally designed in terms of engine power rather than thrust, we next convert the constraint diagram from thrust to weight ratio into an installed power requirement by specifying a propulsive efficiency $\eta = 0.6$ (note that un-supercharged piston engine power varies with altitude so we also allow for this in the conversion using the Gagg and Ferrar model (see Gudmundsson [15]) with $\text{Power}_{SL} = \text{Power}/(1.132\sigma - 0.132)$ where σ is the air density ratio):

In [65]: PropEff = 0.6

In [66]:
```
WSlistCVT_Pa = np.linspace(Start_Pa,8500,Resolution)
PlistCVT_kW = []
i = 0
for WS in WSlistCVT_Pa:
    TW = q_cruise_Pa*(CDmin/WSlistCVT_Pa[i]\
                + WSlistCVT_Pa[i]*k*(n_cvt_/q_cruise_Pa)**2)
```

```
            P_kW = 9.81 * TW * DesignGrossWeight_kg\
            * CruisingSpeed_mpsTAS / PropEff \
            / (1.132*CruisingAltDens_kgm3/SeaLevelDens_kgm3-0.132)/1000
            PlistCVT_kW.append(P_kW)
            i = i + 1
        WSlistCVT_kgm2 = [x*0.101971621 for x in WSlistCVT_Pa]

In [67]: WSlistROC_Pa = np.linspace(Start_Pa,8500,Resolution)
        PlistROC_kW = []
        i = 0
        for WS in WSlistROC_Pa:
            TW = RateOfClimb_mps/ClimbSpeed_mpsCAS\
            + CDmin*q_climb_Pa/WSlistROC_Pa[i] \
            + k*WSlistROC_Pa[i]/q_climb_Pa
            P_kW = 9.81 * TW * DesignGrossWeight_kg\
            * ClimbSpeed_mpsCAS / PropEff \
            / (1.132*ClimbAltDens_kgm3/SeaLevelDens_kgm3-0.132)/1000
            PlistROC_kW.append(P_kW)
            i = i + 1
        WSlistROC_kgm2 = [x*0.101971621 for x in WSlistROC_Pa]

In [68]: WSlistGR_Pa = np.linspace(Start_Pa,8500,Resolution)
        PlistGR_kW = []
        i = 0
        for WS in WSlistGR_Pa:
            TW =(TakeOffSpeed_mpsCAS**2)/(2*9.81*GroundRun_m) + \
            q_TO_Pa*CDTO/WSlistGR_Pa[i]\
            + muTO*(1-q_TO_Pa*CLTO/WSlistGR_Pa[i])
            P_kW = 9.81 * TW * DesignGrossWeight_kg\
            * TakeOffSpeed_mpsCAS / PropEff \
            / (1.132*TakeOffDens_kgm3/SeaLevelDens_kgm3-0.132)/1000
            PlistGR_kW.append(P_kW)
            i = i + 1
        WSlistGR_kgm2 = [x*0.101971621 for x in WSlistGR_Pa]

In [69]: WSlistCR_Pa = np.linspace(Start_Pa,8500,Resolution)
        PlistCR_kW = []
        i = 0
        for WS in WSlistCR_Pa:
            TW = q_cruise_Pa*CDmin*(1.0/WSlistCR_Pa[i])\
            + k*(1/q_cruise_Pa)*WSlistCR_Pa[i]
            P_kW = 9.81 * TW * DesignGrossWeight_kg\
            * CruisingSpeed_mpsTAS / PropEff \
            / (1.132*CruisingAltDens_kgm3/SeaLevelDens_kgm3-0.132)/1000
            PlistCR_kW.append(P_kW)
            i = i + 1
        WSlistCR_kgm2 = [x*0.101971621 for x in WSlistCR_Pa]

In [70]: WSlistAPP_kgm2 = [WS_APP_kgm2, WSmax_kgm2, WSmax_kgm2\
                        , WS_APP_kgm2, WS_APP_kgm2 ]
        PlistAPP_kW = [0, 0, Pmax_kW, Pmax_kW, 0 ]
```

```
In [71]: figCOMP = plt.figure(figsize = (10,10))
         PlotSetUp(0, WSmax_kgm2, 0, Pmax_kW, '$W/S\,[\,kg/m^2]$'\
                 , '$P\,[\,kW]$')
         axCOMP = figCOMP.add_subplot(111)

         ConstVeloTurnPoly = ConstraintPoly(WSlistCVT_kgm2,PlistCVT_kW\
                                           ,'magenta',0.1)
         axCOMP.add_patch(ConstVeloTurnPoly)

         RateOfClimbPoly = ConstraintPoly(WSlistROC_kgm2,PlistROC_kW\
                                         ,'blue',0.1)
         axCOMP.add_patch(RateOfClimbPoly)

         TORunPoly = ConstraintPoly(WSlistGR_kgm2,PlistGR_kW,'green',0.1)
         axCOMP.add_patch(TORunPoly)

         CruisePoly = ConstraintPoly(WSlistCR_kgm2,PlistCR_kW,'red',0.1)
         axCOMP.add_patch(CruisePoly)

         AppStallPoly = ConstraintPoly(WSlistAPP_kgm2,PlistAPP_kW,'grey',0.1)
         axCOMP.add_patch(AppStallPoly)

         axCOMP.legend(['Turn','Climb','T/O run','Cruise', 'App Stall'])

         textstr = '\n                        The feasible aeroplane\
          lives\n                        in this white space'
         axCOMP.text(0.05, 0.95, textstr, transform=axCOMP.transAxes\
                    , fontsize=14, verticalalignment='top')

Out[71]: <matplotlib.text.Text at 0x77c7748>
```

The resulting combined constraint diagram is shown in Figure 10.6. The code used to carry out this analysis can also be downloaded from https://aircraftgeometrycodes.wordpress.com/uavbook/.

10.5   The Combined Constraint Diagram and Its Place in the Design Process

Let us consider Figure 10.6, as generated by the parametric constraint analysis report of Section 10.4. It clearly shows the boundaries of the feasible space, wherein the first-order values of W/S and P will have to reside. At this stage, the bottom right-hand corner of the feasible domain is generally a good choice, as this will minimize the required power (and therefore the size of the engine) and minimize the wing area (which is likely to yield the lowest mass and lowest cost solution). However, the availability of a discrete set of engines and other practical constraints may mean that the designer will have to exercise some level of judgment within the bounds identified by the constraint analysis. For the aircraft in this example, it suggests a small gasoline or medium-sized glowplug engine of around 25 cc with a wing span of 3 m, given the design MTOW of 15 kg and aspect ratio of 9. These results

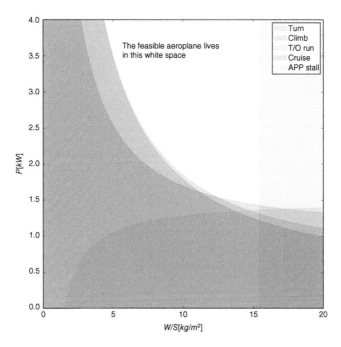

Figure 10.6 Typical constraint diagram. Each constraint "bites" a chunk out of the P versus W/S space; whatever is left is the feasible region, wherein the design will have to be positioned.

accord with the data presented by Gundlach [18], where he shows that typical power to weight ratios for small UAVs lie between 0.05 and 0.13 horsepower per pound of take-off weight, or an installed power for this design of between 1.2 and 3.2 kW. Our Decode-1 aircraft has an MTOW of 15 kg, an OS Gemini FT-160 26.5 cc glowplug engine capable of delivering 1.66 kW and a wing loading of 15.5 kg/m^2, that is, it lies exactly where one would expect it on the diagram.

The engine power and wing loading values thus selected will serve as the starting point of the subsequent analysis as we progress to more refined estimates of the key flight physics parameters of the aircraft – this is the subject of the next chapter.

11

Spreadsheet-Based Concept Design and Examples

Having finished the outline constraint analysis, a more complete concept model can be constructed. Typically a finished concept design will be specified in around 100 numbers, of which perhaps 20 will relate directly to geometry – if greater detail is entered into, there is a danger that the design team will lose understanding of what parameters are controlling which aspects of the design. There does, however, need to be sufficient detail to enable simple sketches of the concept to be generated. The resulting design also needs to be *balanced*. Balance implies, for example, that lift is equal to weight (in level flight or a turn), available thrust is equal to drag (again in level flight or during takeoff or climb), the center of gravity lies close to the center of lift, and fuselage volume is sufficient for fuel, avionics, and payload, and so on, without being over-sized. Achieving balance almost always requires an iterative process to adjust key dimensions, something that can generally be specified in the form of an optimization problem and which can be set up and solved in a spreadsheet or similar environment. Constraint analysis provides the starting point for this process, enabling the design team to specify likely initial wing sizes and powerplant choices.

It is normal at the concept level to *avoid* using physics-based analysis codes to assess a design geometry, particularly during the iterative steps needed to balance the project. Rather, various design rules and equations will be invoked as in the previous chapter, see also, for example, the texts by Raymer [11] or Gudmundsson [15]. The problem with a physics-based approach is the need for more complete geometry data than is commonly available and the much longer calculation times involved. Research is progressing in this area but such tools tend to be limiting in the restrictions they impose on the design team. Thus CFD and FEA are not commonly used in concept design balancing work. They can, however, be invoked once a balanced design has been achieved to check whether the assumptions used in the design are still valid for the final concept. If they are not, assumptions can be adapted and a rebalancing carried out. It cannot be stressed too strongly that getting a design correctly balanced is of critical importance if the final product is to be satisfactory. It is clearly highly undesirable if the aircraft when finally built has an incorrect location of the center of gravity (CoG) and

Small Unmanned Fixed-wing Aircraft Design: A Practical Approach, First Edition.
Andrew J. Keane, András Sóbester and James P. Scanlan.
© 2017 John Wiley & Sons Ltd. Published 2017 by John Wiley & Sons Ltd.

ballast has to be added – this might sound obvious but some quite major engineering projects have suffered because basic issues of balance and trim had not been dealt with correctly at the outset. When comparing configurations, it is again critical that all designs have been worked up properly into balanced possibilities to allow a fair decision. If it is intended to run more complex calculations, these should be completed and any subsequent rebalancing carried out before decisions are taken. In all cases, some measure of design merit will be needed to support the choices being made. This will have to take account of a range of possible performance and cost measures in some statement of overall *value*.

For the purposes of this chapter, we will begin with unmanned air vehicles (UAVs) with a dry take-off weight less than 40 kg. Initially, these will be taken to be powered by a single pusher propeller and piston engine system. Twin carbon tail booms and a U-shaped tail will be adopted. The fuselage can be nylon or glass-fiber-covered laser-cut foam with additional longitudinal stringers if desired. Reinforcement in laser-cut plywood or laser-sintered nylon will be possible. Molds and fiber layout will not be considered so as to reduce tooling and manpower costs. This information represents our topology and manufacturing choices and becomes embedded in the rules used to estimate weights for structural and fuselage components.

The designer starts with a given payload mass as well as landing and cruise speed combination (i.e., a design brief): here we begin with a payload of 2 kg, and landing, take-off, and cruise speeds of 15, 16, and 30 m/s, respectively. Maximum endurance is desired within an all-up maximum take-off weight (MTOW) – including payload and fuel – of 15 kg as used in the previous chapter.

The following sections set out the basic principles we use for sizing such a small, propeller-driven, piston-engine UAV. Most of the approaches we use are covered in Anderson [19] and Gudmundsson [15] with some additional data from Raymer [11], together with engine data sheets and propeller data sheets based on manufacturers' data and some results from our own experimental work. An extensive collection of independent propeller data can be found on the UIUC Propeller Data Site as compiled by John Brandt, Robert Deters, Gavin Ananda, and Michael Selig.[1] The Web-based program javaProp[2] can also be used to further check propeller behavior.

11.1 Concept Design Algorithm

The basic steps of the UAV platform concept design process used here are as follows:

1. Decide target payload mass, MTOW, and cruise speed (at a maximum cruise ceiling, for us typically 400 ft as set by UK CAA regulations) – we aim to maximize range within this MTOW budget and speed setting – other choices of start point are possible.
2. Choose the overall aircraft configuration (canard, monocoque, tail booms, etc.) – here taken as a tail boom pusher as already noted.
3. Choose the landing speed and take-off speed (or set take-off speed to be the same as landing speed), which govern wing loadings if we assume we do not have high lift devices – typical values are in the 12–18 m/s range. Higher landing speeds give faster,

---

[1] http://m-selig.ae.illinois.edu/props/propDB.html.
[2] http://www.mh-aerotools.de/airfoils/javaprop.htm.

Table 11.1 Typical fixed parameters in concept design.

Item	Value
Aspect ratio (span²/area)	9
Reserve fuel fraction	0.1
Minimum (wing) coefficient of drag at cruise speed, C_{Dmin}	0.0143
Viscous drag constant, k_{visc}	0.0329
Pressure and induced drag constant, k_i	0.0334
Minimum drag (wing) coefficient of lift at cruise speed, C_{Lmin}	0.1485
Coefficient of parasitic drag at cruise speed	0.0375
C_l/C_d at landing speed	5
Coefficient of parasitic drag at landing speed	0.05
C_l/C_d at take-off speed	5
Coefficient of parasitic drag at take-off speed	0.035
Equivalent rolling coefficient of friction	0.225
Propulsive efficiency at cruise speed	0.6
Tail-plane aspect ratio (span²/area)	4
Fin aspect ratio, ((twice fin height)²/area) assuming two fins	3
Bank angle in level turn	60°
Minimum rate of climb from cruise	5m/s
Percentage extra thrust desired to start ground roll	0%

smaller aircraft with greater range but which are more likely to have accidents on landing!

4. Choose values for various fixed parameters – many of these are aerodynamic, all are nondimensional or do not depend on the airframe size and here are based on typical aircraft, operational speeds, previous designs and simple calculations (such as $C_d = C_{Dmin} + k_{visc}(C_l - C_{Lmin\_drag})^2 + k_i(C_l)^2$), see, for example, Table 11.1.

5. Place some limits on what is possible or acceptable for the design under consideration – examples are given in Table 11.2.

6. To start the design process, choose the total main wing area from the constraint diagram along with likely coefficients of lift at cruise and landing (these are based on the chosen wing loading and the expected aerodynamic performance, including allowance for flaps or other landing aids if fitted), together with the fuel weight and longitudinal positions of tail-plane spar (to control static pitch stability) and payload (e.g., assumed to be forward of the front bulkhead and hence fixed by the location of this bulkhead) – ultimately we adjust these to *balance* the design and also ensure that they do not go outside the limits just set.

7. Choose the installed power needed from the constraint diagram or try and estimate the likely lift and drag of the aircraft at cruise, level turn, climb, takeoff, and landing for the given MTOW, so as to try and see what kind of installed power will be needed. If MTOW is fixed, this can be done without having to estimate the mass of the vehicle from its components, which is a distinct advantage of starting from a fixed MTOW.

Table 11.2   Typical limits on variables in concept design.

Item	Value
Maximum wing C_l/C_d at cruise speed	16
Maximum wing lift to total drag at cruise speed (allows for all drag elements)	8
Maximum coefficient of lift at landing speed	1.3
Minimum allowable rotation at take-off before grounding happens	17°
Minimum static margin (expressed in fractions of main wing chord)	0.1

Table 11.3   Estimated secondary airframe dimensions.

Item	Value	Units
Fuselage depth	200	mm
Fuselage width	150	mm
Nose length (forward of front bulkhead)	150	mm
Diameter of main undercarriage wheels	100	mm
Longitudinal position of engine bulkhead	−200	mm
Vertical position of base of fuselage	−110	mm
Vertical position of tail-plane	0.0	mm
Vertical position of engine	60	mm
Vertical position of center of main undercarriage wheels	−300	mm

8. Choose an engine/propeller combination and some likely dimensions for the aircraft fuse-lage (we set the datum on the center-line in way of the main wing spar, which is taken to lie at the quarter chord point). These dimensions can be used to sketch the aircraft; for the 15 kg Decode-1 aircraft used as the first example in this chapter, which has twin tail booms and conventional U-shaped rear control surfaces, they are as shown in Table 11.3. For other designs, they will need setting to different but appropriate values, based on prior experience or educated guesses.

9. Assess the likely empty weight based on the dimensions now available and the structural construction philosophy adopted, together with the components required to operate the aircraft (such as avionics, undercarriage, servos, and so on, that is, an outline list of component parts) plus the maximum fuel weight (without reserve). Ultimately MTOW includes

- payload weight
- structural weight
- avionics, systems, and servo weight
- propulsion system weight
- undercarriage and miscellaneous weight
- fuel weight including reserve.

When summing these, note that the dry weight does not include payload or the main fuel, but does include the fuel reserve.

10. Check if the following conditions have been met:

- the aircraft can rotate by at least the minimum angle specified at takeoff without fouling any part of the structure,
- the static margin is greater than the minimum required without being excessive,
- if a nose wheel is not fitted, the pitching moment caused by the propeller thrust and wheel drag (set equal and opposite to the thrust) will not cause the aircraft to pitch nose down into the ground, given the position of the CoG,
- there is enough weight difference between the MTOW, the structural weight, and the payload to carry some fuel!
- the wing C_l/C_d ratio at cruise is below the maximum chosen,
- the wing lift to total drag ratio at cruise is below the maximum chosen,
- the C_l at landing is below the permitted maximum,
- the installed static thrust is enough to start the aircraft rolling and to permit takeoff (assume that static thrust does not change until after the aircraft leaves the runway – in fact, it is likely to rise slightly for a well chosen propeller),
- the installed power is sufficient to achieve the cruise speed and to carry out acceptably banked level turns and climbs,
- the calculated aircraft weight including fuel is no more than the target MTOW (if it is less, more fuel is carried to simply increase the range),

11. Adjust the guessed inputs to maximize the range while meeting the constraints.

When this process is completed, the overall geometry of the balanced platform can then be used to begin the next stage of the design process. To carry out the above calculations, information is needed on a number of other aspects as described in the following sections.

11.2 Range

Range is given by the Breguet equation (for piston engined aircraft): Range $= (3600/9.81) \times (\eta/\text{SFC}) \times (\text{Lift/Drag}) \times \ln(\text{MTOW}/(\text{MTOW} - M_{\text{fuel}}))$ where SFC is in kg/kWh and the resulting range is in km. η is the propulsive efficiency at the cruise speed, typically around 0.6 but clearly varies with propeller choice and also speed. Note that here M_{fuel} does not include the fuel reserve.

11.3 Structural Loading Calculations

To size spars and booms, and hence estimate their weights, some form of maximum g to design against will be needed. This will come from gust or maneuver loads or some arbitrary choice such as $4g$. Then simple beam theory can be used alongside design decisions on the likely form and materials to be selected for these items (typical materials are carbon-fiber tubes, solid or box-section plywood beams, or aluminum rods or tubes). Note that the FAA regulations (FARS 14, CFR, part 25) state that the maximum maneuver load factor is normally to be 2.5 but if the airplane weighs less than 50 000 lbs, the load factor is to be given by $n = 2.1 + 24\,000/(W + 10\,000)$, though n need not be greater than 3.8; while UK JAR 25 says: "the limit load maneuvering load factor n for any speed up to V_n may not be less than $2.1 + 24\ 000/(W + 10\ 000)$ except that n may not be less than 2.5 and need not be greater than 3.8, where W is the design MTOW." Here the unit of weight is lbs. JAR-VLA (the equivalent standard for very light aircraft) simply says the limits should be between 1.5 and 3.8.

11.4 Weight and CoG Estimation

To estimate the total weight of the aircraft, a list must be made of all items carried (such as individual components of the avionics, including the wiring harness, undercarriage components, fuel tanks, servos, engine, etc.) and the locations of each item relative to the chosen datum noted, typically on a separate weights sheet (a relatively detailed example of a weights table is given in Chapter 15). For wing and fuselage structure, some construction method must be chosen and then estimates made using projected areas and lengths. Any spars or booms must be suitably sized with simple beam theory calculations as noted above. Monocoque areas subject to significant structural loads are difficult to estimate weights for accurately during concept sizing, but typically a surface area and thickness plus allowance for any rib stiffening has to suffice. Typical materials for stressed monocoques are carbon-fiber lay-up (possibly using foam sandwich techniques) or selective laser sintered (SLS) Nylon – in this chapter we restrict designs to SLS nylon or glass-fiber-covered hot-wire-cut foam. Tables 11.4 and 11.5 list some of the variables that might be used to scale weights and the items which might be scaled from them, respectively. Table 11.6 lists items whose weights are rather less easy to estimate, but which nonetheless may vary with aircraft size and must be allowed for.

11.5 Longitudinal Stability

Static longitudinal (pitch) stability calculations are always needed, which require suitable downwash data for the elevator surfaces. Here we take the main wing quarter chord point as the datum and center of lift of the main wing:

static margin $= L_{\text{CoG}}/\text{Chord} + ((L_{\text{CoG}} - L_{\text{tail}})/\text{Chord}) \times (A_{\text{tail}}/A_{\text{wing}})$
$\times(1 + 2/(3AR/4))/(1 + 2/(3AR_{\text{tail}}/4)) \times (1 - d_\eta/d_\alpha)$

where L_{CoG} is the longitudinal position of the center of gravity forward (positive) of the main wing quarter chord point, L_{tail} is the longitudinal position of the tail-plane quarter chord behind (negative) the main wing quarter chord point. The value of 2 used twice is from theoretically perfect inviscid two-dimensional thin airfoil theory of 2π for the lift curve slope – in practice, a value of 1.9 is more likely. The value of 3/4 used twice is the Oswald span efficiency and this is on the pessimistic side, 0.85 might be more likely. However, since both the perfect slope value and the span efficiency are applied to both wing and tail terms, the errors tend to cancel; if the main wing and tail-plane aspect ratios are equal they cancel completely. The terms essentially penalize low-aspect-ratio tail-planes slightly. Also in the downwash term d_η/d_α can be estimated from data provided in Raymer [11] (p. 482) and depends on wing aspect ratio (span^2/area); wing semispan (assuming a rectangular wing); vertical position of tailplane compared to the main wing; longitudinal position of tail-plane quarter chord point behind wing quarter chord point; tail aspect ratio; $r =$tail-plane longitudinal position/semi-span; $m =$tail-plane vertical position/semispan. We leave consideration of

Table 11.4   Variables that might be used to estimate UAV weights.

Name	Long name/definition	Typical value	Unit
Awing	Total wing area	1.54	m^2
AR	Aspect ratio (span^2/area)	9.00	—
Thick	Aerodynamic mean thickness	62.1	mm
Atail	Tailplane area	226 222	mm^2
Afin	Fin area	203 600	mm^2
y_ tail_ boom	Horizontal position of tail booms	271.5	mm
Span_ tail	Tailplane span	951.3	mm
Chord_ tail	Tailplane mean chord	237.8	mm
Height_ fin	Fin height (or semispan) for two fins	390.8	mm
Chord_ fin	Fin mean chord	260.5	mm
Vmax_ C	Maximum cruise speed	30.0	m/s
x_ main_ spar	Long position of main spar	0.0	mm
x_ fnt_ bkhd	Long position of front bulkhead	240.3	mm
x_ tail_ spar	Long position of tailplane spar	−1127.9	mm
x_ rear_ bkhd	Long position of rear bulkhead	−200	mm
x_ mid_ bkhd	Long position of middle bulkhead	20.2	mm
Depth_ Fuse	Fuselage depth	250	mm
Width_ Fuse	Fuselage width	190	mm
Len_ Nose	Nose length (forward of front bulkhead)	200	mm
Len_ Engine	Length of engine	125	mm
Mengine	Engine mass	2.072	kg
Dprop	Propeller diameter	494	mm
DTop	Design topology	3	—

dynamic stability until more detailed analysis is to take place, and instead rely on sensible tail volume coefficients to ensure a reasonable starting point has been chosen.

11.6   Powering and Propeller Sizing

By examining data from propeller manufacturers and their recommendations, it is possible to derive a simple regression curve that allows probable propeller diameter to be derived from engine capacity and likely pitch from diameter:

$$D(\text{in.}) = 4.239\text{Cap}(\text{cu. in.})^2 + 11.096\text{Cap}(\text{cu. in.}) + 14.616$$
$$P(\text{in.}) = 0.516D(\text{in.})$$

Table 11.5 Items for which weight estimates may be required and possible dependencies.

Item	Typical no. per a/c	Possible dependency
Wing spars (incl. center spars)	2	Wing area, spar length, speed
Wing ribs+ nylon parts	6	Wing cross-sectional area
Wing foam	2	Wing box volume
Wing covers	4	Wing area
Ailerons	2	Wing box volume
Flaps	2	Wing box volume
Tail boom	2	Tail-plane area, spar length, speed
Tail fin	2	Individual fin area
Tailplane	1	Tail-plane area
Rudder	2	Individual fin area
Elevators	1	Tail-plane area
Engine	2	List of engine weights
Muffler	2	List of engine weights
Propeller	2	Prop dia.
Generator	2	Engine mass/?fixed
Ignition unit	2	Engine mass/?fixed
Fuel tank	0	Engine mass/?fixed
Aileron servos	2	Wing area
Flap servos	0	Wing area
Throttle servo	1	Engine mass/?fixed
Rudder servo	2	Individual fin area
Elevator servo	1	Tail-plane area
Wheel steering servo	2	Wing area
Linkages and bell-cranks	8	Total wing area/?fixed

Also, one can estimate the likely engine capacity from engine power by looking at a range of small engines to get

Cap (cu. in.) = 0.0189(Power (kW))^2 + 0.9288Power (kW).

Then the (uncorrected) Abbott equations[3] link power, thrust, pitch, diameter, and rotational speed as:

$$Power(W) = P(in.) \times D(in.)^4 \times rpm^3 \times 5.33 \times 10^{-15}$$
$$StaticThrust(oz.) = P(in.) \times D(in.)^3 \times rpm^2 \times 10^{-10}.$$

---

[3] http://www.rcgroups.com/forums/archive/index.php/t-1217933.html

Table 11.6 Other items for which weight estimates may be required.

Item	Typical no. per a/c
R/C receiver	1
Batteries	2
Autopilot	2
Misc. avionics	1
Wiring and aerials	1
Engine covers	2
Main fuselage	0
Rear nacelle fuselage	2
Mid-wing skin & fuel tank	1
Rear wing box	1
Nose	2
Front bulkhead	2
Main undercarriage structure	2
Main wheels	2
Undercarriage mounting point	1
Catapult mounting	1
Tail undercarriage structure	2
Nose undercarriage structure	0
Tail wheels	2
Nose wheel	0

Therefore, we can eliminate rpm to link thrust to power:

StaticThrust(oz.) = $(D(\text{in.}) \times P(\text{in.}))^{1/3} \times (Power(\text{W})/5.33E - 15)^{2/3} \times 10^{-10}$.

So, given an engine power, we can estimate its likely static thrust when matched to a suitable propeller. Since we can also estimate the required thrust for cruise, banked turns, climb, and takeoff as described in the previous chapter, it is then possible to make an engine selection given the overall aircraft weight. For example, the estimated thrust needed for takeoff is given by

$$\frac{T}{W} = \frac{V_{\mathrm{L}}^2}{2g\,d_{\mathrm{GR}}} + \frac{q\,C_{\mathrm{D}}^{\mathrm{TO}}}{W/S} + \mu_{\mathrm{TO}}\left(1 - \frac{q\,C_{\mathrm{L}}^{\mathrm{TO}}}{W/S}\right)$$

where the three terms represent the kinetic energy required during the ground roll, the mean aerodynamic drag on the runway, and the mean rolling resistance on the runway (recall that here q is the dynamic pressure at 70.71% of the take-off speed V_{L}). The estimated thrust for climb is given by

$$\frac{T}{W} = \frac{V_V}{V} + \frac{q}{W/S}C_{Dmin} + k\frac{1}{q}\frac{W}{S}$$

where V_V is the vertical velocity in the climb and k is the lift-induced drag factor.

Next, using the UIUC Propeller Data Site,[4] variations in propeller thrust with airspeed for a given diameter and rotational speed can be deduced by regressing plots of the thrust coefficient C_T versus the advance J. For example, for thin electric propellers with 12 in. pitch, the thrust coefficient can be approximated from the advance ratio as $C_T = 0.4876J^4 - 0.6571J^3 + 0.0629J^2 + 0.0101J + 0.0900$. It is also possible, using JavaProp (and a suitable base propeller design operating at fixed torque), to calculate static thrust and to derive regression curves for thrust and required power as the forward velocity changes. In either case, these can be used to simulate runway roll-out and initial climb to check the results from the above equation and also to check whether the assumed propulsive efficiency is sensible. However, when the design is close to balance, we need to recognize that only existing engines can be specified and one should then switch to data for actual engines. Figure 5.3 given previously shows the powers and thrusts of typical UAV engine/propeller combinations.

11.7 Resulting Design: Decode-1

As part of the DECODE program mentioned in the introduction, we designed several aircraft adopting the principles set out in this book. The first of these, Decode-1, was a simple single-engine pusher aircraft that we used to gain information on performance and weights for subsequent design work. A pusher design was selected so as to give an uninterrupted view forward for the payload, but it does mean the fuselage length has to be sufficient to allow the payload mass to balance out the rear-mounted engine mass so as to yield an acceptable center of gravity. The aircraft was also designed to allow a range of different experimental wings to be trialed and was sized so as to fit inside our largest wind tunnel without significant modification. Figures 11.1 and 11.2 show this aircraft with conventional wings attached.

The basic design brief for the aircraft is given in Table 11.7. If this brief, together with the data from Tables 11.1–11.3, is fed into our spreadsheet system, which follows the algorithm previously outlined, the design details given in Tables 11.8 and 11.9 result; see also Figures 11.3–11.5, which show the inputs, results, and simplified layout sketches we create inside the spreadsheet during design studies[5]. We find that seeing the aircraft in planform and side view in this way is very helpful when making design judgments early on in the design process; it is immediately apparent how similar this is to the layout of the final aircraft.

Note that three of the constraints set out in Table 11.2 are actively limiting the design: the peak wing C_l/C_d at cruise speed is 16; the allowable rotation at takeoff before grounding happens is 17°; and the static margin (expressed in fractions of main wing chord) is 0.1. Of these, it is the C_l/C_d value that is most fundamental aerodynamically, the other two largely dictating the fuselage and tail-boom length. The maximum value of C_l/C_d is set by the wing technology being deployed and depends on the detailed choice of planform, sections, twist,

---

[4] http://m-selig.ae.illinois.edu/props/propDB.html.
[5] Note the highly structured spreadsheet layout we adopt, which also always show the formulae being used alongside the results calculated – the geometry page is additionally structured to allow a simple cut-and-paste into the python code we use to invoke the AirCONICS CAD package.

Figure 11.1   Decode-1 in the R.J. Mitchell wind tunnel with wheels and wing tips removed and electric motor for propeller drive.

Figure 11.2   Decode-1 in flight with nose camera fitted.

Table 11.7 Design brief for Decode-1.

Item	Value	Units
Payload mass	2	kg
Maximum take-off weight set by design	15	kg
Maximum cruise speed	30	m/s
Landing speed	15	m/s
Take-off speed	16	m/s
Length of ground roll on take-off	60	m
Runway altitude	0	m
Cruising height	121.92	m

Table 11.8 Resulting concept design from spreadsheet analysis for Decode-1.

Item	Value	Units
Maximum fuel weight (without reserve)	1.04	kg
Total wing area	0.966	m^2
Long position of front bulkhead	640	mm
Long position of tail-plane spar	−995	mm
Coefficient of lift at cruise speed	0.280	—
Maximum range at cruise speed (Breguet)	106.9	km
Lift at cruise speed	147.15	N
Drag at cruise speed	28.93	N
Endurance at cruise speed	60.0	min
Limit load factor	3.80	g
Turn rate	68.7	°/s
Longitudinal position of MTOW CoG fwd of main spar	42.8	mm
Maximum possible rotation at takeoff	17.0	°
Engine selected	OS Gemini FT-160	—
Specific fuel consumption at cruise speed	0.75	kg/kWh
Maximum required power	1.45	kW
Maximum installed power	1.49	kW
Coefficient of lift at landing speed	1.11	—
Wing C_l/C_d at cruise speed	16.0	—
Lift to drag ratio at cruise speed	5.09	—
Static margin	0.10	—
Thrust to weight ratio	0.361	—
Wing loading	15.53	kg/m^2

The first five entries here are manipulated by the solver tool within the spreadsheet to maximize the range.

Table 11.9 Design geometry from spreadsheet analysis for Decode-1 (in units of mm and to be read in conjunction with Tables 11.3 and 11.8).

Item	Value
Total wing span (rect. wing)	2948.3
Aerodynamic mean chord	327.6
Propeller diameter	435.0
Tail-plane span	797.4
Tail-plane mean chord	199.3
Fin height (or semi-span) for two fins	293.0
Fin mean chord	195.3
Long position of middle bulkhead	220.0
Horizontal position of tail booms	239.2

and wing tip treatment. The value of 16 used here is intimately related to the wing loading discussed in the previous chapter, and for operations at sea level and 30 m/s gives a wing loading of 15.5 kg/m^2. The value of C_l/C_d used is something we know we can achieve in practice using the construction methods we have adopted; higher values are possible but these tend to make the wings more difficult and expensive to make. For example, we chose to taper our wings linearly and avoid twist when working with hot-wire-cut foam; twisted or elliptic planform wings would give better control of induced drag, but instead we often make use of quite sophisticated 3D-printed wing tips to limit induced drag, see again Figure 11.2.

Once one is happy that the concept is sufficiently well developed to warrant further effort, the basic geometry details can then be used to generate a more realistic and fully three-dimensional outer envelope for the aircraft. To do this, we use the AirCONICS suite of programs,[6] which leads to the design shown in Figure 11.6. This can then be used for further analysis or to start the process of building a detailed CAD model suitable for generating manufacturing drawings. Note that the AirCONICS geometry includes wing taper, an aerodynamically shaped circular fuselage, faired in-tail booms, and slab-sided undercarriage legs. These are just working assumptions at this stage, but the resulting model is sufficiently detailed for reasonable CFD and FEA analysis. No attempt at this stage has been made to add control surfaces to the wings, tail, or fins, so any CFD results would be solely for the cruise configuration. These can be added later using AirCONICS, as will be seen in Chapter 13. The final design shown in Figure 11.2 has greater wing taper, wing tips, and a triangular section fuselage, but otherwise is broadly similar to this initial AirCONICS model.

11.8 A Bigger Single Engine Design: Decode-2

The next aircraft we consider is an enlarged variant of Decode-1 that has 50% greater payload capability and endurance, a larger margin on installed power (a shorter ground roll is

---

[6] https://aircraftgeometrycodes.wordpress.com/airconics/.

	A	B	C	D	E	F	G
1	Name	Long Name/ Definition	Group	Value	Formula	Units	Notes
2	Mpayload	payload mass	input	2.00	2	kg	
3	MTOW_de	max take-off weight set by design	input	15	15	kg	fixed by CAA reqt or design
4	DTop	Design Topology	input	1	1	–	single_pusher=1, single_tractor=2, twin_tractor=3
5	AR	aspect ratio (span^2 / area)	input	9	9	–	chosen for wing shape in use
6	Vmax_C	maximum cruise speed	input	30.0	30	m/s	the main operational speed
7	V_L	landing speed	input	15.0	15	m/s	this speed is a key design driver
8	V_T	take-off speed	input	16.0	16	m/s	estimated at the rotation point
9	H_TO	runway altitude	input	0	0	m	
10	Phi	bank angle in level turn	input	45	45	deg	chosen for adequate handling
11	ROC	rate of climb at 80% cruise speed	input	5	5	m/s	chosen for adequate handling
12	Gnd_Roll	length of ground roll on take-off	input	60	60	m	chosen for adequate handling
13	hn_minus	minimum required static margin	input	0.1	0.1	–	chosen for adequate handling
14	theta_TO_	minimum rotation angle required for take-of	input	17	17	deg	chosen for adequate handling
15	PCntFuel_	reserve fuel percentage	input	10.0	10	%	chosen as a safety factor
16	PCntEXThr	the percentage extra thrust desired for take-	input	0.0	0	%	chosen as a safety factor, use a negative number on twin engine designs where full one engine out performance is not needed
17	zeta_C	propulsive efficiency at cruise speed	input	0.600	0.6	–	this is a key design driver but 60% is conservative
18	CL/CDma	maximum wing Cl/Cd cruise speed	input	16.00	16	–	chosen for wing shape in use; Fp suggests this can be as high as 18 (was 12)
19	LD/max_C	maximum airframe L/D at cruise speed	input	8.00	8	–	chosen from wind tunnel data at 20 m/s, 2deg A0A
20	CLmax_L	max coefficient of lift at landing speed	input	1.30	1.3	–	this is a key design driver, we achieved 1.6 in tunnel tests with very large flaps; FP suggest 1.516 at 16 deg A0A
21	x_main_sp	long position of main spar	input	0.0	0	mm	chosen datum - only cange this from zero to explore impact of moving wing on CoG
22	x_fuel	long position of fuel tank if fitted	input	200.0	200	mm	set by design style
23	ARtail	tailplane aspect ratio (span^2 / area)	input	4.0	4	–	chosen for tailplane shape in use
24	ARfin	fin aspect ratio ((twice fin height)^2 / area)	input	3.0	3	–	chosen for fin shape in use, assumes two fins
25	Depth_Fus	fuselage depth	input	200	200	mm	
26	Width_Fus	fuselage width	input	150	150	mm	
27	Len_Nose	nose length (forward of front bulkhead)	input	150	150	mm	
28	Dia_Whee	diam of main undercariage wheels	input	100	100	mm	
29	Len_Engir	length of engine	input	125	125	mm	
30	x_rear_bkh	long position of rear bulkhead	input	-200	–200	mm	
31	z_fuse_ba	vert position of base of fuselage	input	-110	–110	mm	
32	z_tail_boo	vert position of tailboom	input	0	0	mm	
33	z_engine	vert position of engine	input	60	60	mm	
34	z_uncarria	vert position of centre of main undercarriage	input	-300	–300	mm	
35	x_payload	long position of payload	input	640	–D42	mm	needs to either be fixed or set equal to one of the moving bulkhead locations
36	Htvol	horizontal tailplane volume coefficient	input	0.5	0.5	–	value of 0.5 taken from Raymer (reduce by 0.7071 for V tail)
37	Vtvol	vertical tailplane volume coefficient	input	0.04	0.04	–	value of 0.04 taken from Raymer (reduce by 0.7071 for V tail)
38	H_C	max cruising height	input	121.92	=400*12*25.4/1000	m	400 foot CAA regs

Figure 11.3 Decode-1 spreadsheet snapshot – inputs page.

	A	B	C	D	E	F	G
39							
40	Mfuel	max fuel weight (without reserve)	solver	1.04	1.035865549811185	kg	set by solver to maximise range within overall take-off weight
41	Awing	total wing area	solver	0.966	0.9658300565835414	m^2	typically set by solver to size wings (implies choice of angle of attack) but limited by max CL/CD set in solver; FP suggests CL = 0.0884 AoA + 0.1132 at 24 m/s, AoA=(CL-0.1132)/0.0884 =1.66 degs
42	x_fnt_bkh	long position of front bulkhead	solver	640	640.305106180148	mm	set by solver to ensure CoG lies near main spar
43	x_tail_spa	long position of tailplane spar	solver	-995	-995.29345480668...	mm	should be set by solver to minimize tail mass, but have to iterate by hand to avoid cra...
44	CL_C	coefficient of lift at cruise speed	solver	0.280	0.2796282296597...	-	set by solver to size wings (implies choice of angle of attack) but limited by max CL/CD set in solver; FP suggests CL = 0.0884 AoA +0.1132 at 24 m/s, AoA=(CCL-0.1132)/0.0884 =1.66 degs
45	range	max range at cruise speed (pregaut)	goal	1069	=(3600/9.81)*(D17/D47)*(D48/D49)*LN(D3/ABS(D3-D40))	km	derived from aero sheet
46							
47	SFCS	specific fuel consumption at cruise speed	engine	0.75	=Engines!$D$18	kg/kw hr	estimated from engine data sheet
48	L_C	lift at cruise speed	aero cruise	14715	=WingAerodynamics!$D$19	N	max take-off weight
49	D_C	drag at cruise speed	aero cruise	28.93	=WingAerodynamics!$D$34	N	cruise speed, total wing area and Cd +F Cdp
50	enduranc	endurance at cruise speed design	design	0.99	=((1000*D45)/D6)/3600	hr	maximized by solver
51	Lfact	limit load factor	design	3.80	=MAX(1/COS(PI()*D10/180),2.5, MIN(3.8,2.1+24000/(10000-D3/0.453	g	based on regulations and desired bank angle
52	turn Rate	turn rate	design	68.7	=57.3*9.81*SQRT(D51^2-1)/D6	deg/sec	
53	DeltaAwir	error in wing estimate	design	5.39249E-05	=ABS(9.81*D3/(0.5*D44*D65*D62)-D41)	m^2	controls error in Solver solution
54	MTOW	calculated value of MTOW that solver must	weight	15.000	=D2+D40*(100+D/15)/100-Weights!$D$37	kg	derived from weights sheet
55	MTOW er	difference in MTOW values being reduced	weight	0.00025	=ABS(D3-D54)	kg	
56	LCoG	Longitudinal position of MTOW CoG fwd of	weight	42.82	=((1000*Weights!$D$37*Weights!$D$38+D40*D22+_D2*D35)/D3	mm	set by solver constraint to be near 1/4 chord
57	theta_TO	max possible rotation at take-off	geometry	1700	=Geometry!$D$36	degs	set by undercarriage height, tail boom length and tail boom height but kept to at least 12 degs by solver
58	Pmax	max required power	engine	1.45	=D49*D6/D17/1000	kw	thrust, prop effcy and speed
59	Pinst	max installed power	engine	1.49	=Engines!$D$17	kw	set to be the next biggest engine compared to Pmax in the look-up table, factor 2 allowing for one-engine failure on twin engine aircraft
60							
61	CL_L	coefficient of lift at landing speed	aero landi	1.11	=WingAerodynamics!$D$45	-	limited by solver, wing area, landing speed and required lift
62	CL/CD_C	Cl/Cd at cruise speed	aero cruis	16.00	=WingAerodynamics!$D$24	-	CL and CD, but limited by solver to user specified upper limit
63	L/D_C	lift to drag ratio at cruise speed	aero cruis	5.09	=WingAerodynamics!$D$37	-	L and D, but limited by solver user specified upper limit
64	hn_Minus	static margin	geometry	0.1000	=((D56/1000)/AORT(D41/D5)-0.25)+(0.25+(D56-D43)/1000*D23/1000000/D41*SQRT(D41/D5))*(0.75*D23/(0.75*D)23+2)/(0.75*D5*D5+2))*(1-DownwashData!B34)	-	uses Oswald span efficiency of 0.75 and expressed in fractions of main chord, first part of expression allows for datum to be at 0.25 chord paint, limited by solver to be > 0.25
65	rho_C	Density at cruise height	aero cruis	1.21	=WingAerodynamics!$D$40	kg/m3	
66							
67	T/W	thrust to weight ratio	engine	0.361	=Engines!$D$23	-	
68	W/S	wing loading	aero cruis	15.53	=D3/41	kg/m2	
69							
70	x_mid_bk	long position of middle bulkhead	geometry	220	=AVERAGE(D42,D30)	mm	
71							
72	**Type**	**Constraint**	**Lower Lit**	**Value**	**Upper Limit**	**Units**	**Comments**
73	Search B	0.01 <Mfuel <40	0.01	1.04	40	kg	limits search range and requires positive fuel load
74		0.01 < CL_C < 2	0.01	0.280	2	-	limits search range and restricts cruise Cl to realistic value
75		0.01 <Awing < 8	0.01	0.966	8	m^2	limits search range and restricts wing areas to realistic value
76		200 <x_fnt_bkhd <3000	200	640	3000.00	mm	limits search range and ensures sensible fuselage geometry
77		-2000 x_tail spar <-600	-2000	-995	-600	mm	limits search range and ensures sensible tail geometry
78	Aero Con	CL_L < CLmax_L		1.11	1.30	-	limits cruise Cl to realistic value
79		CL/CD_C < CL/CDmax_C		16.00	16.00	-	limits cruise Cl/CD to realistic value
80		L/D_C < L/Dmax_C		5.09	8.00	-	limits cruise L/D to realistic value
81	Geom Co	0.1 <static margin	0.100	0.100		-	requires positive static margin
82		17 <theta_TO	17	17.00		deg	ensures adequate rotation clearance on takeoff
83	Solution	(Abs(MTOW-MTOW_Est)		0.00025	0.00025	kg	ensures worksheet is converged
84		error in Awing < setting accuracy		0.00005	0.0001	m^2	ensures worksheet is converged

Figure 11.4 Decode-1 spreadsheet snapshot – results summary page.

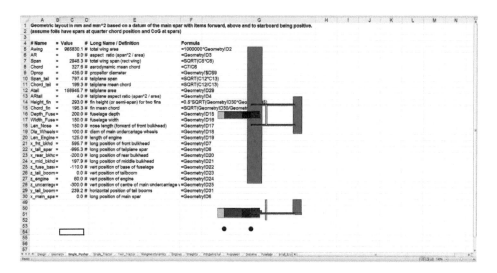

Figure 11.5   Decode-1 spreadsheet snapshot – geometry page.

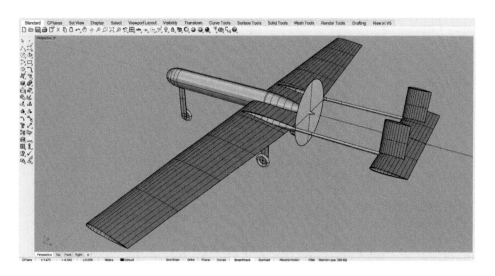

Figure 11.6   Decode-1 outer geometry as generated with the AirCONICS tool suite.

specified), and usefully lower landing speed, all made possible by the use of an inverted V tail and large flaps. The inverted V tail is allowed in the design process by reducing the tail volume coefficients needed by 30%, which gives a smaller, lighter tail and easier control of the static margin. The flaps allow a greater lift coefficient in the landing configuration and hence lower approach speed. Decode-2 is also a single pusher design with a nose-mounted camera system. The design brief is set out in Table 11.10 and the secondary dimensions in Table 11.11. The wing loading is slightly lower (this is now controlled by the choice of maximum landing $C_{l\,land}$ of 1.55) but a slightly less stringent rotation at takeoff of 14° is chosen. The other parameters given in Table 11.2 are held fixed as before.

Table 11.10   Design brief for Decode-2.

Item	Value	Units
Payload mass	3	kg
Maximum take-off weight set by design	23.5	kg
Maximum cruise speed	30	m/s
Landing speed	12.5	m/s
Take-off speed	16	m/s
Length of ground roll on take-off	45	m
Runway altitude	0	m
Cruising height	121.92	m

Table 11.11   Estimated secondary airframe dimensions for Decode-2.

Item	Value	Units
Fuselage depth	225	mm
Fuselage width	250	mm
Nose length (forward of front bulkhead)	200	mm
Diameter of main undercarriage wheels	125	mm
Longitudinal position of engine bulkhead	−250	mm
Vertical position of base of fuselage	−110	mm
Vertical position of tail-plane	0.0	mm
Vertical position of engine	60	mm
Vertical position of center of main undercarriage wheels	−325	mm

Table 11.12   Resulting concept design from spreadsheet analysis for Decode-2.

Item	Value	Units
Maximum fuel weight (without reserve)	0.40	kg
Total wing area	1.554	m^2
Long position of front bulkhead	688	mm
Long position of tailplane spar	−1407	mm
Coefficient of lift at cruise speed	0.272	—
Maximum range at cruise speed (Breguet)	37.3	km
Lift at cruise speed	230.54	N
Drag at cruise speed	46.17	N

(*continued*)

Table 11.12 (*Continued*)

Item	Value	Units
Endurance at cruise speed	20.7	min
Limit load factor	3.80	g
Turn rate	68.7	°/s
Longitudinal position of MTOW CoG fwd of main spar	42.1	mm
Maximum possible rotation at takeoff	14.00	°
Engine selected	Saito FG-57T	—
Specific fuel consumption at cruise speed	0.5	kg/kWh
Maximum required power	2.31	kW
Maximum installed power	3.04	kW
Coefficient of lift at landing speed	1.55	—
C_l/C_d at cruise speed	15.99	—
Lift to drag ratio at cruise speed	4.99	—
Static margin	0.1000	—
Thrust to weight ratio	0.431	—
Wing loading	15.12	kg/m^2

The first five entries here are manipulated by the solver tool within the spreadsheet to maximize the range.

The resulting design is detailed in Tables 11.12 and 11.13 along with the screenshot in Figure 11.7. The aircraft is shown in flight in Figure 11.8 and the outer shape modeled with AirCONICS in Figure 11.9. Note that our spreadsheet sketch does not distinguish between an H and an inverted V tail (simple horizontal and vertical projections being provided); this level of detail has to be worked up after the initial spreadsheet-based concept has been defined, which is straightforward to create within AirCONICS.

11.9 A Twin Tractor Design: SPOTTER

Our third example is the twin tractor aircraft SPOTTER. This is bigger again and is designed to have a high level of redundancy to ensure better operational resilience. The aircraft is designed to carry an interchangeable underslung payload pod of up to 5 kg weight and makes use of an integral fuel tank. It carries dual on-board charging systems powered from the main engines. The design brief is set out in Table 11.14 and the estimated secondary dimensions in Table 11.15. The wing loading is now slightly lower (this is again controlled by the choice of maximum landing $C_{l\,\text{land}}$ of 1.55), and a rotation at takeoff of 15° is used. The other parameters given in Table 11.2 are held fixed as before. Here the engines are sized so as to be

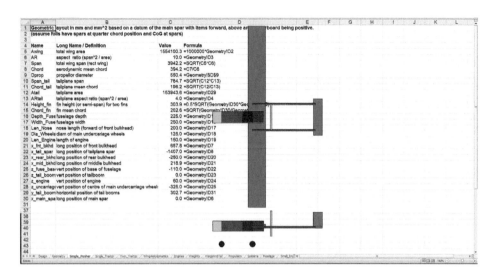

	Name	Long Name / Definition	Value	Formula
1	Geometric layout in mm and mm^2 based on a datum of the main spar with items forward, above and starboard being positive.			
2	(assume foils have spars at quarter chord position and CoG at spars)			
4	Name	Long Name / Definition	Value	Formula
5	Awing	total wing area	1554100.3	=1000000*Geometry!D2
6	AR	aspect ratio (span^2 / area)	10.0	=Geometry!D3
7	Span	total wing span (rect wing)	3942.2	=SQRT(C5*C6)
8	Chord	aerodynamic mean chord	394.2	=C7/C6
9	Dprop	propellor diameter	550.4	=Geometry!$D$9
10	Span_tail	tailplane span	784.7	=SQRT(C12*C13)
11	Chord_tail	tailplane mean chord	196.2	=SQRT(C12/C13)
12	Atail	tailplane area	153943.6	=Geometry!D29
13	ARtail	tailplane aspect ratio (span^2 / area)	4.0	=Geometry!D4
14	Height_fin	fin height (or semi-span) for two fins	303.9	=0.5*SQRT(Geometry!D30*Geo
15	Chord_fin	fin mean chord	202.6	=SQRT(Geometry!D30/Geomet
16	Depth_Fuse	fuselage depth	225.0	=Geometry!D1
17	Width_Fuse	fuselage width	250.0	=Geometry!D16
18	Len_Nose	nose length (forward of front bulkhead)	200.0	=Geometry!D17
19	Dia_Wheels	diam of main undercarriage wheels	125.0	=Geometry!D18
20	Len_Engine	length of engine	150.0	=Geometry!D19
21	x_fnt_bkhd	long position of front bulkhead	687.8	=Geometry!D7
22	x_tail_spar	long position of tailplane spar	-1407.0	=Geometry!D8
23	x_rear_bkhd	long position of rear bulkhead	-250.0	=Geometry!D20
24	x_mid_bkhd	long position of middle bulkhead	218.9	=Geometry!D21
25	x_fuse_bas	vert position of base of fuselage	-110.0	=Geometry!D22
26	z_tail_boom	vert position of tailboom	0.0	=Geometry!D23
27	z_engine	vert position of engine	60.0	=Geometry!D24
28	z_uncarriag	vert position of centre of main undercarriage wheel	-325.0	=Geometry!D25
29	y_tail_boom	horizontal position of tail booms	302.7	=Geometry!D31
30	x_main_spa	long position of main spar	0.0	=Geometry!D6

Figure 11.7 Decode-2 spreadsheet snapshot.

Figure 11.8 Decode-2 in flight with nose camera fitted.

Table 11.13 Design geometry from spreadsheet analysis for Decode-2 (in units of mm and to be read in conjunction with Tables 11.11 and 11.12).

Item	Value
Total wing span (rect. wing)	3942.2
Aerodynamic mean chord	394.2
Propeller diameter	550.4
Tail-plane span	784.7
Tail-plane mean chord	196.2
Fin height (or semi-span) for two fins	303.9
Fin mean chord	202.6
Long position of middle bulkhead	218.9
Horizontal position of tail booms	302.7

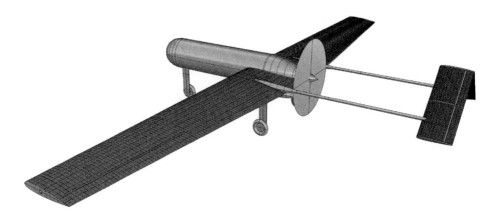

Figure 11.9 Decode-2 outer geometry as generated with the AirCONICS tool suite.

Table 11.14 Design brief for SPOTTER.

Item	Value	Units
Payload mass	5	kg
Maximum take-off weight set by design	30	kg
Maximum cruise speed	30	m/s
Landing speed	12.5	m/s
Take-off speed	16	m/s
Length of ground roll on takeoff	60	m
Runway altitude	0	m
Cruising height	121.92	m

Table 11.15   Estimated secondary airframe dimensions for SPOTTER.

Item	Value	Units
Length of engine	150	mm
Fuselage depth	125	mm
Fuselage width	125	mm
Nose length (forward of front bulkhead)	200	mm
Diameter of main undercarriage wheels	150	mm
Longitudinal position of engine bulkhead	580	mm
Vertical position of base of fuselage	0.0	mm
Vertical position of tailplane	0.0	mm
Vertical position of engine	60	mm
Vertical position of center of main undercarriage wheels	−350	mm

Table 11.16   Resulting concept design from spreadsheet analysis for SPOTTER.

Item	Value	
Maximum fuel weight (without reserve)	1.18	kg
Total wing area	1.957	m^2
Long position of front bulkhead	580	mm
Long position of tailplane spar	−1370	mm
Coefficient of lift at cruise speed	0.276	—
Maximum range at cruise speed (Breguet)	77.7	km
Lift at cruise speed	313.92	N
Drag at cruise speed	62.46	N
Endurance at cruise speed	41.8	min
Limit load factor	3.80	g
Turn rate	66.5	°/s
Longitudinal position of MTOW CoG fwd of main spar	47.94	mm
Maximum possible rotation at take-off	15.00	°
Engine selected	Twin OS GF40	—
Specific fuel consumption at cruise speed	0.5	kg/kWh
Maximum required power	2.98	kW
Maximum installed power	5.60	kW

(continued)

Table 11.16   (*Continued*)

Item	Value	
Coefficient of lift at landing speed	1.55	—
C_l/C_d at cruise speed	15.87	—
Lift to drag ratio at cruise speed	5.03	—
Static margin	0.1000	—
Thrust to weight ratio	0.489	—
Wing loading	16.36	kg/m²

The first five entries here are manipulated by the solver tool within the spreadsheet to maximize the range.

Table 11.17   Design geometry from spreadsheet analysis for SPOTTER (in units of mm and to be read in conjunction with Tables 11.15 and 11.16).

Item	Value
Total wing span (rect. wing)	4196.3
Aerodynamic mean chord	466.3
Propeller diameter	493.6
Tail-plane span	1154.1
Tail-plane mean chord	288.5
Fin height (or semispan) for two fins	424.1
Fin mean chord	282.7
Long position of middle bulkhead	165.0
Horizontal position of tail booms	271.5

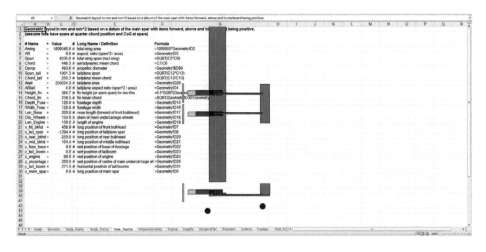

Figure 11.10   SPOTTER spreadsheet snapshot.

Figure 11.11 SPOTTER in flight with payload pod fitted.

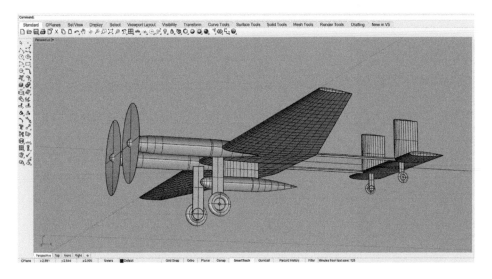

Figure 11.12 SPOTTER outer geometry as generated with the AirCONICS tool suite.

able to maintain flight with one engine nonoperational but not so as to be able to complete a take-off run with only one engine with full fuel and payload. The resulting design is detailed in Tables 11.16 and 11.17 along with the screenshot in Figure 11.10. The aircraft is shown in flight in Figure 11.11 and the outer shape modeled with AirCONICS in Figure 11.12.

12

Preliminary Geometry Design

Once the spreadsheet analysis has been completed and a set of preliminary dimensions has been found, a more realistic series of analyses can be undertaken before the project is passed on for detailed design. The purpose of these analyses is to check that a number of the key assumptions made when using spreadsheets to size the aircraft will be borne out in practice. The primary checks that need to be completed before proceeding on to detail design are typically the following:

- that the lift and drag estimates for the aircraft in cruise, turn, takeoff, and landing configurations are sufficiently accurate that the desired aircraft performance will be achieved.
- that the flight stability of the aircraft will be acceptable.
- that acceptable stress levels in the primary aircraft structure can be achieved within the allowable weight budget.
- that the aircraft can be made sufficiently stiff to avoid control surface operational and aeroelastic problems within the allowable weight budget.

The majority of the calculations underpinning these checks depend on the quality of the geometrical description available, so we begin by first building a much more realistic model of the airframe. Such models can be generated in a variety of ways but, if they are to include more than just the lifting surfaces, they almost always depend on some form of computer aided design (CAD) program. Here we continue to use the AirCONICS system[1] that drives the Rhino CAD platform.[2] Typical outputs from this process have already been shown in the previous chapter; here we give more details on how this works and the steps involved in producing realistic wetted surfaces for the airframe, including control surfaces and a series of components sufficient for preliminary structural analysis.

Once this definition is in place, more detailed experimental aerodynamic data and computational fluid dynamics (CFD) approaches can be used to check the lift and drag performance and stability of the airframe against the assumptions already made in the spreadsheet analysis. If necessary, the spreadsheet process can be refined in the light of such information and the

---

[1] https://aircraftgeometrycodes.wordpress.com/airconics/.
[2] http://www.rhino3d.com/.

Small Unmanned Fixed-wing Aircraft Design: A Practical Approach, First Edition.
Andrew J. Keane, András Sóbester and James P. Scanlan.
© 2017 John Wiley & Sons Ltd. Published 2017 by John Wiley & Sons Ltd.

geometry updated: generally any deficiencies in the aerodynamic characteristics will lead to changes in the overall planform shape. Then attention can be turned to the internal definition of the airframe structure and more refined hand calculations, backed up by finite element analysis (FEA), applied to check stressing and stiffness, including aeroelastic effects. When an adequate structural model is also in place, a relatively rigorous weight and center of gravity check can be made. Again, if the spreadsheet assumptions are inadequate, refinements can be made and the whole process updated – structural problems typically lead to increases in the estimated weights, which will adversely impact on wing loading and thus may require modifications to the planform shape. Finally, and only when all is well, the detailed design that ultimately leads to the final definitions needed for manufacture can begin. Sometimes a series of experimental models will be built and tested before final production is committed to; these can be used to validate any computations carried out as well as providing data for future design studies.

12.1   Preliminary Airframe Geometry and CAD

Because it may very well be necessary to iteratively build and analyze the airframe geometry model several times before a fully accurate and balanced design is achieved that passes the checks just noted, it is of enormous importance that the process for generating the airframe geometry is as automated and robust as possible. This is precisely the rationale behind the AirCONICS suite and why we find its use in this stage of design so important. It is, however, not the only way to produce initial CAD definitions; almost all modern CAD packages support some form of parametric geometry capability that can be programatically controlled. Essentially, this involves first deciding the basic topology of the airframe and then linking as many geometry variables as possible to each other so that only a small number need to be specified as the airframe is changed during sizing and balancing. For the examples studied here, the full geometry is typically fully defined by around 20 master dimensions, see, for example, those specified in Tables 11.9, 11.3, and 11.8.

To generate the geometry used for analysis, we first need an outer wetted surface for the aircraft. As already noted, our preferred tool is Rhino since this can be readily controlled with Python scripts and quickly produces clean and closed wetted surfaces. Since all aircraft contain a range of similar features, most noticeably airfoil-like lifting surfaces, the AirCONICS suite of Python scripts generates these very efficiently. The basic components of the airframe are then built from wing, taiplane, and fin lifting surfaces, linked together via tubular spars with a range of "pods" and cylinders created as elongated bodies of revolution to represent fuselages, junctions, and cowlings. All these elements are joined together to form a single closed "polysurface" that can be exported directly to the analysis tools used for aerodynamic and structural analysis. Simple undercarriage elements can be included by using toroidal tires and fin-like structures for the legs (although it is normal to omit these while carrying out initial CFD validations of drag). Payload pods can also be added at this stage if required. Figure 12.1 shows a typical Rhino airframe as built using AirCONICS, here for a single tractor engine, twin-boom, H-tail design. Note that a tapered wing planform has been adopted but with straight leading edge – this simplifies the structural design by allowing adequate space for a straight wing spar.

The Python scripts used to create our geometries essentially consist of three distinct sections: first, a short section of code is used to define the variables taken from the spreadsheet as a set

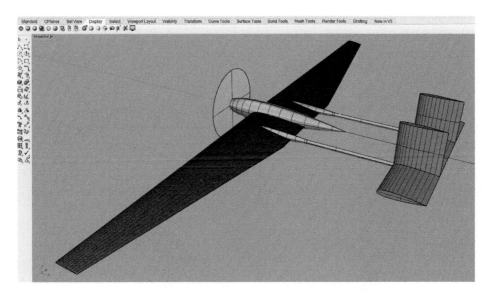

Figure 12.1 Basic AirCONICS airframe geometry for a single tractor engine, twin-boom, H-tail design.

of controlling parameters, see, for example, Figure 11.5. Second, all the main functions are defined that will be used in the geometry creation. Typically this would include the following:

- A set of functions that vary with span to define each lifting surface (generally different for the wings, tail and fins):
  - *DihedralFunction*. A user-defined function describing the variation of dihedral as a function of the leading-edge spanwise coordinate.
  - *TwistFunction*. A user-defined function describing the variation of twist as a function of the leading-edge spanwise coordinate.
  - *ChordFunction*. A user-defined function describing the variation of chord as a function of the leading-edge spanwise coordinate.
  - *SweepAngleFunction*. A user-defined function describing the variation of sweep angle as a function of the leading-edge spanwise coordinate.
  - *AirfoilFunction*. A user-defined function of the previous functions to set up the variation of cross-section as they vary with the leading edge spanwise coordinate.
- A small group of operators to define pods as bodies of rotation created from airfoil sections with optional parallel middle sections.
- A small group of operators to define landing gear elements created from cylinders, struts, and a torus for the tire.

Following this, the main body of the script invokes the various functions plus a few inbuilt objects, such as simple cylinders, to build all the aircraft elements and union them together into a single closed "polysurface" solid that can be exported either as an STL file to a meshing tool or as a Parasolid, ACIS, STEP, or IGES file for use with other CAD packages. We also provide options in the script to include control surfaces and to subdivide the whole airframe into parts suitable for preliminary structural analysis.

12.2   Designing Decode-1 with AirCONICS

If the dimensions for the Decode-1 design given in the previous chapter are used in this way, the Rhino geometry already shown in Figure 11.3 can be generated. Note that a circular cross-section fuselage has been adopted but with airfoil-like leading and trailing parts; parts at the rear would need to be cut away to accommodate the engine in the actual design, but at this stage it is not practical to model the full details of the engine nor would this be desirable for preliminary aerodynamic calculations. A propeller disk and hub can be modeled, however, by

Figure 12.2   AirCONICS model of complete Decode-1 airframe with control surfaces, undercarriage, and propeller disk.

extending the fuselage cone down to a point and including an actuator disk model in the CFD analysis. In the actual design, the entire middle body of the aircraft between the tail booms and including their junctions with the wing are to be manufactured as a set of SLS nylon parts. It will therefore be possible in the detailed design stage to add fillets and fairings between the various items. Whether or not such items matter during preliminary design is not completely obvious – they tend to make little difference to the CFD drag predictions but they will help reduce drag in reality. However, since drag prediction using CFD is notoriously difficult, this may not be worth the effort at this stage and any drag savings are likely to be swamped by losses arising from items like the unfaired engine and various unrepresented items on the undercarriage legs and wheels, aerials, and so on.

The complete Python script for producing the Decode-1 AirCONICS model is given in Appendix B. This script enables the outer mold surface to be generated and the resulting body to be divided into its major structural parts, see Figure 12.2. This is sufficient for aerodynamic and stability analysis and the structural sizing of the main spars and booms. To carry out more precise stressing or to prepare parts for manufacture, further effort is required. Some of these steps can be fully automated with very little further designer input, as will be seen later, while other aspects will require the selection of various additional subsystems such as engines and undercarriage, servos, and spar clamping arrangements. All these aspects should ideally just require selections to be made from libraries of pre-existing parts, which are then joined together with the outer mold line definitions. Generally, the addition of such parts lies in the scope of detailed design, see Chapter 17.

13

Preliminary Aerodynamic and Stability Analysis

A more precise analysis of the aerodynamic flows around the unmanned air vehicle (UAV) can be used to obtain better estimates of the lift, drag, moments, or stall behavior than can be gained by scaling from the fixed coefficients and angles used in the spreadsheet analysis. It may also be desirable to estimate local pressure forces on the lifting surfaces for full stability analysis or subsequent structural analysis. There is generally no point attempting to further size and optimize wings and powerplant without more realistic aerodynamic data. Ideally it would be possible to swiftly take the entire air vehicle geometry, as created in Rhino, select an angle of attack (AoA) and flight speed, and run a series of calculations to gain the information required. It is, however, still not easy to accurately predict the aerodynamic performance of airframes by calculation, so a range of approaches should be adopted to suit the project in hand.

At the most basic level, the many tabulated sets of experimental results in Abbott and von Doenhoff [20] or Hoerner [9, 10] can be used to estimate forces on individual airframe elements, but it is difficult to allow for the interference effects between the multiple elements that make up a complete air vehicle. To proceed further, computational fluid dynamics (CFD) methods are generally used. The main task in CFD-based aerodynamic analysis, given a suitable outer wetted surface model, is the generation of an appropriate panel or mesh representation that can be used to drive the chosen CFD solver. The choice of solver to use will depend on the type of results required. The simplest reliable physics-based computational tools are generally panel codes with coupled empirical boundary layer models. The most well known of these that are freely available are probably XFoil,[1] which can analyze two-dimensional airfoil sections, and XFLR5,[2] which combines results from XFoil to allow wings and combinations of wings to be studied (sometimes with simple fuselages). Neither code will predict stall with complete accuracy, but they are good at understanding lift at reasonable angles of attack and drag at such angles for lifting surfaces. When used properly, it should be possible to allow

[1] http://web.mit.edu/drela/Public/web/xfoil/.
[2] http://www.xflr5.com/.

Small Unmanned Fixed-wing Aircraft Design: A Practical Approach, First Edition.
Andrew J. Keane, András Sóbester and James P. Scanlan.
© 2017 John Wiley & Sons Ltd. Published 2017 by John Wiley & Sons Ltd.

for the interference of the flows between the main wings and the tail and to estimate local pressures on the lifting surfaces for use in stability and structural analysis. To go beyond this, some form of mesh-based approach is currently the aerospace industry standard, typically in the form of a Reynold's averaged Navier–Stokes (RANS) solver such as Fluent$^{®}$,[3] Star-CD$^{®}$,[4] or OpenFoam$^{®}$.[5]

During the earliest stages of design for a small UAV, the costs of very detailed RANS-based CFD analysis can rarely be justified, so panel code results should definitely be used to first check that the primary lifting surface lifts and drags at moderate angles of attack are broadly as expected. If sensible lift and drag coefficients have been used in the spreadsheet analysis, this should always be the case. If the use of panel codes reveals significant differences in the low-AoA lift and drag, something will be wrong and this should be resolved before proceeding. If consistent results are being generated, these may be all that is required before carrying out a full stability analysis and then moving on to carry out the structural analysis. Sometimes, however, it is desirable to deal with more detailed geometrical descriptions or more complex vortex and boundary layer flows, such as those occurring near wing-tip devices and in slotted flaps. If such information is required, work can begin on RANS-based analysis, which can deal with arbitrarily complex geometries. Common to such solvers is the need for an appropriate mesh to describe the surface of the airframe and extending into the fluid domain. Given such a mesh, and provided suitable solver choices are made, it is possible to attempt predictions of the flow around complete air vehicles at quite high angles of attack.

Unfortunately, getting accurate results from RANS codes, particularly for separated flow, is by no means straightforward; any designer new to such methods is well advised to treat their results with great caution unless and until they have been able to produce results for a few known configurations that agree with experiments. When carrying out validation tests, the NASA technical memorandum server[6] is a goldmine of valuable data as is the NASA Turbulence Modeling Resource.[7] The turbulence modeling resource gives examples of using CFD to establish the behavior of several well-studied airfoils showing what may be expected from the very best quality meshes. Even their studies are unable to accurately predict everything that one might wish for. RANS-based CFD meshing and solution is, of course, a huge topic of research and there are major conferences devoted to the subject of drag prediction of aerodynamic surfaces, for example. Clearly, in a text such as this, only a very simple overview of good practice can be given. Consequently, the approaches offered here are aimed at small teams with limited resources; in no way are they meant to represent the best practice of large aerospace companies.

13.1 Panel Method Solvers – XFoil and XFLR5

XFoil[8] is a panel-based method that uses full potential methods to calculate the bulk flows around two-dimensional sections which can then be corrected for the presence of the

---

[3] http://www.ansys.com/Products/Simulation+Technology/Fluid+Dynamics/Fluid+Dynamics+Products/ANSYS+Fluent.
[4] http://www.cd-adapco.com/products/star-cd%C2%AE.
[5] http://www.openfoam.com/.
[6] http://naca.larc.nasa.gov/search.jsp.
[7] http://turbmodels.larc.nasa.gov/index.html.
[8] http://web.mit.edu/drela/Public/web/xfoil/.

boundary layer. The code is aimed at dealing with viscous analysis of airfoils, allowing for forced or free transition, transitional separation bubbles, limited trailing edge separation, and lift and drag predictions just beyond C_{Lmax}. Since the corrections used are based on well-established experimental data, very good approximations of lift and drag can be predicted as long as the boundary layer remains attached to the airfoil. Once separation has begun, leading on to stall, panel codes are less reliable, which is a fundamental limitation of such methods. Figure 13.1 shows the kinds of data provided by XFoil, here at a high AoA where separation has begun. To obtain this result, the following XFoil steps are used:

- The desired foil is selected with the "NACA 0012" command.
- The panel nodes are set directly from the foil with the "PCOP" command.
- The direct operating point menu is entered with the "OPER" command.
- Viscous analysis is selected with the "Visc" command.
- The Reynold's number (here 4.4 million) is entered with the "$Re4\,400\,000$" command.
- The airfoil behavior at the desired AoA (here 16°) is calculated with the "ALFA 16" command.

By dividing the lifting surfaces into stream-wise strips, it is possible to take the results from XFoil and compute the behavior of whole wings – this is what XFLR5 does,[9] while also allowing for simple streamlined fuselage elements and multiple combinations of wings. Again, it will not deal so well with heavily separated or stalled flows and, although interference

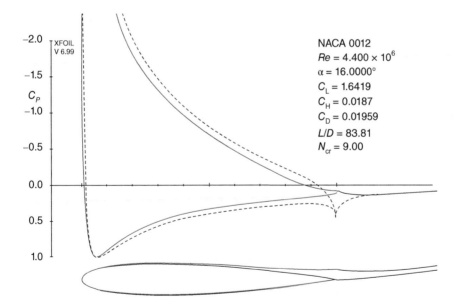

Figure 13.1 C_p and streamline plot for the NACA0012 foil at 16° angle of attack as computed with XFoil.

---

[9] http://www.xflr5.com/.

effects between widely spaced lifting elements (such as wings and tails) can be dealt with, such methods cannot accurately predict the benefits of slotted flaps and other boundary layer control systems. XFLR5 does, however, readily permit calculation of the aircraft's dynamic stability, which can be used to check whether elevator and fin sizes and positions are acceptable given the likely mass, inertias, and flight speeds. To analyze a simple wing with XFLR5, the following steps are used:

- The desired foil section to be used in the wing design is chosen along with the operational Mach number for the wing; the section is loaded into XFLR5 via the "Direct Foil Design" menu. AirCONICS provides a good selection of basic foils, the UIUC database has many more.
- The foil section is then analyzed with the "XFoil Direct Analysis" menu in batch mode; analysis type 1 is used to specify a range of section operational Reynold's numbers and an increment in these along with a sweep through likely section angles of attack. This builds an internal database of results using XFoil from which the wing analysis can be generated (typical values are Reynold's numbers from 100 000 to 6 million in steps of 100 000 with AoA sweep from $-4°$ to $16°$. The system simply aborts any combinations that cannot be converged by XFoil; such a sweep typically takes 2 or 3 min. Figure 13.2 shows the results for the NACA 64–210 section at Mach 0.17).
- The wing is loaded under the "Wing and Plane Design" menu by reading an XML file as created by the AirCONICS system for the wing of interest; this comprises the locations and sizes of sections along the wing and references the previously chosen airfoil section. (Note: if this step is carried out before the chosen section has been loaded, a simple planform diagram is created instead.)
- Alternatively, the wing is created directly in XFLR5 with manual inputs, in the "Plane Define a new plane" area, where a title and description can be set before defining the main wing; each section requires values for y, chord, (leading edge) offset, dihedral, twist, section name (from the drop-down of sections built earlier), and the number and distribution of panels (typically between 10 and 30 in cosine or sine pattern).
- The desired analysis is defined by using type 1 again but now with the chosen operational speed and one of the four available analysis types (Lifting Line Theory, Horseshoe Vortex, Ring Vortex, or 3D Panels; the first of these does not allow viscous affects to be included, so we prefer to use the 3D Panel approach as the computational costs are still slight, but this is possible only if the tail and fin are not included. 3D panels have to be used to generate full pressure maps for structural analysis, however).
- The computed reference area, span, and chord length should be checked against what is expected.
- If desired, inertia terms, nonstandard atmospheric properties, and extra drag items can next be added.
- The sequence of angles of attack can then be specified and analyzed and checked for any errors; the most likely error is that one of the sections along the wing goes outside the previously computed set of polars for a given wing AoA (a local section lift coefficient is requested that is above what can be achieved for the section at the given operating point). If this happens, modify the chosen sweep of angles to ensure that all angles can be fully

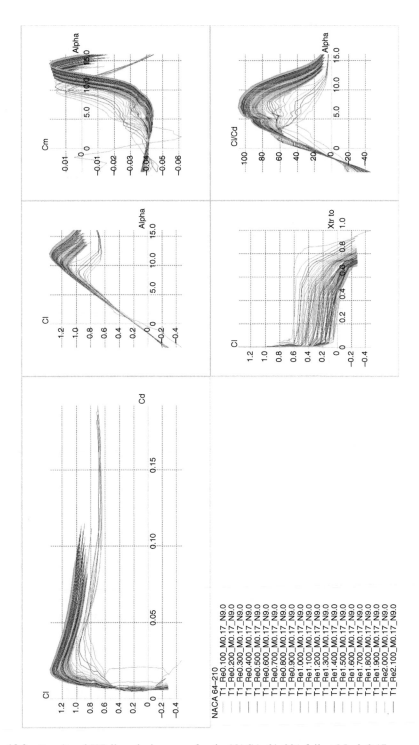

Figure 13.2 Results of XFoil analysis sweep for the NACA 64–201 foil at Mach 0.17 as computed with XFLR5.

dealt with.[10] Figure 13.3 shows the results of a typical sweep through the available angles of attack. These results can be exported to a text file for subsequent analysis or comparison with other results.

XFLR5 also allows tail fins and elevators to be added and also a second wing (typically for bi-plane configurations). These are all defined as for the main wing and, in the case of fins, can be either single- or double-sided. All the surfaces can have dihedral, sweep, and twist as required, and the code allows for the flow fields around the individual items to impact on each other in terms of downwash but not wake effects. It is also possible to add fuselage-like elements to link the lifting surfaces together but these are not correctly modeled aerodynamically. Dynamic stability analysis can readily be carried out to give the natural frequencies, damping of modes, and so on. To demonstrate what these codes can achieve during design, they are used in the subsequent subsections to predict the behavior of simple 2D airfoil sections, simple wings, and a built-up airframe.

13.2   RANS Solvers – Fluent

There is a range of well-established RANS-based CFD solvers, many of these provide student editions and some, such as OpenFOAM,[11] are free to download. We tend to use the Fluent system supplied by Ansys Inc.[12] It is a very well regarded code with many years of development behind it. It also offers a range of boundary-layer modeling capabilities suitable for UAV lift and drag prediction, see Table 13.1. The basic settings we adopt are the following:

- pressure-based, steady-state, 3D, double-precision MPI parallel solver;
- viscous Spalart–Allmaras or k-ω SST turbulence model with standard air for the fluid domain;
- symmetry boundary conditions to either side of the aircraft (or to one side and on the center plane for half models);
- pressure outlet boundary conditions above and behind the model (for positive angles of attack – the upper and lower boundary conditions are swapped for negative angles);
- velocity inlet boundary conditions in front and below the model (for positive angles of attack, the ratio of velocity components sets the AoA, and for zero AoA we make both upper and lower boundaries symmetry planes);
- a reference area equal to the wing planform, a reference length equal to the root chord, and air density and viscosity at the take-off height;
- the SIMPLE solution method with second-order discretization and default solution controls (as noted below, it may be necessary to start with first-order methods or reduce relaxation controls to gain convergence);
- standard initialization (so that all cells start with the free-stream velocities);

---

[10] The length of time needed to carry out the sweep will be directly related to the number of angles analyzed and the number of sections used to define the wing. If a complex wing is being chosen with as many as 100 sections, this can take up to an hour. If this is not acceptable, remove some of the intermediate sections in the XML file to simplify things.

[11] http://www.openfoam.com/.

[12] http://www.ansys.com/Products/Simulation+Technology/Fluid+Dynamics/Fluid+Dynamics+Products/ANSYS+Fluent.

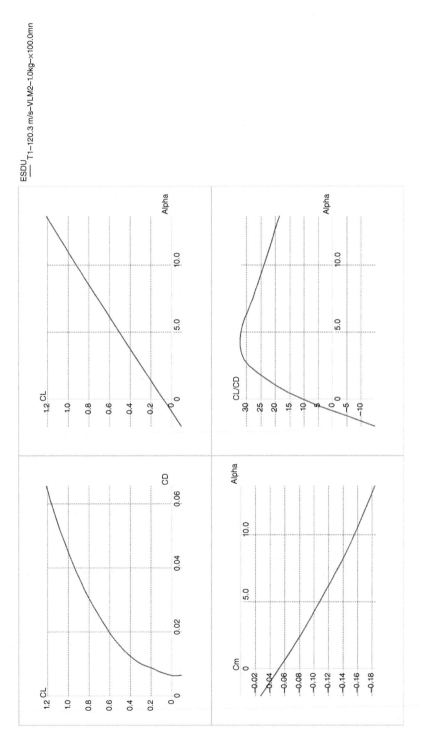

Figure 13.3 Results of XFLR5 analysis sweep for a wing generated from the NACA 64–201 foil sections at Reynold's number of 4.4 million and Mach 0.17.

Table 13.1   A summary of some of the Fluent turbulence models based on information provided in Ansys training materials.

Model	Behavior and usage
Spalart-Allmaras	Primarily intended for aerodynamic and turbomachinery problems that are not strongly three dimensional. It does not work very well in heavily separated flows. This model is relatively inexpensive and is often the best place to start analysis for airfoils, wings and fuselages. It will deal with transonic and supersonic conditions.
Standard $k - \epsilon$	A particularly robust and relatively efficient model that is good for starting simulation runs but does not work very well if there are adverse pressure gradients, significant streamline curvature, or separation
RNG $k - \epsilon$	Designed for flows with transition, moderate swirl, and vortices such as behind bluff bodies and where there is boundary layer separation
Realizable $k - \epsilon$	Similar to RNG $k - \epsilon$ but often converges more easily and is sometimes more accurate
Standard $k - \omega$	This model is good at dealing with complicated boundary layer flows with adverse pressure gradients and separation such as airfoils and wings at high angles of attack, though it tends to predict transition as occurring too early and the separation as being too strong
SST $k - \omega$	An improved but more expensive version of standard $k - \omega$ with a better three-equation (γ SST) transition model that more accurately predicts the transition location. This model is commonly the best end point for RANS studies of airfoils and wings, although even then getting the correct transition point is not completely straightforward. It is even more difficult to get the correct transition behavior on fuselages

- solution for 2000 up to 6000 iterations even if standard solution convergence is achieved before this, since solutions often change considerably after the basic convergence tests have been met (see Figure 13.4). Note that, as the onset of stall approaches, a steady-state solution may not be achievable, especially with lower resolution meshes and it is then not unusual to find the lift and drag values wander as the point of separation moves backwards and forwards, with flow even re-attaching after separation (Fluent allows lift and drag coefficients to be defined and monitored during solution). If this happens, the best that can be done is to average over a very large number of iterations, but even so such results should be treated with extreme caution);
- when using the k-ω models, we generally begin with a Spalart–Allmaras model to initially converge the model before switching to the k-ω model, often checking the nondimensional first cell height, y+, and adapting the mesh at the change-over point to refine any boundary layer cells that are too coarse. It is important that second-order models are used in the final stages to get accurate answers, but these can be harder to converge so we often start with first-order upwind models before gradually switching to second-order models, usually leaving momentum until last (and sometimes with reduced relaxation to aid stability).

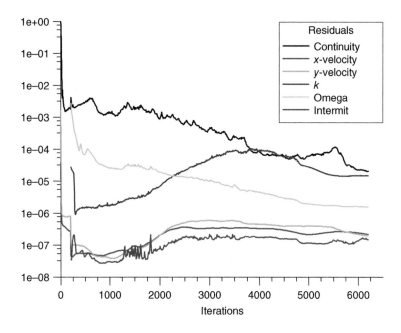

Figure 13.4   Convergence plot of two-dimensional k-ω SST RANS-based CFD analysis.

Once the solution has completed, as well as y+, we also examine the pathlines around the airframe and static pressure plots on the wetted surface, see Figure 13.5. We look to make sure that there are no, or only very limited, areas of reverse flow or separation (unless deliberately studying stall effects). If all is well, we then extract the surface-integrated forces in the vertical and horizontal directions so as to calculate the lift and drag coefficients. Note that the lift and drag are defined relative to the freestream velocity vector and not the chord, etc. Using the axes definition in Figure 13.5, for example, and for a positive AoA α, the lift vector is $[-\sin(\alpha), 0, \cos(\alpha)]$ instead of vertical, that is, $[0, 0, 1]$, and the drag vector is $[\cos(\alpha), 0, \sin(\alpha)]$ instead of horizontal, that is, $[1, 0, 0]$. Thus the lift force (L) is given from the horizontal (H) and vertical (V) force components by $L = -H\sin(\alpha) + V\cos(\alpha)$ and the drag force (D) by $D = H\cos(\alpha) + V\sin(\alpha)$.

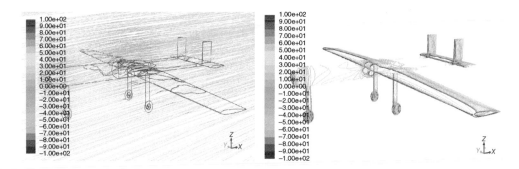

Figure 13.5   Pathlines and surface static pressure plot from Fluent RANS based CFD solution.

By stepping through a range of AoAs, achieved by varying the velocity inlet boundary condition component speeds, lift versus drag polar plots can be generated and compared with the lift and drag coefficients that were used in the initial spreadsheet design balancing process. It is then also possible to get local estimates of aerodynamic loads on the airframe elements for structural analysis by storing static pressure map files. After the design has been completed, these predictions can also be compared with those measured in the wind tunnel if such measurements are made.

It should be possible to get the lift values to agree quite tightly with experiment – generally to within 5%; drag predictions are harder to reconcile but the correct trends should be observed – absolute values within 50% of those measured should be obtainable – though RANS-based CFD results tend to be pessimistic on drag prediction when compared to hydraulically smooth models in low-turbulence wind tunnels, in our experience. Getting stall angles correct to within a degree or two, and thus a precise value of maximum lift coefficient, is much harder and is often beyond the capabilities of steady-state RANS analysis unless very refined and highly structured meshes are used; and these can be very time consuming to build and solve. It is also possible to switch to unsteady RANS models or to adopt large eddy simulation methods, but both these lie outside the scope of this book and are currently only rarely used in design work, and then only by the most sophisticated aerodynamic teams.

13.2.1   Meshing, Turbulence Model Choice, and y+

Producing a mesh for RANS CFD involves taking a representation of the solid geometry (and any nearby solid boundaries) and constructing a discretization of the surrounding fluid domain. There are many tools that can be used to do this. In this section, we will show the results for meshes produced with ICEM,[13] Harpoon,[14] and dedicated codes produced at NASA for CFD turbulence model validation.[15] Other popular meshing tools include Tgrid,[16] Boxer,[17] Gridgen,[18] and Centaur.[19] All meshing tools have their strengths and weaknesses and often the final choice will rest on familiarity and ready availability.

Here meshes are considered at several levels of resolution, aimed at two different types of results: first, lift and drag at low AoAs in the absence of separation, and, second, highly nonlinear flows with stall and separation. We always aim to adopt meshing approaches that give reasonable results with limited effort, even if this results in slightly larger meshes and therefore longer solver run times. For example, the fast hex-core mesher Harpoon can generally create a CFD mesh around an aircraft containing some 5 million cells in a minute or so, including the generation of a tightly controlled boundary layer mesh. Refinement zones around the lifting surface edges and wing tips can easily be added if desired, but these will rapidly cause the cell count to rise. We tend to use five or more expansion layers in our main mesh plus a boundary layer mesh (possibly extended into the wake) for more accurate work. If possible, we exploit symmetry and run half models. We find that, when trying to recover test results for airfoil

---

[13] http://resource.ansys.com/Products/Other+Products/ANSYS+ICEM+CFD.
[14] http://www.sharc.co.uk/.
[15] http://turbmodels.larc.nasa.gov/naca0012_grids.html.
[16] http://resource.ansys.com/Products/Other+Products/ANSYS+TGrid.
[17] http://www.cambridgeflowsolutions.com/en/products/boxer-mesh/.
[18] http://www.pointwise.com/gridgen/.
[19] https://www.centaursoft.com/.

sections, the far-field boundaries being used must be at least 20 chord lengths from the foil but, for dealing with whole airframes, these can be reduced somewhat (for airfoils alone, the drags are so low that very careful meshing is required, while for complete airframes with induced parasitic and interference drag, a less conservative approaches can suffice). Once a basic solution is in place, we sometimes add the effects of propellers by using actuator disk models.

We typically start with a coarser mesh and the simple Spalart–Allmaras turbulence model where a full airframe model can typically be built with perhaps 2 000 000 cells while a refined boundary layer model will typically have 10–20 times as many cells, although these can still be solved in affordable times if a reasonably powerful multicore computer with sufficient memory is available. Whether the cost is justified will depend on the sophistication of the overall project. As already noted, simple meshes will allow reasonable estimates of lift to be produced; good estimates of drag will require more expense, while attempts at prediction of separation and stall will need very finely controlled boundary layer and trailing-edge meshes.

The number of cells in the boundary layer area depends on both the results required and the type of turbulence model to be used: when studying stall and drag at higher angles of attack, we use the $k - \omega$ SST method, while for simpler work we use the Spalart–Allmaras approach and opt for a rather coarser mesh that draws on wall functions. The fineness of boundary layer meshes is characterized by the nondimensional cell height parameter y+. The desired first cell height in the boundary layer can be estimated from the target y+ using chord-based Reynold's number (Re), free-stream velocity (V), air density (ρ), and air viscosity (μ), together with a flat-plate approximation as follows[20]:

$$C_f = (2log(Re) - 0.65)^{-2.3},$$
$$\tau_{\text{Wall}} = 0.5\rho V^2 C_f,$$
$$u^* = \sqrt{\tau_{\text{Wall}}/\rho},$$
$$\text{FirstCellHeight} = y+_{\text{desired}}\mu/(\rho u^*).$$

For the best results, which require viscous sublayer models, y+ should ideally be below 1. Some references suggest that keeping y+ below 5 is sufficient, but Fluent advises that average values of 0.8 should be aimed for, with finer meshes being preferable. To achieve a fine y+, a boundary layer mesh will be needed together with a series of expansion stages moving away from the boundary, typically with expansion ratios varying from 1.15 to 1.4, while for coarser meshes using wall functions, a main domain mesh plus a few well-positioned refinement zones are all that may be needed. Very fine expansion ratios are needed if flow separation is to be correctly predicted (Fluent recommends 1.15 or less). For approximate results, the wall function approach can be used by adopting mean values of y+ of 60, and mostly lying in the range 30–150 (it is in fact very difficult to get an ideal distribution of y+ without detailed intervention in the meshing stage; so a compromise between accuracy and effort must be made unless significant expertise is available).

The y+ parameter is reported by the solver after convergence, and it is good practice to always plot this quantity out for the surface of the aircraft, both as a histogram and surface

---

[20] Note, however, that since most finite volume CFD codes give velocities at the cell centers, using this estimate for cell height will give a y+ value that is somewhat less than the desired value (for Fluent, a simple flat-plate simulation using a cell height calculated from the formula for y+ = 1, results in actual values of 0.54). There are a number of online calculators that offer to work out cell heights for given y+ values and they mostly ignore this aspect of finite volume codes. The actual y+ values achieved should always be checked at the end of the CFD run. At the time of writing, the calculator at http://www.cfdyna.com/CFDHT/Y_Plus.html correctly allows for such aspects.

contours, before studying lift and drag data. Some CFD solvers (including Ansys Fluent) allow mesh adaptation to try and match the y+ values to those desired once the solution has begun. Note also that y+ changes with the flight speed, height, and AoA. According to Ansys Fluent, the main choice in boundary layer modeling is between resolving the viscous sublayer where

- the first grid cell needs to yield y+ less than 1 but
- this adds significantly to the mesh count although it allows models such as $k - \omega$ SST to be used to resolve transition and separation so that
- generally speaking, if the forces on the wall are key to the simulation (aerodynamic drag, turbomachinery blade performance), this is the approach to take

and using a wall function where

- the first grid cell needs to be $30 < y+ < 300$ (too low, and model is invalid; too high and the wall is not properly resolved) together with
- a wall function, and high Reynold's number turbulence model such as Spalart–Allmaras, and
- generally speaking, this is the approach to take when one is more interested in the mixing in the middle of the domain, rather than the forces on the wall.

In all cases, meshes with y+ between 5 and 30 should be avoided, the range known as the buffer layer, because of the large variation of various turbulence source terms in this layer where existing models cannot handle the wall treatment well when the first grids lie there.

Figure 13.6 shows a typical Harpoon mesh cross-section, while Figure 13.7 shows a plot of y+ for this mesh at a Reynold's number of 4.4 million and designed to work with a Spalart–Allmaras approach. Note the mean value is around 60 as desired, but there is a spread from 40 to 80. Reducing this range further would take significant effort with the meshing tool and this is rarely warranted during the earliest stages of design. Note also that the octree mesh is deliberately stretched in the direction of flow; here an aspect ratio of 2 is set for far-field cells and this significantly reduces the total cell count needed in the model. Figure 13.8 shows a plot of y+ for a refined mesh designed to work with the $k - \omega$ SST approach. Note the mean value is now around unity as desired with very few cells above two. In all cases, the first layer

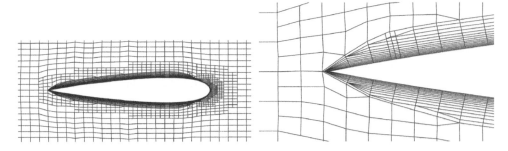

Figure 13.6 Section through a coarse-grained 3D Harpoon mesh for typical Spalart–Allmaras UAV wing model and close-up showing a boundary layer mesh.

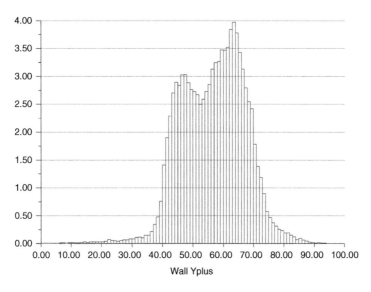

Figure 13.7 Histogram of y+ parameter for typical boundary layer mesh using the Spalart–Allmaras one-parameter turbulence model.

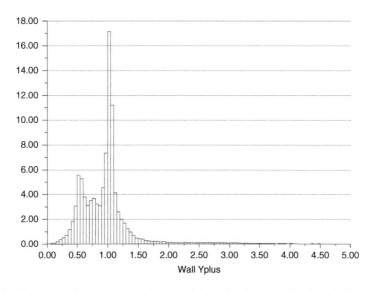

Figure 13.8 Histogram of y+ parameter for typical boundary layer mesh using the $k - \omega$ SST turbulence model.

cell height is estimated using air speed, Reynold's number, density, and viscosity and then revised after a short initial CFD run (for an aircraft with a wing planform area of $1\,m^2$, and flying at $20\,m/s$ in standard air at ground level, a first cell height of $0.03\,mm$ will lead to an average y+ of around unity).

13.3 Example Two-dimensional Airfoil Analysis

The simplest aerodynamic analysis that can be considered is flow past a two-dimensional sym-metric airfoil. We begin with the NACA0012 airfoil, which has been extensively studied and for which good-quality experimental data is available. First we show what can be achieved for this airfoil using XFoil; the NASA four-digit sections are built in to this toolkit directly so that a drag polar can be produced very rapidly, see Figure 13.9. Notice that very good agreement to experimental data is given until the section is highly separated and stall begins, when XFoil overpredicts the lift that is possible.

Next, to show the impact of meshing, we compare three two-dimensional RANS meshes with experimental results and tabulated data reported on the NASA turbulence modeling Web site. Figures 13.10 and 13.11 illustrate two NASA meshes, one with $y+$ ranging up to 2 and containing around 4000 cells, the second with $y+$ held below 0.35 and 57 000 cells. The third mesh is a block-structured model produced with ICEM and with $y+$ averaging 0.75 and 60 000 cells, see Figure 13.12. Note that the ICEM mesh achieves a worse $y+$ for a similar cell count because it is much less highly optimized to the section. The NASA meshes are the result of very careful placement of the mesh control points. Figure 13.13 shows the results from using these meshes with the Fluent $k - \omega$ SST turbulence model and 6000 solver iterations (this yields a very highly converged solution and allows the use of a relaxation setting on momentum set down to 0.2 to aid stability of solution). Also included are three results tabulated on the NASA Web site from their CFD studies as well as the experimental results provided there.

Notice that for low AoAs (up to 6°), all the RANS meshes give similar lift and drag results and these agree well with experimental ones. At higher angles, where the onset of separation begins, none of the meshes correctly resolves the drag, overestimating this by up to 100% and

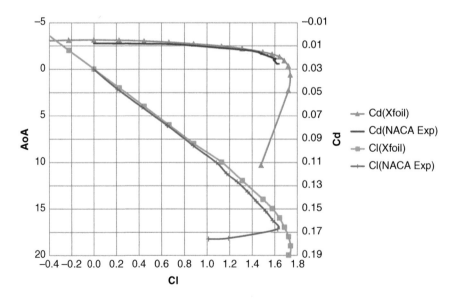

Figure 13.9 Lift versus drag polar for NAC0012 airfoil from XFoil and experiments. Note that when plotted in this way, both lift and drag coefficients may be found at a given angle of attack or, for a given lift coefficient, drag coefficient and angle of attack may easily be read off.

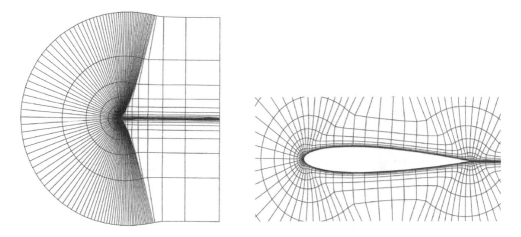

Figure 13.10   Low-resolution NASA Langley 2D mesh around the NACA0012 foil.

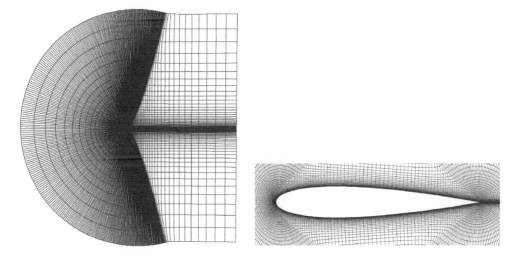

Figure 13.11   Middle-resolution NASA Langley 2D mesh around the NACA0012 foil.

being not better than the much less expensive XFoil panel analysis. Also, none of the meshes accurately resolves the correct peak of the lift curve though the two finer meshes are within two degrees. The NACA0012 section is difficult for RANS-based CFD to resolve because (a) it has a low thickness to chord ratio, meaning that the zero AoA drag is very small; (b) when very smooth, the section is capable of attached flows up to 17° after which the stall is rather sudden; and (c) the studies carried out by NACA were for speeds where the boundary layer is thin requiring large meshes to resolve the viscous sublayer. This level of (dis)agreement is typical for RANS-based CFD near the onset of separation, even using high-quality validation meshes, since a separation bubble on the low-pressure side of the airfoil has to be correctly predicted from first principles if accurate results are to be generated, see Figure 13.14.

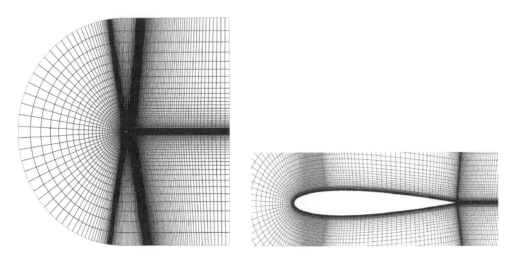

Figure 13.12   ICEM 2D mesh around the NACA0012 foil (courtesy of Dr D.J.J. Toal).

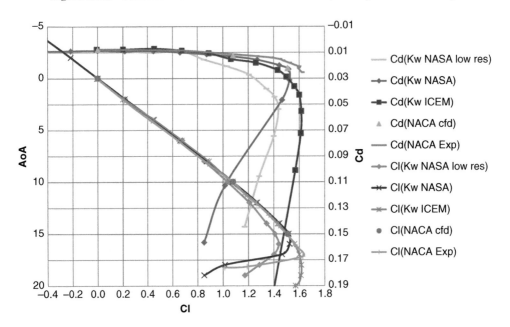

Figure 13.13   Experimental and 2D computational lift and drag data for the NACA0012 airfoil (using the $k - \omega$ SST turbulence model). Adapted from Abbott and von Doenhoff 1959.

13.4   Example Three-dimensional Airfoil Analysis

Next we consider a three-dimensional slice of the NACA0012 airfoil that is 0.05 chord lengths wide and sandwiched between a pair of symmetry boundary conditions. In this case, Harpoon meshes are used and again compared to the 2D data available on the NASA Web site. Results are calculated using both simple meshes ($30 < y+ < 150$ and the Spalart–Allmaras turbulence

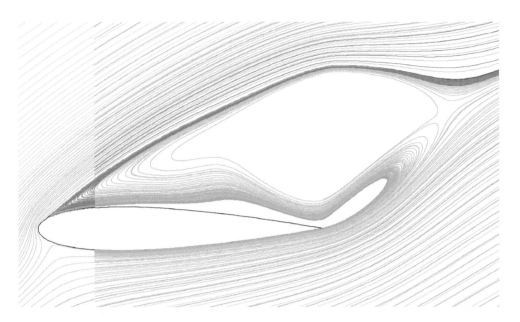

Figure 13.14   Computed two-dimensional flow past the NACA0012 foil when almost fully stalled.

model) and fine boundary layer models ($y+ < 1$ and the $k - \omega$ SST turbulence model). The meshes used are those already shown in Figure 13.6 and Figure 13.15. The Harpoon meshes have 548 000 and 5 344 000 cells, respectively. The $y+$ plots for these meshes are those already given in Figures 13.7 and 13.8.

The resulting lift and drag data are plotted in Figure 13.16. Note that at low AoAs, both the meshes again correctly predict the lift, while the finer $k - \omega$ SST turbulence model mesh is better at predicting the drag, as would be expected. Once separation begins, the approach to stall the quality of the mesh critically impacts on the ability to predict the angle of maximum lift and the drag. Here the lower resolution Spalart–Allmaras mesh gives drag values that are more than 100% in error and stall much too early. Neither of the three-dimensional meshes is as good at predicting high AoA behavior as the two-dimensional meshes considered in the previous subsection. In summary, neither of these models is completely accurate, again making clear the limitations of RANS-based analysis. Of course, the mesh quality is as much related

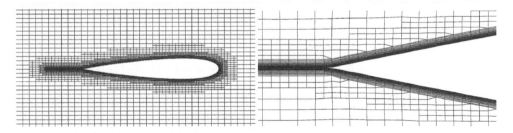

Figure 13.15   Section through a fine-grained Harpoon 3D mesh around the NACA0012 foil suitable for the $k - \omega$ SST turbulence model. Note the wake mesh extending from the trailing edge.

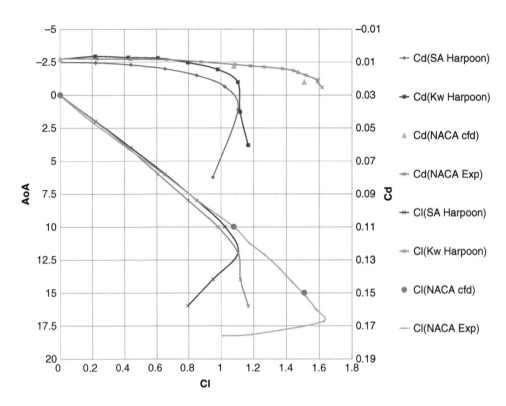

Figure 13.16 Experimental and 3D computational lift and drag data for the NACA0012 airfoil (using the Spalart–Allmaras and $k - \omega$ SST turbulence models). NASA.

to the effort and experience of the user as the underlying tool, but these results indicate that octree-based meshing tools are not as well suited to this analysis as the block-structured ICEM mesh or the very high quality NASA test meshes, though they are much easier and faster to use when meshing complicated three-dimensional airframes.

13.5 3D Models of Simple Wings

For low-speed, low-drag wing design, the NACA experimental report by Sivells and Spooner [21] is a useful source of results. This gives the lift, drag, and moment values for two slender, tapered, unswept wings of aspect ratio 9 at moderately high speeds ($Re = 4.4 \times 10^6$ and Mach= 0.17) with a range of trailing-edge high-lift devices, with and without leading-edge roughness. If one is able to recover some of the details of this study with CFD approaches, then confidence can be had that results for less slippery wings with thicker boundary layers will be sensible.

    This NACA study is based on the NACA 64–210 and 65–210 sections. Results for two-dimensional tests of the NACA 64–210 section can be found in Abbott and von Doenhoff [20] and also, both dry and in heavy rain, on the NASA server (Bezos *et al.* [22]). Interestingly, even for the dry cases, these two sources of experimental data differ, particularly with regard to the section drag coefficients. We show both these sets of data in Figure 13.17 together

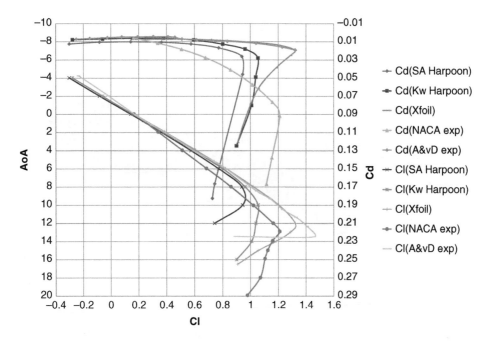

Figure 13.17 Experimental [20, 22] and computational lift and drag data for the NACA 64–210 section. Source: Adapted from Abbott 1959.

with those calculated using meshes from the octree-based tool Harpoon, again with both lower resolution Spalart–Allmaras and higher resolution $k - \omega$ SST turbulence models and appropriate meshes. Also shown are results from XFoil, which predict the Abbott and von Doenhoff data quite well but are some way from the later Bezos et al. data. The Harpoon meshes have 460 000 and 4.97 million cells, respectively, using 3D models, again with spans of 0.05 chord length between symmetry planes.

The XFoil and $k - \omega$ SST results predict the low AoA lift and drag behavior well but, again, neither gives the ultimate stall angle that accurately, although the errors are less than the differences between the two sets of experimental data! Figure 13.18 shows a RANS solution when the airfoil has begun to stall; in fact, wind tunnel data suggest that the onset of such stalls may well be delayed at this angle for very smooth airfoils in low-turbulence wind tunnels.

Turning next to the complete Sivells and Spooner wing, Figure 13.19 includes two sets of experimental results and shows data calculated using XFLR5 and Fluent with Spalart–Allmaras and $k - \omega$ SST turbulence models with Harpoon meshes. Figure 13.20 shows the XFLR5 model. First, the NACA 64–210 airfoil is loaded and analyzed from Reynold's numbers from 10^5 to 6×10^6 and at angles of attack from $-4°$ to $16°$. This builds a database of results that can then be used for the wing analysis. At the Reynold's number used in the reported work, XFLR5 gives lift and drag results from $-3°$ to $13.5°$; outside of this range, some of the sections stall and the results cannot be computed for the whole wing. For the Fluent models, just the octree-based Harpoon meshing tool has been used. The Spalart–Allmaras Harpoon mesh has 1.225 million cells, giving $y+$ values between 30 and 100. The $k - \omega$ SST Harpoon mesh has 13.6 million cells including a trailing-edge wake mesh, giving mean $y+$ values 1.3, that is, not quite as fine as before.

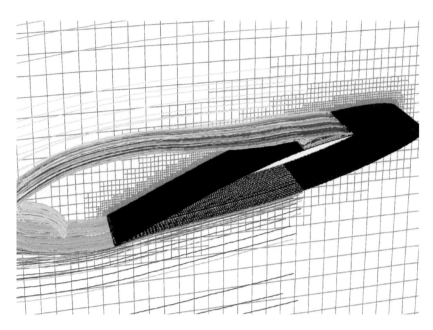

Figure 13.18 Pathlines from a RANS $k - \omega$ solution for the NACA 64–210 airfoil at $12°$ angle of attack. Note the reversed flow and large separation bubble on the upper surface.

Here there is little to choose between the RANS-based solutions: both give good estimates of lift at low AoA but overpredict the drag compared to the experiments with the smooth wing, while the XFLR5 results underpredict the drag. In addition, the behavior at maximum lift still cannot be accurately resolved using octree-based meshes. Even the XFLR5 model fails to solve some $2°$ before the smooth wind tunnel model stalls. If the degree of mesh resolution is increased for the $k - \omega$ SST Harpoon mesh, using 76 million cells so that $y+$ can be brought mainly below 1, better results can be obtained for the drag, see Figure 13.21. Figure 13.22 illustrates the computed flow around the wing for this refined mesh at $11°$ AOA, the point of maximum lift predicted in this case. Note that some of the pathlines show the onset of stall. Sivells and Spooner [21] give a sketch for this wing at $11.5°$, albeit with a roughened leading edge, that shows separation to begin at just this location in the span-wise direction (the pathlines oscillate and the static pressure suction at the front of the wing collapses). In summary, with a very refined boundary layer, the Fluent model behaves at separation rather like the experimental wing with roughened (and thus more realistic) leading edge. Such results need to be treated with caution but clearly are of use in understanding the behavior of the wing over a range of operational conditions.

13.6   Example Airframe Aerodynamics

Having established what may be achieved with CFD codes for sections and simple wings, we next apply these to a whole airframe, in this case for the Decode-1 UAV outer mold line already created using AirConics. We show the use of panel codes for aerodynamics and stability and RANS approaches for aerodynamics.

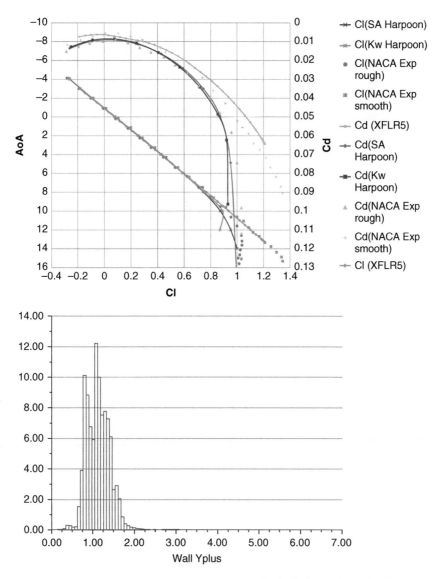

Figure 13.19 Experimental and computational lift and drag data for the Sivells and Spooner [21] wing and $y+$ for the $k - \omega$ SST Harpoon mesh. Source: NASA.

13.6.1 Analyzing Decode-1 with XFLR5: Aerodynamics

When using XFLR5, only the lifting surfaces can be analyzed for lift and drag (though deflected control surfaces can be included by adding small trailing-edge distortions to the shapes of the airfoil sections where the controls are assumed to be[21]). While the fuselage

---

[21] We find that flap deflections beyond around ±5° cause XFoil to fail and this unfortunately limits the utility of this capability.

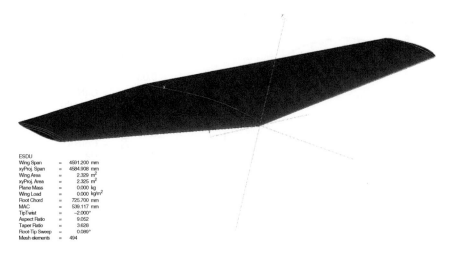

ESDU		
Wing Span	=	4591.200 mm
xyProj. Span	=	4584.908 mm
Wing Area	=	2.329 m²
xyProj. Area	=	2.325 m²
Plane Mass	=	0.000 kg
Wing Load	=	0.000 kg/m²
Root Chord	=	725.700 mm
MAC	=	539.117 mm
TipTwist	=	–2.000°
Aspect Ratio	=	9.052
Taper Ratio	=	3.628
Root-Tip Sweep	=	0.089°
Mesh elements	=	494

Figure 13.20   XFLR5 model of the Sivells and Spooner wing.

can be included, the guidance supplied with the code advises against it; the tail booms and undercarriage details must be omitted completely. They can, however, be allowed in the drag calculations by being added as extra drag coefficients with associated areas. These can be taken from the previous spreadsheet analysis or from texts such as Hoerner [9]. Figure 13.23 shows the resulting model built directly from the AirCONICS system and imported to XFLR5 via the medium of an XML file, while Figure 13.24 shows a set of polar plots for three flight speeds, for a mass of 15 kg and a fixed center of gravity (CoG) position 42.8 mm forward of the main wing quarter chord position (as taken from the previous spreadsheet analysis and with the CoG input to XFLR5 as 39.1 mm from the main wing leading edge). The plot of the imported geometry also shows the lift distribution in the cruise condition. Such data can be used to inform subsequent structural analysis such as sizing of the main spars.

Note that the previous spreadsheet-based approach to aerodynamic design was not based on any given airfoils: rather, typical airfoil lift and drag performance were assumed. By using XFLR5 and XFoil, it is possible to begin to understand the impact of actual section choices on the aircraft. Here, a NACA 23 012 section has been used for the main wing and 0212 for the elevator and fins. The main wing has zero setting angle and zero washout at the tips while an elevator setting angle of $-2.85°$ has been chosen. Figure 13.24 shows that, in this configuration, at 30m/s, the aircraft has to fly at an AoA of $2.53°$ to achieve the required cruise Cl of 0.28 where the lift to drag ratio is comfortably greater than the value of 16 assumed in the spreadsheet. The chosen elevator setting angle allows the aircraft to fly stably at this AoA as can be seen by the fact that Cm is then zero. Moreover, the Cm value is positive at the zero lift AoA and the Cm curve has a negative slope, indicating that the design is stable in pitch, as expected.

This is not, however, an efficient configuration, as the fuselage would not be aligned with the direction of flight in the cruise condition, leading to extra drag. It is general practice to adjust the main wing and tail setting angles to produce a condition where both the required Cl and a zero Cm are achieved with the aircraft horizontal. To do this, it is useful to conduct a series of fixed-speed, fixed-AoA, and stability runs in XFLR5 with varying wing and elevator settings to achieve the required conditions. The stability calculation is particularly useful in this regard,

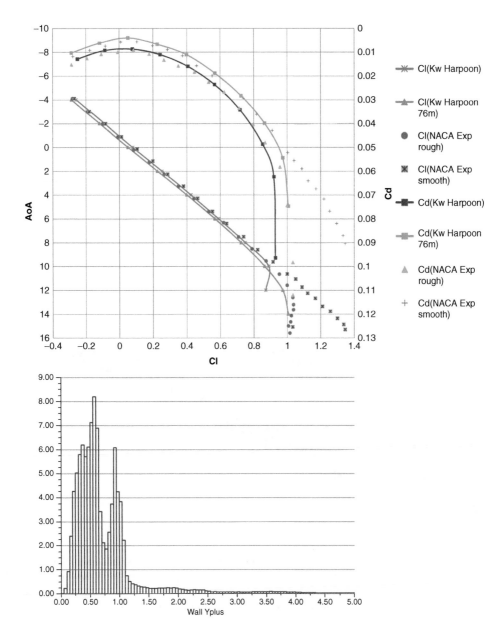

Figure 13.21 Experimental and computational lift and drag data for the Sivells and Spooner [21] wing with enhanced $k - \omega$ SST Harpoon mesh of 76 million cells and y+ for the enhanced mesh. Source: NASA.

as it automatically searches for the speed and AoA where Cm is zero and reports these in the operating point.[22] To begin this process, we first set the main wings to a setting angle of 2.53°

---

[22] Further details on setting up and interpreting all the information generated by the XFLR5 stability calculation are given in the next section.

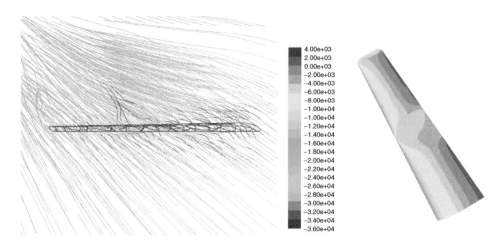

Figure 13.22 Pathlines and static pressure around the Sivells and Spooner [21] wing with enhanced $k - \omega$ SST Harpoon mesh at 11° angle of attack. Source: NASA.

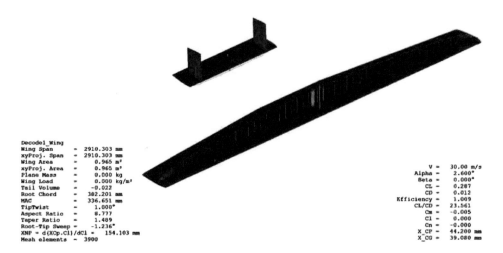

Figure 13.23 XFLR5 model of Decode-1 airframe as generated by AirCONICS with main wing setting angle of 0° and elevator setting angle of −2.85°, at an angle of attack of 2.6° and 30 m/s. Note the use of cambered sections for the main wing and symmetrical profiles for the elevator and fins. The green bars indicate the section lift, with the tail producing downforce to ensure pitch stability.

(i.e., the AoA where a Cl of 0.28 was previously achieved) and reduce the tail setting angle by a similar amount. XFLR5 then suggests that the stable flight point would be at 30.08 m/s and at an AoA of −0.06°. A further iteration to the tail setting angle of −0.34° yields a flight point at 29.98 m/s and an AoA of −0.04°, which is the desired condition to three significant figures. Figure 13.25 shows the lift, drag, and moment polar plots of this final configuration.[23]

---

[23] These results are compared with Fluent calculations and experimental data in the section on Fluent analysis of Decode-1.

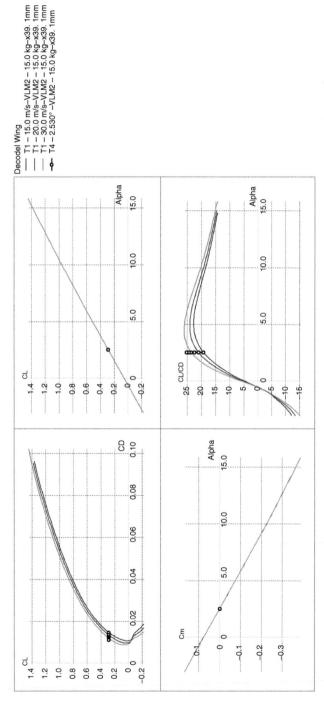

Figure 13.24 XFLR5-generated polar plot for Decode-1 airframe as generated by AirCONICS with main wing setting angle of 0° and elevator setting angle of −2.85°, showing speed variations from 15 to 30 m/s. The black circles indicate flight at an angle of attack of 2.53° at which Cm is zero.

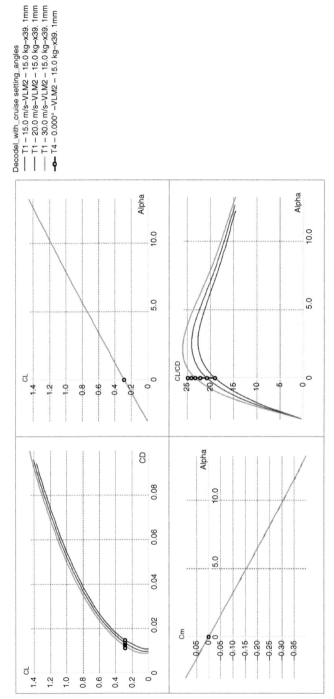

Figure 13.25 XFLR5-generated polar plot for Decode-1 airframe as generated by AirCONICS with main wing setting angle of 2.53° and elevator setting angle of −0.34°, showing speed variations from 15 to 30 m/s. Note that Cl is 0.28 and Cm is zero at an angle of attack of 0° as required in the cruise condition.

Note here that at the assumed landing speed of 15m/s, the required Cl is 1.11, which is achieved at an AoA of approximately 9.25° where the lift/drag ratio is 17; XFLR5 will not solve above 12.2° for this configuration and flight speed, indicating that the stall angle is around this value, thus giving a stall margin during landing of some 25%. This margin may seem quite large, but in fact XFLR5 tends to be optimistic with regard to stall angles and lift/drag ratios, so this is a sensible value to aim for.

In Chapter 11 on spreadsheet-based design, care was taken to ensure static stability was obtained, and estimates of the impact of the wake of the main wing on the tail were used to do this. As just shown, with XFLR5 it is possible to check this analysis by studying the pitching moment coefficient curve. Note that the curve can be adjusted by altering the position of the CoG (this alters the slope), by changing the setting angle for the elevator (this moves the curve up and down on the plot), by moving the longitudinal position of the elevator relative to the main wing (and thus changing its lever arm, altering both slope and position of the curve), or by altering its vertical position (and thus changing the impact of the main wing downwash on the elevator). Of course, one is not free to set these arbitrarily; clearly, the longitudinal CoG must be realistic given the design in hand and large setting angles for the elevator will increase drag, both by directly increasing the drag of the elevator and also by requiring greater lift from the main wing. If an acceptable position for the CoG in front of the neutral point is not immediately achieved, it may be necessary to reposition heavy items within the aircraft; in the extreme case it may be necessary to lengthen the fuselage of the aircraft forward to do this. Alternately, if the design uses simple carbon-fiber spars for the tail booms, it may be easier to adjust the longitudinal position of the tail. Hopefully a sensible tail volume coefficient will have been adopted at the outset so that significant changes are not required. Figure 13.26 shows a set of polar plots for different CoG positions, tail length, and elevator setting angle for Decode-1.

13.6.2   Analyzing Decode-1 with XFLR5: Control Surfaces

At this point, it is also possible to estimate the effectiveness of movable control surfaces on the design for small control deflections. In general, control surfaces are sized by their span-wise and chord-wise extents: as already noted, we typically opt for ailerons that extend over the outboard 50% of the main wing with flaps over the inboard parts of the wing (although these are sometimes omitted on simpler designs). For rudders and elevators, we opt for full length surfaces (and for very slow speed flying at more extreme angles of attack, full moving elevator surfaces). For hinged control surface, we design these to extend across the rearmost 30% of the airfoil section. Hoerner [10] suggests that there is little to be gained from extending a plain flap control surface beyond this and that 20% is often a good trade-off in terms of performance, structural penalties, weight, and so on. We find for small UAVs it is generally worthwhile going as far as 30% and that this can be easily accommodated in the structural design approach we adopt.

The flap hinge moment coefficient for a control surface, defined in terms of flap area and flap mean chord, $C_H = H/(0.5\rho V^2 S_f \bar{c}_f)$, typically varies fairly linearly with the flap deflection and the overall lift coefficient, with $dC_H/d\delta$ being -0.0075 per degree and dC_H/dC_L around -0.05 for 30% chord ratio flaps (both being negative because of the normal convention that a positive control surface deflection produces a negative moment). $dC_H/d\delta$ decreases in magnitude (becomes less negative) as the chord ratio gets greater than this, while dC_H/dC_L gets

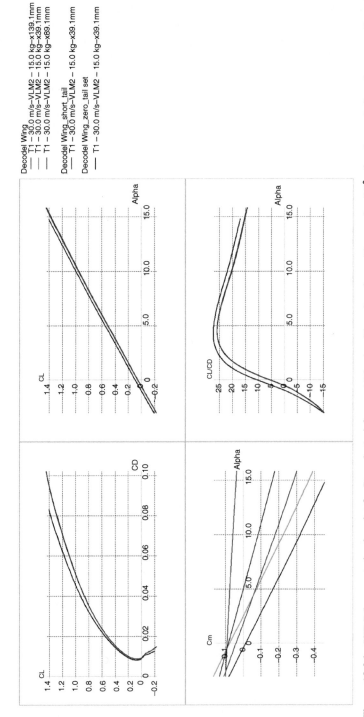

Figure 13.26 XFLR5-generated polar plot for Decode-1 airframe at 30 m/s with main wing setting angle of 0°, showing variations in center of gravity position by 100 mm, reduction in tail length by 300 mm, and elevator set at an angle of 0°.

bigger in magnitude (more negative). Notice that, even when not deflected, an aileron or flap (and to a lesser extent an elevator) will have a noticeable hinge moment simply induced by the fact that the overall section is creating lift and thus there is a pressure differential on the flap itself and that this effect increases as the flap chord ratio rises. Likely maximum hinge moments should be used to select appropriate servos to control the surface, typically assuming a one-to-one mechanical advantage ratio in the linkage (this is slightly conservative since servos typically have a greater angular range than control flaps and this can be used to gain slight mechanical advantage by suitable selection of the servo arm and control horn lengths). Figure 6.9 shows how servo weights typically vary with the required torques.

XFLR5 deals with control surfaces by allowing automated edits of airfoil section shapes to approximate the behavior of deployed simple trailing edge flaps. Note, however, that no account is taken of the gaps that are inevitable around the control surface or the changes to the flows that these give rise to, nor will it deal with very large control surface deflections: it is necessary to move on to a RANS-based analysis if a more complete treatment is required. The basic calculations that XFLR5 permits are those of roll, pitch, and yaw moments at a given flight speed and AoA. These can be used to estimate steady-state roll, pitch, and yaw rates. For example, if on Decode-1, ailerons extending 50% of the span and 30% of the chord are deflected by $\pm 5°$ at 2.53° AoA and 30 m/s, the corresponding roll, pitch, and yaw moment coefficients are -0.0355, -0.0012, and -0.0004, respectively; note that there is very little pitch or yaw coupling as desired. When deflected in this way, assuming the previous values for the hinge moment slopes and a wing $C_L = 0.287$, we find that the resulting hinge moment coefficient for the ailerons is $C_H = -0.0075 \times 5 - 0.05 \times 0.287 = -0.0519$. Taking the aileron area to be 0.04 m^2 with a mean chord of 0.1 m, the required hinge moment torque is 0.114 Nm $= 16.2$ oz. in. In practice a much larger deflection might occur in flight and at dive speeds, so rather higher torques should be specified; if instead of a 5° deflection a value of 30° is used with a dive speed of 45 m/s, the required hinge moment torque coefficient becomes $C_H = -0.0075 \times 30 - 0.05 \times 0.287 = -0.239$, leading to a torque requirement of 0.94 Nm (133 oz.in.) at 45 m/s. Using the equation for typical servo weights given in Figure 6.9 reveals that the servo will likely weigh around 2 oz. or 56 g, a perfectly reasonable value for a wing servo on an aircraft of this size.

13.6.3 Analyzing Decode-1 with XFLR5: Stability

Having set out a basic aerodynamic analysis with XFLR5, it is then possible to use the code to more fully investigate the stability of the design by locating the eight natural flight modes of oscillation on a root locus plot. These can be compared with the tabulated acceptable values provided by the U.S. military. In a linearized analysis, these modes consist of four longitudinal modes, namely a pair of symmetric phugoid modes and a pair of symmetric short-period modes, and four lateral modes, namely one spiral mode, one roll damping mode, and a pair of Dutch roll modes (free directional oscillations): that is, five distinct types of behavior. XFLR5 does this by using finite differencing in the six degrees of freedom of flight to calculate the first-order stability derivatives, which can then be used to calculate the various eigenmodes (changes are made in wind velocities by 0.01 m/s and angular velocities by 0.001 rad/s). Because the code is so fast, these additional runs are not too significant an overhead in the calculation. In general, the results are conservative because XFLR5 tends to underestimate the various damping terms in the analysis because it tends to underestimate drag.

XFLR5 follows the analysis presented by Etkin and Reid [23] and, in particular, solves equations (4.9,18 and 4.9,19) presented there. To do this, it derives the linearized stability derivatives by finite differencing, and these are then used to solve and store the eigenmodes and 18 nondimensional derivatives. The derivatives are returned as CXu, CLu, Cmu, CXa, CLa, Cma, CXq, CLq, Cmq, CYb, Clb, Cnb, CYp, Clp, Cnp, CYr, Clr, and Cnr. The first nine of these are of horizontal and vertical forces and roll moment with respect to changes in forward speed, of AoA at constant pitch, and of varying pitch at constant AoA (note these last two are different since the direction of instantaneous flight may not be horizontal). They are defined as follows:

$$CXu = (X_u - \rho u_0 S C_{w0} sin(\theta_0))/(\rho u_0 S/2) = C_{Xu};$$

$$CLu = -(Z_u + \rho u_0 S C_{w0} cos(\theta_0))/(\rho u_0 S/2) = -C_{Zu};$$

$$Cmu = M_u/(\rho u_0 \bar{c} S/2) = C_{mu};$$

$$CXa = X_w/(\rho u_0 S/2) = C_{Xa};$$

$$CLa = -Z_w/(\rho u_0 S/2) = -C_{Za};$$

$$Cma = M_w/(\rho u_0 \bar{c} S/2 = C_{ma});$$

$$CXq = X_q/(\rho u_0 \bar{c} S/4) = C_{Xq};$$

$$CLq = -Z_q/(\rho u_0 \bar{c} S/4) = -C_{Zq};$$

$$Cmq = M_q/(\rho u_0 \bar{c}^2 S/4) = C_{mq}.$$

The last nine are the side force derivative, the dihedral effect, the weathercock stability, the side force due to rolling, the damping in roll, the yaw due to roll cross derivative, the side force due to yaw, the roll due to yaw, and the damping in yaw. They are defined as follows:

$$CYb = Y_v u_0/(qS) = C_{Y\beta};$$

$$Clb = L_v u_0/(qSb) = C_{l\beta};$$

$$Cnb = N_v u_0/(qSb) = C_{n\beta};$$

$$CYp = Y_p(2u_0)/(qSb) = C_{Yp};$$

$$Clp = L_p(2u_0/b)/(qSb) = C_{lp};$$

$$Cnp = N_p(2u_0/b)/(qSb) = C_{np};$$

$$CYr = Y_r(2u_0)/(qSb) = C_{Yr};$$

$$Clr = L_r(2u_0/b)/(qSb) = C_{lr};$$

$$Cnr = N_r(2u_0/b)/(qSb) = C_{nr}.$$

Here, ρ is the air density, u_0 is the reference flight speed, θ_0 is the reference angle of climb, S is the wing planform area, $\bar{c}$ is the mean aerodynamic chord, and b is the span. The various terms such as X_u and N_r are the derivatives, that is, the rates of changes of forces and moments due to changes in velocities and angular velocities.

Phugoid oscillation is a macroscopic mode of exchange between kinetic and potential energies (forward and vertical velocities) and is normally slow, lightly damped, and may be stable or unstable. A very simple estimate of the phugoid frequency is given by Lanchester's approximation as $9.81/(\sqrt{2}\pi V_0)$, where V_0 is the aircraft's speed in m/s. An unstable or divergent phugoid is mainly caused by a large difference between the incidence angles of the wing and the elevator. A stable, decreasing phugoid can be attained by adopting a smaller elevator on a longer tail, or, at the expense of pitch and yaw static stability, by shifting the CoG to the rear. As the phugoid frequency is generally very low, it is quite possible to fly an aircraft with a divergent phugoid mode, although this is probably not desirable. Its damping is inversely proportional to the lift/drag ratio, so that more aerodynamically efficient designs tend to have less phugoid damping. Since the lift/drag ratio varies with speed, so also does phugoid damping.

The short-period mode is primarily vertical movement and pitch rate in the same phase and is usually of high frequency, stable, and well damped. An aircraft with a low short-period natural frequency will seem initially unresponsive to control input. If the natural frequency is too high, the aircraft will feel too sensitive in maneuvering and too responsive to turbulence. Aircraft with low short-period damping ratios tend to be easily excited by control inputs and turbulence, and the resulting oscillations take longer to disappear. Aircraft with high short-period damping can be slow to respond and sluggish; thus a compromise in damping is necessary. FAA regulations require that the short-period oscillation must be "heavily damped" and that "Any ... phugoid oscillation ... must not be so unstable as to increase the pilot's workload or otherwise endanger the aircraft."

The spiral mode is primarily in-heading, nonoscillatory, slow, and generally unstable, requiring pilot input to prevent divergence (pilots commonly do this without noticing on well-designed aircraft). Roll damping is usually stable because positive dihedral, swept wings, high wings, or a low CoG will have been adopted. Dutch roll is a combination of yaw and roll, phased at $90°$, and is usually lightly damped; it is largely controlled by the size of the tail fin, which should have a sensible value if an appropriate volume coefficient has already been used. Larger fins increase both the Dutch roll frequency and its damping. Dutch roll is not an aircraft deficiency but the inevitable result of having directional stability, that is, the fundamental tendency of a stable aircraft to roll away from, but yaw towards, the velocity vector. Aircraft with greater directional than lateral stability tend to Dutch-roll less, but also tend to be spirally unstable. Normally, the design compromise between Dutch roll and spiral instability is to reduce Dutch roll at the expense of spiral instability because spiral dives begin slowly and are more easily controlled than Dutch roll.

To help decide what values for the various dynamic stability modes are acceptable, the U.S. military have developed a specification for the flying qualities of piloted airplanes: MIL-F-8785C. This defines three levels of flying qualities (1–3, with 1 being the best), three phases of flight (A–C, with A being the most demanding and including air-to-air combat and ground attack), and four classes of aircraft (I, light aircraft; II, medium-weight, medium-maneuverability aircraft; III, heavy, low-maneuverability aircraft; and IV, high-maneuverability fighter types). Then for each combination it gives guidance on the acceptable damped natural frequencies and damping ratios. For the purposes of most small UAV designs, class I and flight phases B or C should be assumed. If the aircraft is to be manually flown, a flying quality of level 1 should be aimed at, but if full autopilot control is assumed, then a quality of level 3 can be acceptable as pilot workload is then not an issue. For the five different modes normally encountered, the specification gives the following guidance for class I aircraft:

- *Short-period mode*. Damping ratio limits for level 1, 0.35–1.3 in phases A and C and 0.3–2.0 in phase B (these can be relaxed to 0.25–2.0 and 0.2–2.0 for level 2 in phases A and C and phase B, respectively, and to just lower limits of 0.15 in any phase for level 3);
- *Phugoid mode*. Damping ratio for level 1 to be at least 0.04 (this can be relaxed to zero for level 2 and to time for the amplitude to double (T_2) of 55 s for level 3, that is, divergence is then acceptable);
- *Roll mode*. Maximum time constant, τ_R,[24] 1 s for level 1, phase A or C and 1.4 s for phase B (these can be relaxed to 1.4 and 3.0 s for level 2, respectively, and to 10 s in any phase for level 3);
- *Dutch-roll mode*. Minimum damping ratio of 0.19 for level 1, phase A and 0.08 for phase B or C (these can be relaxed to 0.02 in any phase for level 2 and zero for level 3); minimum damped natural frequency of 1 rad/s (0.159 Hz) for level 1, phase A or C and 0.4 rad/s (0.064 Hz) for phase B (these can be relaxed to 0.4 rad/s in any phase for levels 2 and 3);
- *Spiral mode*. Time for amplitude to double (T_2) for level 1 to be at least 12 s in phases A and C or 20 s in phase B (these can be relaxed to 8 s in any phase for level 2, and 4 s for level 3).

Phugoid, short-period, and Dutch-roll modes being oscillatory give complex pairs of roots in the root locus plot, while roll damping and spiral modes lie on the horizontal axis. These quantities can then be converted into frequencies, damping ratios, time constants, and times for the amplitude to double. Apart from the aerodynamic quantities of the aircraft, the stability behavior is controlled by the location of the CoG and moments of inertia; all must be input into XFLR5 to carry out a stability analysis (xy and zy cross-moments are ignored, as the airframe is assumed port/starboard symmetric). If one is unsure of the moments of inertia, these can be estimated by assuming that the radius of gyration for Ixx is a fraction of the main semispan, say 22%; that for Iyy is a fraction of half the overall length of the aircraft, say 35%; and that for Izz is a fraction of the average of semispan and half length, say 38% (Raymer [11] gives values for these nondimensional radii of 22–34%, 29–38%, and 38–52%, respectively; our values are based on the UAVs we have built and flown). The cross-moment Ixz is typically small, and can generally be set as zero until better information is available. These may then be used with the overall mass to compute the required inertias. For Decode-1 at 15 kg maximum take-off weight (MTOW), these estimates give Ixx as 1.6 kg m^2, Iyy as 1.2 kg m^2, and Izz as 2.7 kg m^2. Later on in the design, when a full CAD model is being created, these values can be checked and updated.

Table 13.2 shows the eigenvalues for the stability analysis of Decode-1 with final wing and elevator setting angles, using these inertia values and the previously calculated XFLR5 stability derivative data at 30 m/s. The results suggest that the aircraft has quality 1 dynamic stability except in the phugoid mode where it lies on the quality 1/2 boundary. The phugoid prediction is pessimistic because the parasitic drag has not been included in XFLR5; when this is included, clear quality 1 is obtained (at the time of writing, XFLR5 did not include parasitic drag in its internal stability analysis but this can be added in separately using the formulae provided by Phillips [24, 25], although these formulae include a few simplifications not needed

---

[24] The roll mode time constant is the time it takes to achieve 63.2% of the final roll rate following a step input.

Table 13.2 Decode-1 eigenvalues as calculated from XFLR5 stability derivatives using the formulae provided by Phillips [24, 25] and the estimated inertia properties for a flight speed of 30 m/s and MTOW of 15 kg.

Mode	Real part	Imag. part	Damp. Rat.	Class I, Phase B quality	Comment
Short period	−8.31	15.8	0.465	1	High-frequency convergent (2.85 Hz)
pPhugoid	−0.0181	0.417	0.0435^a	1/2	Lightly damped low-frequency convergent (0.07 Hz)
Roll damping	−30.1			1	Heavily convergent ($\tau_R = 0.0332$ s)
Dutch roll	−0.641	5.22	0.122	1	Lightly damped medium-frequency convergent (0.84 Hz)
Spiral	0.0453			1	Slowly divergent ($T_2 = 15.3$ s)

Note that the undamped natural frequency in hertz (Hz) is given by the magnitude of the eigenvalue divided by 2π, while the damping ratio is the negative of the ratio of the real part divided by the imaginary part; $T_2 = \log_e(2)/$eigenvalue (for positive eigenvalues) and $\tau_R = \log_e(1 - 0.632)/$eigenvalue (for negative eigenvalues).
^{a}0.151 if parasitic drag is included and thus clearly quality 1.

by the full eigen analysis used in XFLR5). Note also that Dutch roll damping is increased by the drag caused when the fuselage elements are not aligned with the direction of flight, and these aspects are not generally captured by XFLR5 either (it is possible to approximate the fuselage as a further lifting surface, but clearly most fuselages are not really wing-shaped!). Thus the XFLR5 results are essentially worst case values. This analysis has been confirmed by flight trials; the aircraft is responsive and benign when flown manually, and the phugoid and Dutch-roll modes are easily controlled. Figure 13.27 illustrates the behavior of the modes listed in Table 13.2 (with the added parasitic drag) in the time domain.

13.6.4 Flight Simulators

To go beyond a simple XFLR5 stability analysis, yet without resorting to a full Navier–Stokes solver, it is possible to gain further insights into the controllability of a new design by using one of the better aircraft flight simulator packages. These necessarily allow for the behavior of all lifting and control surface actions, and the more powerful systems will give highly realistic simulations of aircraft flight characteristics. The main difficulties in using such an approach to evaluate a new design are (a) installing all the required geometric information into the simulator and (b) having sufficient piloting skills to be able to evaluate the resulting design in a controlled manner. We have used the X-Plane[25] system to assess aircraft designs for

---

[25] http://www.x-plane.com//.

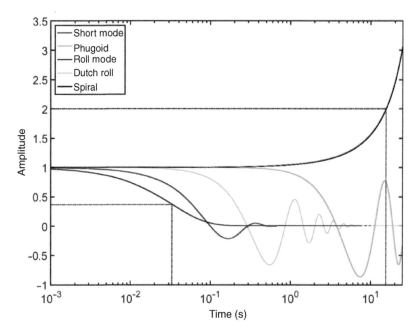

Figure 13.27 Time-domain simulation for XFLR5-generated eigenvalues at 30 m/s taken from Table 13.2 showing τ_R for the roll mode and T_2 for the spiral mode.

a number of years, employing professional test pilots to give us reports on the behavior of our aircraft. We have access to a large-scale aircraft simulator environment in which to do this, see Figure 13.28. X-Plane works with a blade element momentum model and actuator disks to predict the behavior of the aircraft as it flies in an animated 3D world. It is also capable of capturing large amounts of flight data from its simulations. To install a new aircraft into X-Plane, a separate tool called Plane Maker is provided, see http://developer.x-plane.com/?article=creating-a-basic-aircraft-in-plane-maker[26] together with the example provided at http://wiki.x-plane.com/File:A36.zip. Whether or not it is worth the expense and effort of building an accurate simulator model will depend on the project in hand; for most conventional designs, provided the XFLR5 stability analysis is acceptable and reasonable size control surfaces are planned, it is probably not necessary.

13.6.5 Analyzing Decode-1 with Fluent

If a more detailed analysis of the fundamental aerodynamics is required, Fluent is able to model the entire airframe including fuselages, undercarriage elements, wheels, and deployed control surfaces, and even allows for propeller models to be included (e.g., as actuator disks). However, this increased capability comes at a significant extra cost in terms of the effort required to set

---

[26] There are also some good sets of Plane Maker tutorials on YouTube such as those at http://www.youtube.com/watch?v=l4m2wRb7ZBA, http://www.youtube.com/watch?v=TklD4AW2ZGE, http://www.youtube.com/watch?v=uKqIX3ByDHc, http://www.youtube.com/watch?v=coNaY0wivGg, and http://www.youtube.com/watch?v=8AuPIaVIyrs.

Figure 13.28   University of Southampton flight simulator.

up and carry out an analysis. It would be wasteful to commence such studies until a full XFLR5 analysis has been completed and any changes made to the design that might be indicated as being necessary from that analysis. At the very least, the main wing and elevator setting angles should be chosen using XFLR5 before commencing detailed work with Fluent. Note that we generally do not use Fluent to calculate stability derivatives; even at the most basic level, doing this requires six additional runs of the RANS code to carry out finite differencing with respect to the six degrees of freedom of flight and this cost is seldom warranted. Moreover, those working in this field now tend to use fully nonsteady RANS approaches, adjoint sensitivity calculations, or frequency-domain techniques which are even more involved to set up, see, for example, Mader and Martins [26]. If more precise estimates of stability derivatives than can be supplied by XFLR5 are required, it is generally more effective to use experimental-based approximations as available either through the USAF Stability and Control Digital DatCom datasets[27] or the UK ESDU datasets[28].

Because we only use Fluent for dealing with more complex geometries, we build meshes for these using the Harpoon mesher.[29] To get the best results, we try and use the $k - \omega$ SST approach and so require a $y+$ value below 1 for the meshes. However, since our flight speeds are typically 30 m/s or less, the flight Reynold's numbers are lower than for the previously presented validation trials, meaning that a half airframe model including wing, elevator, fin, control surfaces, fuselage, and undercarriage elements can typically be built with between 8 and 9 million cells, although careful $y+$ adaptation around control surfaces can take this up to 2 or even 3 times this. By comparison, a simple wing, elevator, and fin model typically requires less than 4 million cells. Figures 13.29 and 13.30 illustrate a typical mesh that includes fuselage and landing gear, but not control surfaces, together with the equivalent histogram of

---

[27] http://www.pdas.co./datcom.html.

[28] http://www.esdu.com. – the former Technical Department of the Royal Aeronautical Society

[29] http://www.sharc.co.uk/.

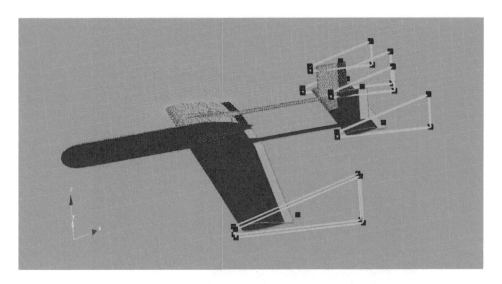

Figure 13.29   Decode-1 mesh shown inside Harpoon along with wake surfaces and refinement zones.

the $y+$ values. Figure 13.31 shows the Fluent convergence history for this model; note the steps in convergence caused by shifts from first to second order and following mesh refinements. Figure 13.32 shows the resulting Fluent lift/drag polars for the aircraft at the cruising speed of 30m/s (as both a full model and one with just the lifting surfaces used in XFLR5), together with results already obtained from the XLFR5 analysis and shown in Figure 13.24. Also shown here are experimental results recorded for the final aircraft in our wind tunnel, but corrected for blockage effects by using CFD results for a model of the aircraft in a model of the tunnel at 6° AoA (discussed further in the section on wind tunnel testing in Chapter 16). The AirCONICS models used to prepare these Fluent polars are shown in Figures 13.33 and 13.34.

Again, a NACA 23012 section has been chosen for the main wing and NACA 0212 for the elevator and fin elements. The setting angle of 2.53° calculated as being necessary for trimmed flight at 30m/s by XLFR5 has also been used in these calculations. Notice that the slopes of the lift curves do not quite agree between the two codes and there are significant differences in the drag values as would be expected from the previous validation studies already described. Also, XFLR5 predicts lift all the way up to 13.2° without any fall-off in the lift slope, beyond which some sections start to stall, and the XFLR5 wing analysis code no longer returns data. For the full aircraft model, Fluent shows a roll-off in lift slope beyond 9° and a drop in lift beyond 12° with a peak Cl of around 0.97, while the measured results gave lift coefficients as high as 1.14 at 11°. As before, it is clear that Fluent is being pessimistic and XFLR5 optimistic in regard to the actual lift performance, which lies somewhere between the two. Interestingly, the Fluent analysis for the wing only gives peak lift values that are very similar to those seen for the whole aircraft in the tunnel.

The measured drag is 50% higher than the Fluent predictions for fully attached flows, even though Fluent is generally pessimistic in drag predictions. This is because the actual aircraft includes a range of additional small drag elements and unknown surface roughness. In the previous spreadsheet analysis, a parasitic drag coefficient of 0.0375 was assumed. If this is

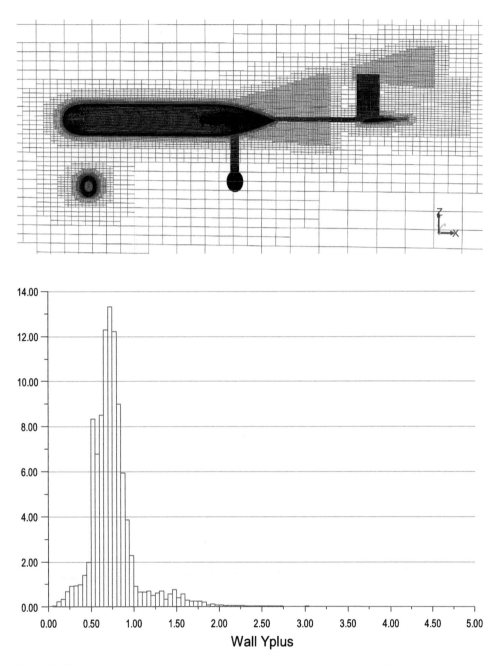

Figure 13.30   Fluent mesh on the center plane for the Decode-1 airframe $k - \omega$ SST analysis at 30 m/s, together with resulting y+ histogram.

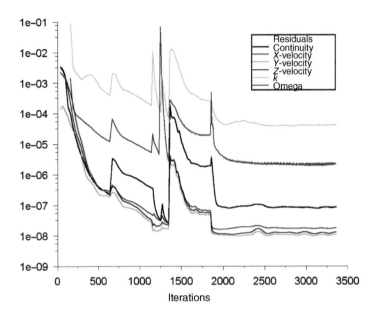

Figure 13.31   Fluent convergence plot for Decode-1 whole aircraft model at 30 m/s.

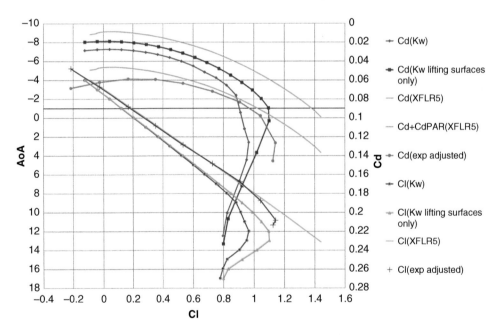

Figure 13.32   Polar plot for Decode-1 airframe at 30 m/s showing both Fluent and XLFR5 results for lift and drag. Those for Fluent include results for just the lifting surfaces and with the complete airframe fuselage, control surfaces, and undercarriage gear; those for XFLR5 show also the impact of adding a fixed parasitic drag coefficient of 0.0375.

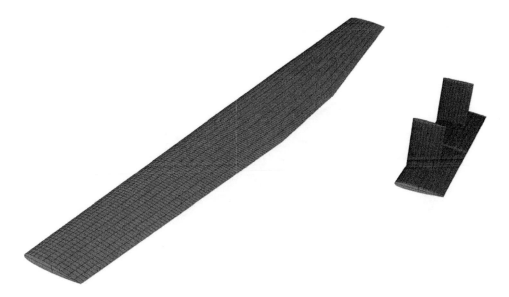

Figure 13.33   AirCONICS model of Decode-1 lifting surfaces.

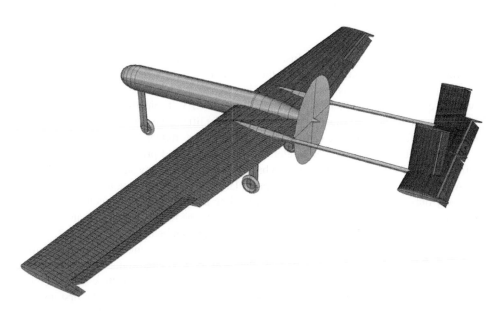

Figure 13.34   AirCONICS model of complete Decode-1 airframe with control surfaces, undercarriage, and propeller disk.

added to the XLFR5 drag coefficient curve, reasonable agreement with experiment is reached, as shown in the figure. A value of only 0.025 is sufficient to correct the Fluent drag coefficient (calculated for the lifting surfaces only) to reach similar levels of agreement. Again, Fluent is somewhat pessimistic in its drag predictions. Fluent does, however, allow investigations of

Figure 13.35 Streamlines colored by velocity magnitude around the complete Decode-1 airframe with deflected ailerons.

aspects such as lift generated by the fuselage; a well-designed fuselage can augment lift by 5% or more, and Fluent studies can reveal these benefits.

The next set of calculations that may be undertaken with Fluent concerns the behavior of the various control surfaces that will be required on the UAV. Typically, moving surfaces are added to give pitch, roll, and yaw control and to augment lift at take-off and landing. On a conventional aircraft, these take the form of elevators, ailerons, rudders, and flaps, respectively. It is clearly useful if the size of these elements can be checked prior to final airframe construction, as initial estimates will have been based on one of the many standard aircraft design texts. As already mentioned, we generally opt for large surfaces to maintain control authority, particularly during the low-speed take-off and landing phases. As already seen, AirCONICS can be used to add the various moving control surfaces to the model, and these can be at arbitrary angles of deflection, and the resulting geometry can then be analyzed in Fluent, though care has to be taken to mesh the zones between the moving elements and the main lifting surfaces in sufficient detail and to ensure that the resulting flow field remains attached to the surfaces; it is unwise to rely on steady-state RANS results for separated flow studies. When carrying out such work, it is often sensible to study just the lifting surfaces in the first instance so as to reduce the mesh sizes being used: for roll moment calculations, a half model is no longer appropriate and so mesh sizes immediately double. Figure 13.35 shows the velocity streamlines past a full Decode-1 model with deflected control surfaces.

It should be noted, however, that the most desirable result from such analysis would be to find the peak lift that can be obtained in the landing configuration with flaps fully deployed. Unfortunately, even a finely meshed model in Fluent is unlikely to be able to do this very accurately. For commercial airliners, it is still common practice to validate the high-lift design using wind tunnel tests. However, whether the effort expended in building complex and highly refined CFD models for small UAVs is really worthwhile remains debatable. We have access to large-scale wind tunnels and rely on these to confirm the high-lift behavior prior to flight trials. Nonetheless, once one becomes familiar with the behaviors of XLFR5 and Fluent (or other equivalent codes) and how best to use them, it is relatively straightforward, albeit sometimes expensive, to validate a good number of the earlier design decisions taken in the spreadsheet analysis prior to detail design and airframe construction.

14

Preliminary Structural Analysis

Having completed the analysis of the wetted surface of the aircraft from a fluid dynamics perspective, attention next turns to the airframe structure. Assessing the strength and rigidity of an airframe is just as important as ensuring that its aerodynamic characteristics are as required. Moreover, building aircraft that are sufficiently but not overly strong is crucial in controlling airframe weight and thus the overall aircraft performance. In our experience, structural analysis can be somewhat of a Cinderella subject in small unmanned air vehicle (UAV) design; it is far from this in commercial aircraft work, and often more engineers will be found working on structures, loading, and weight control than on aerodynamics in large aerospace companies. While much of the work of designing the structure lies in the realm of detailed design, it is very important to have a preliminary structural model as soon as possible so as to better inform calculations on aircraft weight and center of gravity location, also so that any significant shortcomings can be identified before too much further design effort is expended. Essentially the structural design task during preliminary design is to establish the primary dimensions of the main structure that are needed to avoid overstressing the airframe during flight and to prevent any adverse aeroelastic effects stemming from an overly flexible structure. These can then be used to check against the available weight budget used when sizing the aircraft.

Given their low cost, light weight, and ready availability, we always build our main structural elements from carbon-fiber-reinforced tubes. We typically join these together with structural clamps made from 3D printed material (usually selective laser sintering (SLS) nylon) which, where possible, are integral to other parts of the airframe. We also pass them through 3D printed or laser-cut plywood elements to facilitate load transfer from other parts of the aircraft that either generate large forces or entail significant inertias, such as lifting surfaces, engine bearers, servo mountings, undercarriage elements, catapult launch bars, payload items, and heavy avionics components such as batteries. It is, of course, possible to opt for fully mono-coque structures that do not rely so heavily on spars, but since we tend to build our fuselages from fused deposition modeling acrylonitrile butadiene styrene (FDM ABS) or SLS nylon and these materials are not that structurally efficient, we incorporate spars or other reinforcements in our structures even when they are mainly monocoque in layout. Although it is then possible to use basic Euler–Bernoulli beam theory to assess the sizes of the elements involved, it is often simplest to build up an outline model in parametric CAD that is suitable for finite element

Small Unmanned Fixed-wing Aircraft Design: A Practical Approach, First Edition.
Andrew J. Keane, András Sóbester and James P. Scanlan.
© 2017 John Wiley & Sons Ltd. Published 2017 by John Wiley & Sons Ltd.

analysis (FEA) in one of the many readily available analysis packages.[1] Such a geometry definition can easily be set up with the AirCONICS code inside the Rhino environment.

The primary structural cases that need to be considered for small UAVs are the following:

- Maneuver and gust loads on the lifting surfaces;
- Aeroelastic checks (to establish the divergence, control reversal, and flutter onset speeds);
- Inertia and vibration loads, particularly for heavy elements or those near the engines (inertia loads during landing and takeoff can be very severe for small UAVs);
- Landing and take-off loads, which mostly affect the landing and launch bar gear and local supporting structure;
- Powerplant loads, essentially of thrust, torque, inertia, and gyroscopic effects, on the mounting structures.

Only the first two groups here really concern the overall structure and thus need addressing during preliminary analysis, with the other cases mostly being dealt with during detail design (weights for undercarriage, launch and engine mountings, and fittings being estimated by scaling from similar aircraft at this stage). The maneuver and gust load cases lead naturally to the so-called Vn diagram, which may be used to set out the aerodynamic loads on the airframe at various points in the operating envelope. Figure 14.1 shows a typical composite Vn diagram. The key features of this diagram are the positive and negative limits due to stall that set the maximum and minimum lift curves, the maximum and minimum maneuver and gust limits, and the maximum dive speed. These may be used to decide the operating points for which loads need calculating.

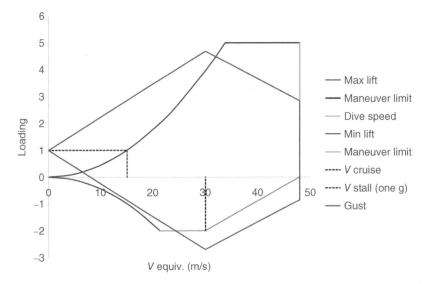

Figure 14.1 Typical composite Vn diagram for gust and maneuver loads on a small UAV (here for Decode-1 assuming maneuver load factors of +5 and −2, 9.1 m/s gust velocity, and a dive speed of 160% of the cruise speed).

---

[1] Note, that although many CAD packages now include a built-in FEA system, we prefer not to use such tools as they rarely have the degree of mesh control and part to part interaction capabilities that we often need.

Typical maximum and minimum maneuver load factors for UAVs are range from $4g$ to $5g$ positive and $-1.5g$ to $-2g$ negative, although these values will depend on the aircraft mission and the regulations being designed to. Combat and aerobatic aircraft are typically designed to higher load factors, and acceptable designs can be built with lower ones. For example, the Federal Aviation Agency (FAA) regulations[2] (FARs 14, CFR, part 23[3] and part 25[4]) state things such as that the maximum maneuver load factor is normally to be 2.5 but that if the airplane weighs less than 50 000 lbs, the load factor is to be given by $n = 2.1 + 24\,000/(W + 10\,000)$, though n need not be greater than 3.8. Loads caused by flying through gusts can be significant drivers in aircraft design. Appropriate load factors can be calculated from the various formulae such as those in FARs 14, CFR, part 23, which states that $n = 1 \pm \dfrac{k_g U_{gE} V_E a \rho S}{2W}$, where k_g the gust alleviation factor is given by $k_g = \dfrac{0.88 \mu_g}{5.3 + \mu_g}$, and μ_g the aircraft mass ratio given by $\mu_g = \dfrac{2W}{\rho c a S g}$. Here, U_{gE} is the equivalent gust velocity and $a = dC_L/d\alpha$ is the slope of the lift curve in radians. A gust velocity of 9.1 m/s has commonly been used for small aircraft up to their cruise speeds, dropping linearly to 4.6 m/s at dive speeds; the FAA regulations use a maximum of 15.2 m/s. For the Decode-1 aircraft with a cruise speed of 30 m/s, the FAA maneuver equation gives $n = 4.49$ and the gust equation gives $n = 1 \pm 3.68$ with a gust velocity of 9.1 m/s; thus a maximum load factor of 5 is quite appropriate and the slightly more modest value of 4 would suffice for many purposes.

The load factors can be taken from the Vn diagram and applied in a number of ways: most obviously by scaling the normal flight loads and applying to the wing spar, for example. Alternatively, elevator and fin loads may be calculated by using maximum C_L values and applying equivalent loads since in small UAVs the structures supporting these elements are less severely loaded by maneuvers and gusts. Note that the angle of attack will influence the direction of aerodynamic loads.

To establish the distribution of loads on the aerodynamic surfaces, a number of approaches can be taken. The simplest and most pessimistic is to apply the full lift and drag forces falling on each surface to the tip of that surface as a point load. This is a very pessimistic approach but one that can be very quickly utilized, particularly if the main structural elements of the aircraft are made up from beam-like elements. Then the simple Euler–Bernoulli beam theory results (found in all standard texts) will allow deflections and stresses to be rapidly and directly calculated. Less pessimistic and only slightly more difficult to apply are uniformly distributed loads (UDLs) of the same total magnitude. More sophisticated approaches seek to distribute the lift and drag forces more accurately, sometimes in both spanwise and chordwise directions. There are a number of semiempirical methods for doing this such as the Schrenk approximation, where the total loading is taken to be an average of a pure elliptical distribution and one that apportions the pressure pro rata to the planform area. If chordwise distributions of loading are to be used with beam models, then local torque distributions will be needed as well. If, however, an XFLR5 analysis has been carried out for the lifting surfaces, it is more sensible to take the pressures calculated there and scale them by the desired load factor to arrive at a load distribution. In either case, the resulting loads can then be broken down into equivalent point or short-span uniform loads and torques and a total load case built by superposing these.

---

[2] http://www.ecfr.gov/cgi-bin/text-idx?c=ecfr&tpl=/ecfrbrowse/Title14/14tab_02.tpl.
[3] Airworthiness standards: normal, utility, acrobatic, and commuter category airplanes.
[4] Airworthiness standards: transport category airplanes.

Alternatively, the load distribution can be carried into a structural FEA code such as Abaqus®[5], either of a spar-based model or of more complete representations of the airframe. If using a relatively complete FEA model, this loading can be applied directly in the form of pressure maps onto the lifting surfaces, which can then provide appropriate force and torque loads on any embedded spars using contact analysis. Our normal practice is to begin with simple, uniform loading calculations on the main spars followed by progressively more detailed FEA models.

14.1 Structural Modeling Using AirCONICS

Starting from the existing wetted surface model already prepared in AirCONICS for aerodynamic analysis, one can begin to build a simple structural model. For most of our aiframes, such as Decode-1, the main wings will be made of foam outboard of longitudinal spars that join the empennage to the main wings. Inboard of this, a 3D printed structure will be adopted, which will directly support the various spars, engine bearer and undercarriage pickup points. In addition, a single rib will commonly be positioned midway along each outer wing where the ailerons start; these carry the inboard ends of the aileron hinges (and sometimes the aileron servos). The wing tips will be 3D printed and also carry the outer ends of the aileron hinges. The elevators and rudders will be dealt with in a similar way with ribs at their ends to support the hinge spars and small 3D printed structural elements to join them to the longitudinal spars, see Figure 14.2.

It can be seen from this that our airframes essentially consist of five types of elements placed inside the wetted surface outer mold line:

- Carbon-fiber reinforced plastic (CFRP) spars (usually of circular cross-section but sometimes other shapes);
- 3D printed fuselage elements (in SLS nylon or FDM ABS);

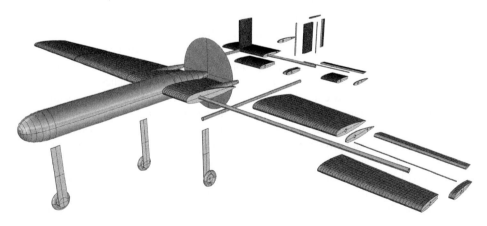

Figure 14.2 Breakdown of Decode-1 outer mold line model into individual components for structural modeling.

---

[5] http://www.3ds.com/products-services/simulia/. Abaqus® is a registered trademark of Dassault Systèmes or its subsidiaries in the US and/or other countries.

- 3D printed or laser-cut ply local load transfer and clamping structures;
- Low-density foam for aerodynamic surfaces;
- Foam cladding in either Mylar film or glass fiber cloth for all the foam parts (in both cases bonded to the foam with appropriate resin-based systems).

To proceed, we begin by defining the acceptable working thicknesses for all elements. For aircraft up to 30 kg in size, we typically adopt the following values:

- Carbon-fiber spars – 0.75–2 mm wall thickness depending on the diameter and selected from readily available stock sizes;
- 3D printed fuselage skins, ribs, and clamps – 1.5–8 mm wall thickness (although it is possible to print thinner structures, we find them too fragile to work with);
- Laser-cut ply ribs and local load transfer structures – from 3 to 10 mm thickness;
- Low-density foam for aerodynamic surfaces – minimum wall thickness of 5 mm, more commonly 10 mm (very fine trailing edges are avoided, either by using a trailing edge radius or by simple truncation);
- Foam cladding in Mylar film – 125 μm thick;
- Foam cladding in glass-fiber cloth – 20 g/m^2 woven cloth, which gives a cladding thickness of 100 μm.

Having made these choices, the next task in partitioning the structure is to decide where the various spar elements should be placed. We generally opt for the principal lifting surface spars to run along the sections at the quarter chord location along with smaller hinge spars at 75–80% of chord when control surfaces are present. These then need to be linked by one or more longitudinal spars or (less commonly) the main hull of the fuselage. We then size the lifting surface ribs and tips that will eventually act as control surface hinge supports as well as helping transfer aerodynamic loads to the spars. Finally, we delineate the extent of the foam elements along the lifting surfaces, fixing where they meet the fuselage and local load bearing structures; all these parameters are defined in the AirCONICS codes we use, see Appendix B for example. Figure 14.3 shows just those parts that will eventually be 3D printed or made from laser-cut ply.

Next the 3D printed parts and foam elements are hollowed out, with the foam parts then being skinned with Mylar or fiber-reinforced plastic. During detailed design, our 3D printed parts are always configured as thin-walled structures locally stiffened against buckling, since this can be accomplished very efficiently during this stage. However, during preliminary design work we often use a thick-walled equivalent structure to represent this approach, and for simple spar analysis just completely solid SLS joining parts.[6] For thick-walled equivalent models, a wall thickness of between 2 and 4 mm is usually reasonable.

---

[6] It is tempting to try and decided on more precise local dimensions of the 3D printed parts by moving to thin-walled models with suitable approximations of the likely internal local stiffening; caution should be taken with such analyses, however, since the final details of local stressing will often be controlled by subtleties such as clearances in contact analysis, local fillets and the positions of any clamping bolts, and so on. It is usually best to delay more precise structural modeling until after the detail design stage has been entered, or even to rely on past practice or experimental testing.

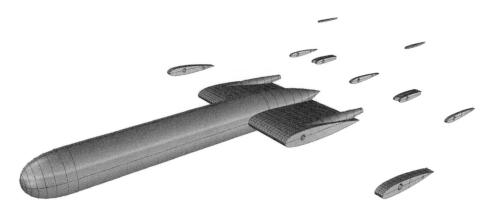

Figure 14.3   Decode-1 components that will be produced by 3D printing or made from laser-cut ply.

Thus the basic steps in converting from the aerodynamic outer mold line to a simplified structural representations are as follows:

- Assume all the moving control surfaces will be solid foam with a Mylar or glass-fiber cloth covering and a full-length hinge pin formed from a small diameter CFRP tube. These items can be dealt with separately and simply need to be partitioned into thin skins and foam cores, which themselves have circular holes cut through for hinge pins – we do not add lightening holes to control surfaces;
- Cut the main wings just outboard of the junction with the longitudinal boom(s) and again at the start (two cuts) and finish of the aileron to create four outer wing parts on each side. The outboard-most of these will become the (3D printed or laser-cut ply) tip, and the other three will become two foam blocks joined together by a rib (of 3D printed plastic or laser-cut ply).
- Cut the elevator and fins off in a similar manner, with end parts forming printed or laser-cut tips and allowing for ribs inboard where the control surface hinges end.
- Cut suitable holes in all parts to accept CFRP spars, boom(s), and hinge pins.
- Hollow out all the foam items forming the main lifting surfaces to reduce weight, leaving suitable wall thicknesses including around spar holes. These can then also be partitioned into thin skins and foam cores.
- Convert the remaining outer mold surface parts into solid or thick-walled structures. These will be the main 3D printed structural elements of the aircraft already illustrated in Figure 14.3.

Once the building blocks have been defined in this way, analysis can begin. We start with simple beam theory assessments before moving on to FEA approaches based on exports from the AirCONICS Rhino model (in the form of a ACIS, STEP, STL, or Parasolid files) and read into whatever FEA system is to be used to carry out the more detailed preliminary structural analysis. In either case, we start with simple UDLs set at the peak load factors already defined in the Vn diagram. We follow this up with distributed loads taken directly from XFLR5 analyses, scaled by load factors from the Vn diagram (we do not try and use loads taken directly from XFLR5 at extreme load factors since the panel solver is not compatible with such conditions). For beam theory calculations, loads are most easily transferred as combinations of point and uniformly distributed forces and torques, while for FEA models it is also possible to work with body forces on the spars or pressure loads on the aerodynamic surfaces.

14.2 Structural Analysis Using Simple Beam Theory

It is relatively straightforward to use simple Euler–Bernoulli beam and torsion theory to analyze a spar that has uniform properties along its length. Since we almost always use uniform section CFRP tubes, this represents the best place to begin the preliminary structural analysis. The key results that are needed are for point and uniform forces and moments applied to cantilever beams with built-in (encastre) ends. These can be obtained from any standard undergraduate text: the key results are summarized in Table 14.1 for convenience. We then use the principle of superposition to sum up the effects of loads along the length of any given spar. If a simple end-point force and moment pair or a completely uniform load is to be applied, this is very straightforward. For example, if we want to place a $4g$ maneuver load on a main spar of 3 m length (wing tip to wing tip), which we assume is clamped into the fuselage at the center plane and ignoring the mass of the wings, then our UDL for a 15-kg aircraft is just $w = 4 \times 15 \times 9.81/3 = 196.2$ N/m. Table 14.1 then shows that for this case the maximum

Table 14.1 Shear forces (Q), bending moments (M), slopes (θ, in radians), and deflections (δ) for Euler–Bernoulli analysis of uniform encastre cantilever beams.

Model	Distribution	Max. value
Concentrated tip force P	$Q(x) = P$	$Q_{max} = P$
	$M(x) = P(x - L)$	$M_{root} = PL$
	$\theta(x) = \frac{P(2Lx-x^2)}{2EI}$	$\theta_{tip} = \frac{PL^2}{2EI}$
	$\delta(x) = \frac{Px^2(3L-x)}{6EI}$	$\delta_{tip} = \frac{PL^3}{3EI}$
Concentrated tip moment q	$Q(x) = 0$	$Q_{max} = 0$
	$M(x) = q$	$M_{max} = q$
	$\theta(x) = \frac{qx}{EI}$	$\theta_{tip} = \frac{qL}{EI}$
	$\delta(x) = \frac{qx^2}{2EI}$	$\delta_{tip} = \frac{qL^2}{2EI}$
Uniformly distributed load w	$Q(x) = w(L - x)$	$Q_{root} = wL$
	$M(x) = -\frac{w(L^2-2Lx+x^2)}{2}$	$M_{root} = \frac{wL^2}{2}$
	$\theta(x) = \frac{wx(3L^2-3Lx+x^2)}{6EI}$	$\theta_{tip} = \frac{wL^3}{6EI}$
	$\delta(x) = \frac{wx^2(6L^2-4Lx+x^2)}{24EI}$	$\delta_{tip} = \frac{wL^4}{8EI}$

Note that for a hollow cylinder of outer diameter d and wall thickness t, the second moment of area $I = \pi(d^4 - (d - 2t)^4)/64$, while for a rectangular cross-section of breadth b, height h, and wall thickness t, the second moment of area $I = bh^3/12 - (b - 2t)(h - 2t)^3/12$.

bending moment occurring at the root (mid-span) of the spar is $196.2 \times 1.5^2/2 = 220.7$ N m. If the spar is made of CFRP with a Young's modulus of 70 GPa and ultimate tensile strength of 570 MPa (see Table 18.1) and inner and outer diameters of 25 and 20 mm, respectively, $M/I = \sigma_{bending}/y$ tells us that the peak bending stress will be $\sigma_{bending\ max} = \frac{220.7 \times 0.025/2}{\pi(0.025^4 - 0.02^4)/64} = 243.7$ MPa, that is, 43% of the ultimate tensile strength of the material. The equivalent tip deflection is $\delta_{tip} = \frac{4 \times 15 \times 9.81/3 \times 1.5^4}{8 \times 70 \times 10^9 \times \pi(0.025^4 - 0.02^4)/64} = 0.1567$ m, an acceptable fraction of the total wing span. If the lift force was not acting directly through the spar but some distance in front or behind it, there would be an additional torque load to be dealt with. If, for example, the lift was acting 20 mm behind the spar centerline, the total torque would be $T = 0.02 \times 4 \times 15 \times 9.81/2 = 5.886$ N m. The maximum stress caused by such a torque $\tau_{max} = \frac{Td}{2J} = \frac{5.886 \times 0.0125}{\pi(d^4 - (d-2t)^4)/128} = 13$ MPa, a negligible extra amount.

For a nonuniform loading, some form of discretization is needed, but a simple hand calculation is still possible. If the load distribution is to be taken from XFLR5, the code conveniently gives section bending moments along each lifting surface that can be applied directly to the spars, station by station along the spans. For a uniform spar, one simply takes the highest bending moment tabulated by XFLR5 for a given spar and again uses $M/I = \sigma/y$ to get the relevant maximum spar stresses. If the tip deflection is also required, one must sum up the deflections section by section along the spar from the root, taking care to allow for the cumulative effect of the slope changes caused by each local section bending moment. To make this clear, consider again the previous case but now assume each spar half is analyzed in just two parts and where the local section lift and drag coefficients as given by XFLR5 (lift coefficient = Cl, viscous drag coefficient = PCd, and induced drag coefficient = ICd) are outer half 0.176423, 0.007079, 0.003350, and inner half 0.303362, 0.007320, 0.000282, and the mean local section chords are 0.2641 and 0.3727, respectively. We take the velocity to be 30 m/s and ρ to be 1.211 kg/m². We first convert the coefficients to an equivalent uniformly distributed force by noting that the drag and lift are perpendicular to each other and also allow for our maneuver load factor of 4 so that the force per unit length are $w = 4 \times 0.5\rho V^2 \bar{c} \sqrt{Cl^2 + (PCd + ICd)^2} = 101.7$ and 246.5 N/m, respectively. (Here, $\bar{c}$ is the local mean section chord and we ignore the slightly different angles these resultant forces have with respect to each other and the airframe.) The deflection and slope at the root are of course zero. At the junction between the two parts, the deflection is due to a combination of the UDL on the inner half of the wing and the force and moment applied at the inner half's tip by the outer half. These quantities are $w = 246.5$ N/m, $P = 0.75 \times 101.7 = 76.3$ N, and $q = 0.5 \times 101.7 \times 0.75 \times 0.375 = 28.6$ N m. Their effects on the inner half of the wing may be deduced from Table 13.1 as a deflection at the junction of $\delta_{junction} = \frac{0.75^2(3 \times 246.5 \times 0.75^2 + 8 \times 76.3 \times 0.75 + 12 \times 28.6)}{24 \times 70 \times 10^9 \times \pi(0.025^4 - 0.02^4)/64} = 0.036$ m and a slope at the junction of $\theta_{junction} = \frac{0.75(246.5 \times 0.75^2 + 3 \times 76.3 \times 0.75 + 6 \times 28.6)}{6 \times 70 \times 10^9 \times \pi(0.025^4 - 0.02^4)/64} = 0.076$ rad. The tip deflection at the end of the outer part is then that for the outer half plus that for the inner half plus that caused by the slope there times the length of the outer half, that is, $\delta_{tip} = \frac{101.7 \times 0.75^4}{8 \times 70 \times 10^9 \times \pi(0.025^4 - 0.02^4)/64} + 0.036 + 0.0076 \times 0.75 = 0.0981$ m, which is somewhat less than predicted from a fully uniform loading, as would be expected. Clearly, such an approach can be extended to multiple subdivisions with the use of a simple spreadsheet.

Following this logic, Figure 14.4 shows how deflection and slope vary for the Decode-1 main spar when flying at 30 m/s and an angle of attack of 2.53° using loading results from XFLR5, again with a load factor of 4. For this 15-kg aircraft, the spar has a total length

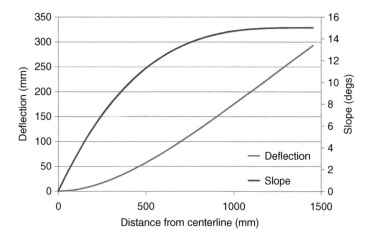

Figure 14.4 Deflection and slope variations for the Decode-1 main spar when flying at 30 m/s and an angle of attack of 2.53° using loading taken from XFLR5, a load factor of 4, and simple beam theory analysis. The spar is assumed to be made from a circular CFRP section of outer diameter 20 mm, wall thickness 2 mm, Young's modulus of 70 GPa, and extending the full span of the aircraft, being clamped on the center plane.

of 2.94 m, an outer diameter of 20 mm and a wall thickness of 2 mm. The resulting tip deflection using simple beam theory is predicted to be 0.292 m. XFLR5 reports a maximum bending moment for this condition of 48.77 N m, which when combined with a load factor of 4 leads to a peak stress in the spar of 420.7 MPa, which is 74% of the yield stress of the CFRP material, and an acceptable margin, bearing in mind that the spar is actually supported by the fuselage over a significant length rather than just at the center plane; this will reduce the stresses and deflections significantly as we will see shortly. To gain further insights, it is simpler to make use of one of the many standard FEA packages for such analyses, an approach we consider next.

14.3 Finite Element Analysis (FEA)

Although it is quite simple to build a finite element model directly from beam elements and continue the kind of analysis set out in the previous section, having already generated a simplified parametric AirCONICS CAD model that includes the main structural spar and boom elements it is usually more straightforward to carry out FEA by importing this CAD model into an FEA package where a suitable 3D mesh of the structure can be built and loads and boundary conditions applied to it. From a structural perspective, it will be clear that the key locations in such a parametric model will be based around the main spar and tail booms and their junctions with the tail and fin spars and the fuselage elements. Figure 14.5 shows these basic components of the structure, here controlled by just the five sets of spar diameters, the junction locations, and spar lengths. Note that the tail and fin spars take the form of the tail shape adopted (H, V, inverted V, etc.). For the simplest forms of analysis, all that need be considered are these spars and the (typically SLS nylon) junction parts. If being analyzed in detail, the internal dimensions of those SLS parts that directly transfer load between the spars clearly need to be parametrically driven, possibly with a thick-walled structure as already mentioned.

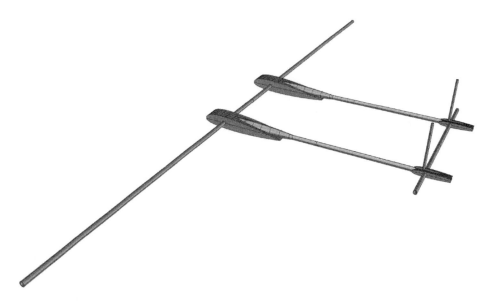

Figure 14.5 Preliminary spar layout for Decode-1. Here the linking parts are taken directly from AirCONICS without being reduced to either thick-walled or thin-walled rib-reinforced structures.

The precise detailing of the junctions would be finalized later on; however, for sizing spars they do not matter greatly and can be replaced by simple fully solid SLS parts to begin with, as in the figure, the aim being merely to provide the supporting boundary conditions. The structure can then be meshed and loaded with the various forces already derived to assess its suitability for the design in question. By using appropriate material properties for the junction parts, more realistic boundary conditions can be imposed than by directly restraining the spars themselves.

14.3.1 FEA Model Preparation

Generally the fastest approach to meshing is to use an unstructured tetrahedral mesh for the SLS nylon, possibly with hex-core conversion if the mesher in use permits it. The spars being simple tubes can always be meshed with quadrilateral plate or hexahedral brick elements. Suitable care must be taken to ensure that sufficient element density is achieved to get reliable results. Unless the user is particularly skilled at FEA meshing, it is generally sensible to start with the simplest set of default meshes the system in use can provide. The level of accuracy required during preliminary designs makes spending large amounts of time on the FEA definition rather a waste. The aim is to ensure that the structure is broadly plausible so that better weight and center of gravity estimates can be made, rather than to attempt detailed stressing of individual components; localized high stress areas will need further study during detail design anyway (note that using a simple mesh will, however, mean stress levels and deflections may differ from fully converged results by 10% or more and this should be borne in mind when considering the results obtained). Suitable material properties must also be entered, see Tables 18.1–18.4, but note that thin glass-fiber claddings do not achieve the

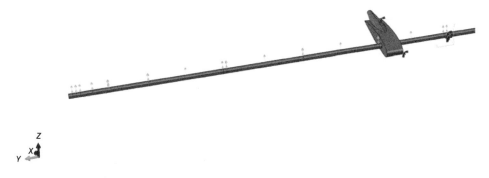

Figure 14.6 Simplified Abaqus® main spar model with solid SLS nylon supports for Decode-1, showing subdivided spar and boundary conditions for a 4*g* maneuver loading.

full material properties detailed there. Loads can again be simple point forces, UDLs, body forces, or complete pressure maps. If densities are included in the model, gravitational affects can be added as well. The interactions between components can be simply modeled as direct links between surfaces or using contact models that allow for sliding and normal forces. While contact models give more realistic results, they can be temperamental to converge, and if this is the case, initial studies can be made where directly linked (tied) surfaces are used. Such models also rapidly allow for all inputs to be checked for any errors (such as quantities being specified in the wrong units or forces in the wrong directions, etc.). Once meshing is complete and loads and boundary conditions are defined, analysis can begin.

An example of such a simplified structural model of the Decode-1 airframe loaded into Abaqus® is shown in Figure 14.6. Note that for this case just one half of the airframe model needs to be loaded and symmetry can then be exploited to save on computational effort. Here, the spars are meshed with hexahedral brick elements, and the SLS parts with tetrahedral elements and a frictionless hard contact model are adopted. The model illustrated has some 85 000 elements in total. It is restrained by clamping the inner faces of the forward SLS parts where they would, in fact, continue onward into the rest of the fuselage structure, and symmetry is imposed on the center plane (this pair of constraints essentially unloads the spar inboard of the nylon parts, perhaps a slightly optimistic model). To begin with, loads are added by simple, uniform body forces on the spars scaled by the overall maximum take-off weight (MTOW) of the aircraft and a load factor such as 4 times gravity to account for maneuver loading. Thus for a first worst case analysis, and as in the previous section, twice the MTOW is applied uniformly to each main spar, and the stresses and deflections are examined, see Figure 14.7.

To build this model in Abaqus, the following steps have been taken (other FEA packages differ a little in the details of such analyses but the same basic information always has to be loaded):

- The geometry created by AirCONICS is exported in a format that Abaqus can easily read; we use the ACIS definitions that create a .SAT file (the airframe is subdivided into its constituent parts first, and each part is further subdivided if divisions are needed for details of loading or meshing).
- Each part is imported into Abaqus, one at a time (those parts that have been internally subdivided – here the main spar – are reassembled on import but preserving the subdivision

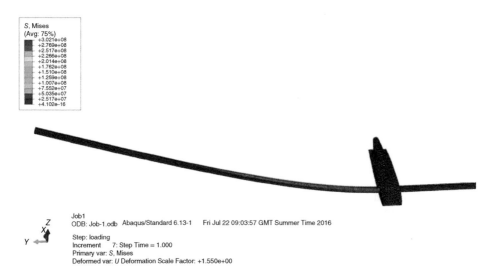

Figure 14.7 Deformed shape and von Mises stress plot for Decode-1 main spar under 4*g* flight loads using a uniform spar load. The tip deflection is 189.7 mm.

boundaries using the "combine into single part" and "retain intersecting boundaries" options on import). This means we have just two parts, the spar and one of the SLS nylon junction parts which supports the spar (and eventually the tail boom).

- It is probably good practice at this stage to convert the imported parts to an Abaqus "precise" geometry using the "Convert to precise" option within the Part editing area of the Geometry Edit tool.
- Next, the material properties are defined and assigned to the parts: the spar is CFRP, and the support SLS nylon with properties taken from Tables 18.1 and 18.2, that is, Young's modulus of 70 and 1.7 GPa, respectively, while Poisson's ratio is taken as 0.3 for the CFRP and 0.4 for the nylon. So here two mechanical elastic property types are created and then linked to the parts (in Abaqus via the creation of 'solid sections', which are assigned to each part in turn).
- The parts are then assembled into a single model by creating an "assembly instance" containing "dependent" parts.
- Each part is separately meshed. To do this, first the mesh control menu is used to decide the type of meshing on each part, and then the approximate global seed size is set before meshing is invoked; here the spar can be sweep-meshed with hexahedral elements, while the more complex shape of the SLS nylon part is meshed with tetrahedral elements, in this case seed sizes of 0.004 and 0.005 are used, which are some three times finer than the suggested default, but still six times larger than would be needed for fully resolved results. This gives a mesh with 14 760 elements in the spar and 71 000 in the nylon part. Clearly, the mesh quality needs to reflect the desired accuracy of results. The chosen values are reasonable, but to gain fully accurate stress results using solid elements, it is normal to have at least three elements spanning the thickness of the spar. This increases the element count substantially, and the simpler model used here is only a few percent different in terms of peak von Mises stress results (the simplest way to assess the mesh being used is to reduce all the mesh seed

values by 10%, which nearly doubles the element count, and see what impact this has on results. If very significant changes occur, the seeds can be reduced in 10% steps until some form of convergence is achieved).

- Following this, the analysis steps and various boundary conditions need to be defined. For this analysis, two steps will be used (initialization and loading): the center plane of the spar and inner face of the SLS nylon part will be constrained (here encastre on the SLS nylon and Y-symmetry on the spar), a contact model is placed between the spar and the nylon part that has zero sliding friction and hard normal contact (finite friction tends to cause convergence problems so is best avoided during preliminary analysis), and lastly body forces will be applied along the length of the spar, either as a uniform loading or based on the section lift and drag estimates given by XFLR5 and used in the previous section, again with a load factor of 4 (only the starboard side of the spar will be loaded).

  - To create the contact pairs (or direct ties), the "find contact pairs" option is used in Abaqus, which creates all likely pairs; those that are to be used are given appropriate properties and the rest are deleted, making sure that the most densely meshed part is the slave and not the master and also that the master part is the one that extends beyond the contact region (if difficulties are experienced in deciding which contact pairs to use, the "Instance" option can be used for the search domain and then hitting the arrow button in the dialog box allows individual parts to be selected for the search to work over; also the separation tolerance can be changed to include extra or exclude pairs being found).
  - The constraint conditions are created under the "loads" menu, where the center plane of the spar and the inner face of the SLS part are set as Y-symmetry and encastre, respectively.
  - Abaqus always provides the first initialization step automatically, and this is where the Y-symmetry and encastre conditions are specified. To enter the applied forces, a second step, here called "loading," is defined that inherits the previous boundary conditions and then applies the desired forces. It is in defining this step that the user selects the type of analysis: here "static, general" (various options are available at this point such as contact stabilization, which may be needed for some more complex analyses).
  - The body force loads are added to the loading step, one for each subpart of the spar and in the vertical direction. For a uniform loading, these will all be the same and can be added at one time, while to model the load variation obtained from XFLR5, individual values will need to be inserted for each subpart. Note that these body forces are in units of N/m^3 if the model is being built in SI units.

- Then, having defined all the required problem information, an Abaqus "job" is created, which allows details of how the available computing resources will be used (memory and parallel options can be defined). This job is then submitted for execution. While execution continues, Abaqus offers a monitor option that allows progress of the job to be observed. If a fine mesh and wide variations in stiffness properties are being used during contact analysis, this job may take some time to stabilize and converge. Simpler models even with many hundreds of thousands of elements will solve in a few minutes.

- Finally, when the job has completed successfully, it will create an "output database" or .ODB file, which contains various standard sets of results such as deflections and stresses that must be loaded for viewing (the results collected during analysis can be changed by the user using the "field output requests" and "history output requests" commands). Generally, the deflections and von Mises stress levels are the quantities of most interest and these are created by default. For the case illustrated in Figure 14.7, the maximum deflection is 189.7 mm and the peak von Mises stress is 302.1 MPa. The reduced deflections and stresses predicted by the FEA stem mainly from the constraints and the additional support provided by the SLS nylon part, meaning that the effective cantilever span is much reduced.[7]

Once the main spar has been checked for overall suitability in this way, more detailed loading models can be considered using the pressure distribution found from XFLR5 to spread loads along each of the spars more realistically, again scaled by a suitable loading factor. This reduces the effective loads so that the maximum deflection for loading of the main spar alone comes down to 148.3 mm and the peak von Mises stress to 255.4 MPa, very comfortably below the likely failure stress of the material. Of course, one does not want the stress to be too low, because this would imply an over-heavy design.

14.3.2   FEA Complete Spar and Boom Model

Having assessed the main spar, we move on to a more complete model using all the main CFRP elements and nylon linking parts shown in Figure 14.5. When building models with multiple spars plus link parts and using contact analysis, care has to be taken to understand possible rigid-body modes in the assembled structure. These can occur because spars are free to rotate inside the nylon sockets created for them, or because a spar may be able to slide out of the joint it has been placed in. Clearly, in reality these behaviors are prevented by the action of clamping forces that will be generated by details that are not yet present during preliminary design. Therefore, additional constraints have to be imposed in the FEA, but care then has to be taken to ensure that such constraints do not provide excessive support or overly restrict the likely deflections; otherwise unrealistic deflections and stresses will result. To do this, we find it is useful to add tie constraints between the nodes on the end faces of spars to the nearest transverse spar in addition to the surface contacts between the spars and the nylon junction parts. Because the spars are all orthogonal to each other, this prevents them from sliding or rotating during analysis (notice that this will locally increase stresses in the spars but that any form of real clamp would also do this; also some spar-to-spar areas may appear not to need tying together to ensure that rotation and sliding of all parts is prevented, but we find convergence is helped by tying all spar ends where possible). We might also revisit how the whole model is being restrained; instead of symmetry and encastre constraints, the center plane of the main spar could simply be restrained in the transverse direction alone and the inner faces of the main forward nylon parts just in the vertical and fore and aft planes – an approach that will be examined later on. Reducing the supports in this way allows the main spar to flex between the nylon parts. Whether this is more realistic would depend heavily on the stiffness of the main fuselage element and the degree to which it might support the spar internally.

In this case, we take the previous 4g loading on the main spar and add pressure loads for the elevators and fins assuming that all the tail sections are operating at a local lift coefficient of unity, clearly an extreme case. Following the previous logic, the resulting mesh has 270 000 elements and 970 000 variables (Figure 14.8). The resulting deformed shape is shown

---

[7] If the spar is analyzed on its own, without the SLS part, using an encastre center plane and the mesh seed reduced to 0.0008 so that around 400 000 elements are used, the predictions of Euler–Bernoulli beam theory given earlier are recovered to within 0.4%.

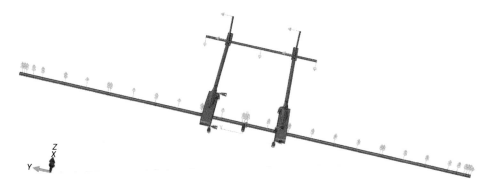

Figure 14.8   Abaqus loading for full Decode-1 spar model under wing flight loads taken from XFLR5 together with a load factor of 4 plus elevator and fin loading based on Cl values of unity.

in Figures 14.9 and 14.10. The various peak stresses and spar deflections are detailed in the first results column of Table 14.2. Note that the peak stress in the nylon is massively higher than the material yield stress: this occurs in the region where the starboard boom emerges from the forward supporting nylon structure. If a refined mesh is used, by reducing the seeds by 10% the whole model then contains some 509 000 elements and has 1.8 million variables in it. Now, Table 14.2 shows that while the various deflections have not changed significantly, the peak nylon stress has increased and the starboard boom stress is now 50% higher, see Figure 14.11. Although this model is almost twice as large, solution times are still much less than those for the RANS-based CFD models of the airframe considered in Chapter 13. Using a desk-side-based machine with 12 cores, the results take around 33 min to compute.

To proceed further, a number of subtleties need to be considered: first, the actual nylon part will not be fully solid; second, as the yield stress of nylon is massively below the peak stresses being indicated, this simple elastic model is no longer valid; and third, the element density in the nylon parts of the model is still rather low. The first thing to review is the mesh density in the region giving concern. In Abaqus this can be controlled locally by using edge seeding: so next a finer set of seeds is placed at the very rear of the nylon part and also along the spar, biased away from a collar placed in this location, see Figure 14.12, although this model requires adaptive stabilization to converge.[8]

When this is done, much more realistic stress results are obtained, although the nylon part still exceeds its yield stress at the very edge of the structure, see column 3 of Table 14.2. Notice that it is the transverse fin loading that is causing these stresses and not the vertical elevator loads, and also that these refinements barely affect the remaining results in the table. If this kind

---

[8] When convergence difficulties are experienced during solution, the user can turn on the Abaqus automatic stabilization mechanisms for the loading step using the "Specify dissipated energy fraction" and "Use adaptive stabilization" boxes when defining the loading step. This does, however, slow down the convergence process considerably and increases significantly the size of output databases, so it is best avoided if at all possible. Convergence problems commonly arise when either the problem is over- or under-constrained, or contact elements have been included where no or very large deformation occurs during the analysis or finite friction has been included. Generally, to avoid such issues it is best to build up the structural model, piece at a time, solving at each stage rather than directly building a fully featured model and attempting to solve it all in one go. During the development of such a model, the use of tie constraints rather than contact pairs simplifies the analysis and speeds up solutions considerably. These can also be used in areas of the model where detailed contact stress results are not required.

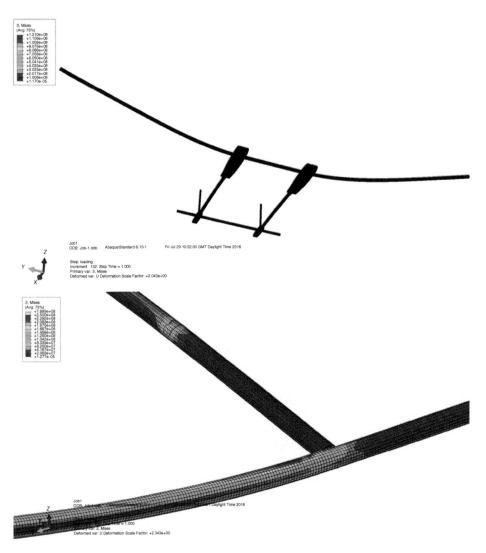

Figure 14.9 Deformed shape and von Mises stress plot for full Decode-1 spar model under wing flight loads taken from XFLR5 together with a load factor of 4 plus elevator and fin loading based on Cl values of unity. The main spar tip deflections are 143.9 mm, the elevator spar tip deflections are 10.8 mm, and the fin spar tip deflections are 11.1 mm.

of refinement is applied throughout, even better results are obtained, although by this stage the solution takes some $3\frac{1}{2}$ h to complete, see column 4 of Table 14.2. For this model, the encastre restraints on the forward nylon parts have been replaced by simple displacement constraints in the vertical and fore and aft planes, while the center of the main spar is just restrained in the transverse direction. This change to boundary conditions significantly reduces the clamping of the model so that the main spar can then flex between the nylon supports, leading to much greater tip deflections, see Figure 14.13. This in turn places higher stresses on the elevator

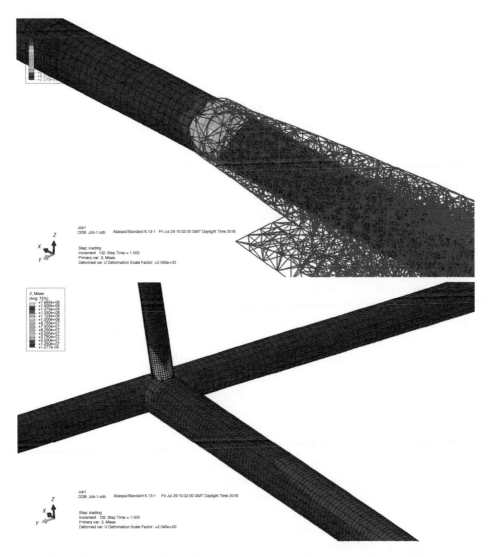

Figure 14.10 Further details of the deformed shape and von Mises stress plot for full Decode-1 spar model under wing flight loads taken from XFLR5 together with a load factor of 4 plus elevator and fin loading based on Cl values of unity.

spar, although the other spar stresses are not changed significantly. Whether this more loosely supported model is more realistic is a matter of conjecture. Clearly, the main fuselage would support the wing spar to some extent; the actual tip flexure would thus probably lie somewhere between the results reported here in columns 3 and 4 of the table. At the same time, the greater mesh refinement reduces the peak stresses in the SLS parts to just 37% over yield, and this again occurs only in the very thin areas of the nylon where local yield would be inevitable in this design.

Table 14.2   A selection of results from various Abaqus models of the Decode-1 airframe.

Model	Units	Basic mesh, encastre/ symmetry	Refined mesh, encastre/ symmetry	Refined boom-biased mesh, encastre/ symmetry	Refined all biased mesh, XYZ pinned
Cell count	—	271 123	508 723	487 627	623 040
Number of variables	—	967 719	1 825 611	1 781 196	2 315 880
Main spar tip defln.	m	0.1490	0.1439	0.1427	0.2540
Main spar max. stress	MPa	262.5	238.0	235.7	235.3
Stbd. boom max. stress	MPa	472.7	676.0	55.5	59.8
Port boom max. stress	MPa	255.4	254.4	48.2	49.5
Elevator spar tip defln.	m	−0.0113	−0.0128	−0.0135	0.0063
Elevator spar max. stress	MPa	15.4	19.2	23.0	66.9
Fin spar tip defln.	m	0.0111	0.0129	0.0134	0.0100
Fin spar max. stress	MPa	115.4	167.4	163.9	180.0
Peak nylon stress	MPa	796.5	937.9	85.9	61.5

Further increasing the mesh refinement and moving to a thick-walled structure will make the Abaqus predictions even more realistic. However, in such analyses it is important to understand the elastoplastic nature of the nylon junction parts. If rather fine areas are adopted in the geometric design of the nylon, as here where the tail booms emerge from the forward nylon supports, it is almost certain that the yield stress of the nylon really will be reached locally. This is generally not a problem since some permanent stretching of these areas of nylon is probably acceptable, although detail design improvements might sensibly be made later on.[9]

Overall, the results in Table 14.2 suggest that the main spar is moderately loaded (to about half yield) while the elevator spar is rather too strong for its role, with the fin spars being somewhere in between. Clearly, it might make sense to reduce the diameter of the elevator spar at this stage, saving mass in a very weight-sensitive area of the aircraft. Reducing the fin spars would be possible from a structural point of view, but they are already of quite small diameter and making them smaller might lead to assembly and maintenance issues. The very high stresses initially found in the booms would have been a cause for concern, but by revising the element density in the contact region, their loading is revealed to be, in fact, quite modest, although reducing their diameters might make the aircraft too flexible from a controllability

---

[9] To model yielding, the material could be treated as an elastoplastic one. This would get around the issue but would also make the solution convergence rather more temperamental and probably to little benefit.

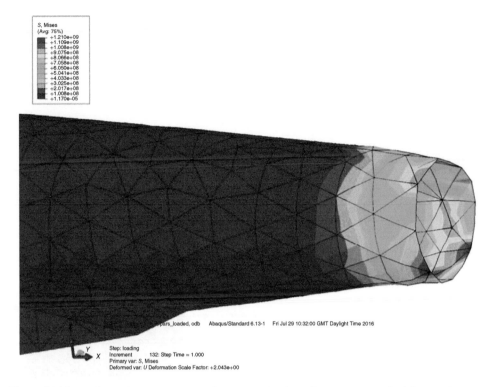

Figure 14.11 Deformed shape and von Mises stress plot for nylon support part in full Decode-1 spar model under wing flight loads taken from XFLR5 together with a load factor of 4 plus elevator and fin loading based on Cl values of unity.

perspective. Accurate stressing of the nylon parts clearly needs more sophisticated models, and as yet no attention has been focused on the main lifting surfaces that are made from foam with a surface cladding.

14.3.3 FEA Analysis of 3D Printed and Fiber- or Mylar-clad Foam Parts

Properly stressing the 3D printed SLS nylon and the fiber- or Mylar-clad foam elements is probably best left until the detailed design stage, but if preliminary results are required for these elements, some steps can be taken without reaching for a fully featured design. For example, to limit design effort and computational cost, the 3D printed nylon parts can be modeled as thick-walled structures. To do this, we simply adopt a specified equivalent thickness having calibrated the resulting weights and stresses against our experience of the performance achievable from a more fully detailed structure; see Figure 14.14 where a 4-mm-thick wall has been created using the shell commands of the CAD package. Although this model is more complicated to build and mesh, the resulting mesh has fewer elements, which can be used to better effect; meshing fully solid parts with a fine mesh rapidly increases the element count. When this approach is used to study the forward SLS nylon boom supports, it reveals that

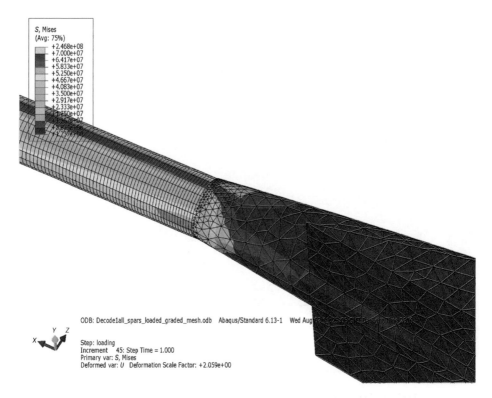

Figure 14.12 Deformed shape and von Mises stress plot for full Decode-1 spar model with locally refined mesh under wing flight loads taken from XFLR5 together with a load factor of 4 plus elevator and fin loading based on Cl values of unity.

the peak contact stress with the tail boom is now predicted to be locally 57.8 MPa but that this falls rapidly away from the fine edges. At the same time, the peak contact stress with the main spar is only 5.6 MPa, while the stresses in the spars are hardly changed from before, see Figure 14.15. Reducing the equivalent wall thickness to 2 mm raises the peak nylon stress in its finest area to 84.6 MPa and that near the main spar to 24.2 MPa, see Figure 14.16, suggesting that 2 mm wall thickness is probably more appropriate for an equivalent stress model in this case. Note, however, that thin-walled structures notoriously suffer from local buckling, and thus far our Abaqus analysis, though dealing with contact stresses, does not allow for buckling (or nonlinear material properties). It is inadvisable to push this kind of simplified structural analysis model too far; the final part will contain rib-stiffening to control buckling and the SLS nylon will undoubtedly suffer from local yielding where it is in contact with the CFRP spars.

To model the aerodynamic surfaces, we need to model both the lightweight foam core and its cladding. Again, a simplified approach is possible, and we proceed as before but join the inner surface of the cladding to the outer surface of of the foam with tie constraints all over. The foam is also created using the shell commands of the CAD package plus suitable fillets, here with a 10 mm wall thickness which is typical of the sort we manufacture. The foam parts are easily meshed with standard 3D brick hex elements, see Figure 14.17. The cladding is created as a very thin shell surrounding the foam, and we use continuum shell elements to model it in

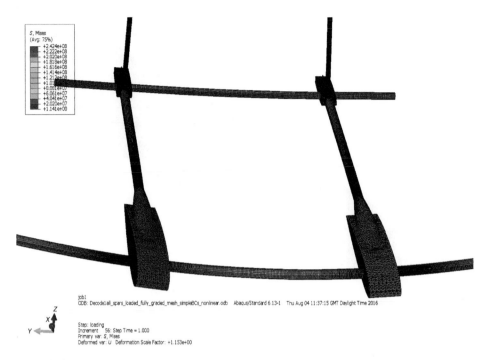

Figure 14.13 Deformed shape and von Mises stress plot for full Decode-1 spar model with fully refined mesh and reduced boundary conditions under wing flight loads taken from XFLR5 together with a load factor of 4 plus elevator and fin loading based on Cl values of unity.

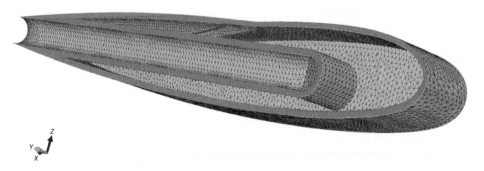

Figure 14.14 Simplified Abaqus thick-walled structural model for Decode-1 SLS nylon part. The mesh for this part contains 25 000 elements.

Abaqus. These allow direct meshing of the solid body produced by AirCONICS despite the very thin nature of the cladding, here 0.1 mm[10] – continuum elements allow for the extreme aspect ratios produced when elements are this thick while being 10 mm and more in the other

---

[10] We find it difficult to make claddings thinner than 0.08 mm, and even this thickness is very hard to achieve evenly over the whole wing surface.

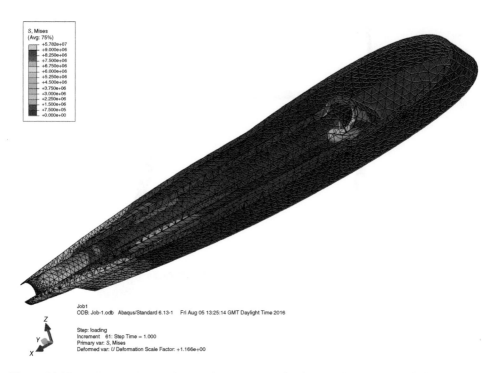

Figure 14.15 Deformed shape and von Mises stress plot for thick-walled nylon part in full Decode-1 spar model with fully refined mesh and reduced boundary conditions under wing flight loads taken from XFLR5 together with a load factor of 4 plus elevator and fin loading based on Cl values of unity.

directions, without the need to switch to traditional surface-based shell models, which are less easily constructed from CAD models. Suitable material properties can be found in Tables 18.3 and 18.4, although we always reduce the Young's modulus value for the glass fiber to allow for the reduced fiber-to-resin ratios achieved in very thin claddings. A value of 25 GPa is a reasonable approximation that is borne out by the results of experimental vibration testing of wings.

It is important when using continuum shell elements that sweep meshing be used to build the mesh and the through-thickness direction is correctly specified as the sweep direction.[11] To do this, we partition the cover into separate parts using the Partition commands in the Tools menu: first, the main upper and lower surfaces are separated at the leading and trailing edges with face partitions; and then, if sharp trailing edges are in use, the upper and lower trailing edges are split off using the "sweep edge along edge" process to further partition off the trailing edges, which are then triangular in cross-section. If a truncated trailing edge is being used (perhaps because a flap has been cut from the section), the part is partitioned at the truncation points. The main upper and lower covers are next meshed using swept medial axis hex meshes with the stack directions defined by selecting the outer faces. If a sharp trailing edges is adopted,

---

[11] If the "Redefine Sweep Path ···" option in mesh controls is used, a small red conical arrow is shown defining the sweep direction. This must be normal to the surface of the cover; selecting the outer surfaces as the top surface in the "Assign Stack Directions ···" option will enforce this.

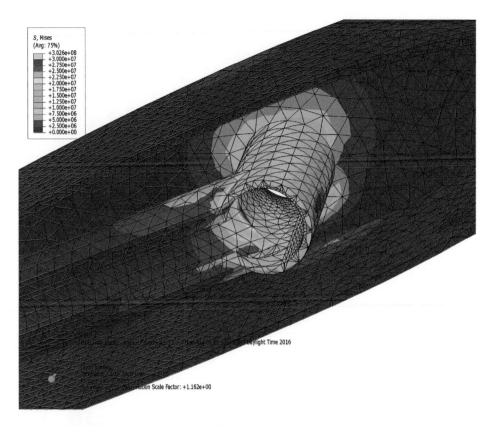

Figure 14.16 Deformed shape and von Mises stress plot for 2 mm thick-walled nylon part in full Decode-1 spar model with fully refined mesh and reduced boundary conditions under wing flight loads taken from XFLR5 together with a load factor of 4 plus elevator and fin loading based on Cl values of unity.

the two resulting wedge sections are meshed using hex wedge elements. These are compatible with the continuum shells, and the fact that we are not using shells in this very small region will make little difference, see Figure 14.18. If a truncated trailing edge is used, this may have to be meshed using tetrahedral solid elements linked with tie constraints.

The foam parts are coupled to the rest of the structure via contact interactions with the spars and also with lugs added to the SLS nylon parts, see Figure 14.19. Pressure loads are then applied to the outer surface of the cladding that surrounds the foam, using values taken from the XFLR5 analysis, and these replace the body forces previously applied to the spars, see Figure 14.20. The pressures are applied using an Abaqus-mapped field created through the "tools - analytical field - create" option, which allows the data from XFLR5 to be entered from a comma separated file with X, Y, and Z coordinates taken directly from the XFLR5 output and the C_p values converted to pressures by multiplying by $0.5\rho V^2$. Note that the data entry dialogue box has a "mapper controls" tab where the tolerances used to do the mapping can be set. Then, when creating a pressure load, the analytical field can be referenced along with a value of 4 for the load magnitude to simulate the $4g$ loading. Obviously, the XYZ pressure

Figure 14.17   Abaqus model of foam core created with CAD shell and fillet commands and meshed with brick hex elements.

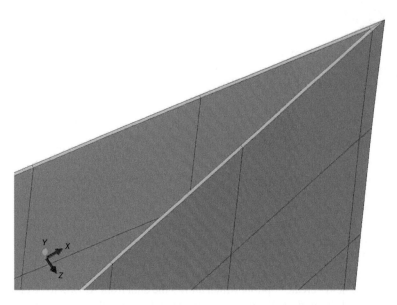

Figure 14.18   Abaqus model of glass-fiber wing cover created with CAD shell commands and meshed with continuum shell hex elements. Note the wedge elements used for the sharp trailing edge.

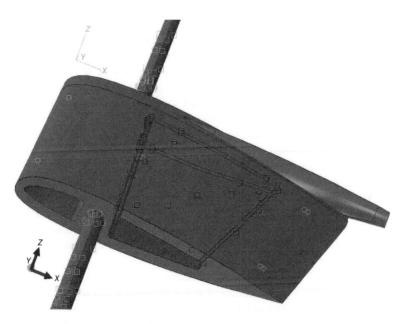

Figure 14.19 Abaqus assembly with foam parts added, highlighting the tie constraint between the foam and the SLS nylon support.

Figure 14.20 Pressure map on Decode-1 foam part under wing flight loads taken from XFLR5.

data must be in the same geometric locations as the AirCONICS CAD model used to create the structural model.

Note that XFLR5 only supplies mid-surface aggregated pressure loads unless a full 3D panel analysis is carried out; such analyses are currently only possible for wing alone configurations in the code. Thus to get the desired pressure loads for the mold line surface, one first sets up the full aircraft with elevators, fins, and so on, to establish the desired flight conditions and then re-analyses the wing alone at the desired speed and angle of attack to generate the pressure loads for the wing (these then being suitably scaled to simulate the desired load cases taken from the Vn diagram). The resulting analytical field inside Abaqus can be used to load multiple parts in the FEA, as Abaqus can correctly work out which part of the field needs to be used for which surface when defining the pressure loads. If pressure loads are required for the elevator, this has to be set as the main wing in XFLR5 and an approximate angle of attack and speed have to be deduced by trial and error so as to get to the same net forces as seen on the elevator for the full configuration. This is a tedious process and rarely warranted for structural analysis of the elevator foam parts, as they typically see smaller stresses than the main wings (though they can flex noticeably if the hinge spars are too small in diameter, something that can adversely affect controllability).

Figures 14.21 and 14.22 show the resulting deflections and stresses in the foam, cover, and supporting SLS nylon part when a section of the main wing foam is analyzed in this way. Here, the peak stresses in the foam are 0.45 MPa, those in the cover 38 MPa, and those in lug region of the nylon 35 MPa. Thus there are small regions of the foam that are close to the material limit in this model, while the nylon is acceptably below its limits and the glass-fiber cladding well below its ultimate strength. Note, however, that the failure mechanisms seen in tensile and compression tests of woven glass-fiber-reinforced plastic specimens are rather different from those we have observed in our airframe parts: ultimate failure of our wing foam parts is caused by delamination of the cover where the adhesive layer between the foam and cover fails locally, followed by crushing of the foam. A much more detailed FEA model would be required to study this kind of behavior since the resin layer between the glass fiber and foam would need to explicitly modeled and a very fine mesh used in the foam element. Moreover, the nonhomogeneous properties of the fiber reinforcement would also need to be allowed for.

Pressure loads can also be extracted from runs of Fluent by using the file export solution data capabilities of the code: an ASCII space delimited file should be created for cell center static pressures on the bodies for which loads are required. This will create a (typically very large) text file with a header row and five columns of data (cell number, x-coordinate, y-coordinate, z-coordinate, and pressure). Before these can be read into Abaqus, the header and any blank trailing rows have to be removed along with the first column.[12] Once read into Abaqus, the size of the original data file does not matter, as Abaqus calculates a loading pattern appropriate to the part the pressure is mapped onto. Because Fluent can be used to study arbitrarily complex geometries, pressures can be applied to all the external surfaces of the Abaqus model in this way, including deflected control surfaces, assuming they both originated from the same AirCONICS (or other) CAD-based geometry definition.

---

[12] Because the files are often very large (perhaps containing over a million lines), this is not completely straightforward. However, a simple Windows batch file can be written to strip the first column. It should contain two lines: @ECHO OFF and for /F εtokens=2-5ε %%a in (%1) DO (echo %%a %%b %%c %%d). Once the data has been stripped in this way, the header line and any blank lines at the bottom should be removed by manual editing before it is read into Abaqus.

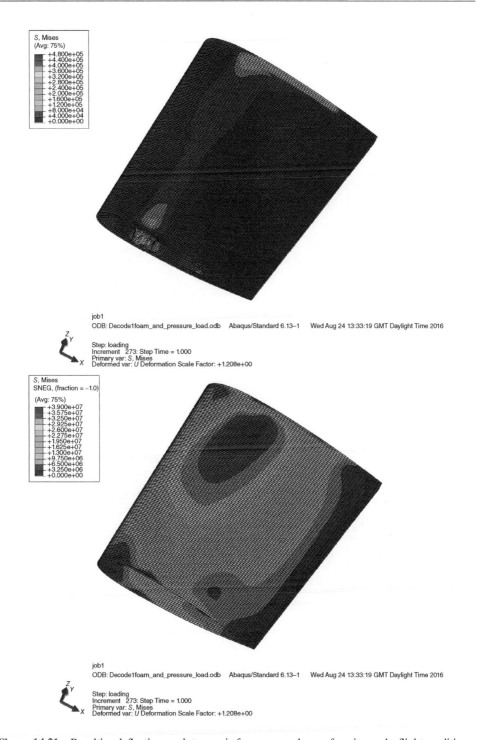

Figure 14.21   Resulting deflections and stresses in foam core and cover for wing under flight conditions.

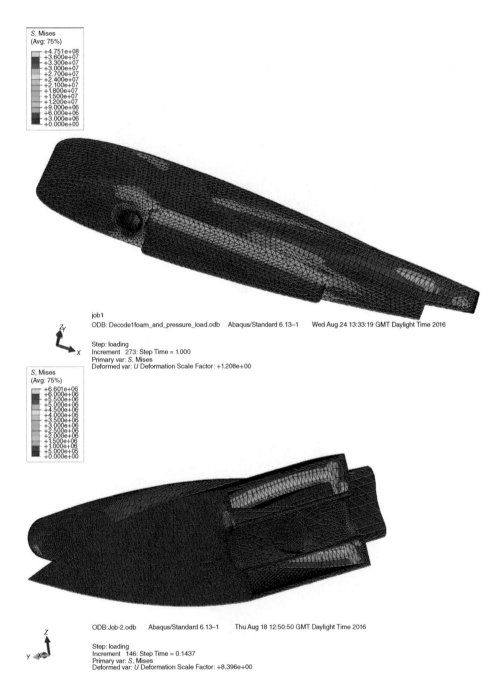

Figure 14.22   Resulting deflections and stresses in SLS nylon part with foam mounting lug for wing under flight conditions.

14.4   Structural Dynamics and Aeroelasticity

Having completed the simple static structural analysis of the preliminary airframe design, finally attention can be given to the topics of structural dynamics and aeroelasticity. It is, of course, well known that aircraft can suffer from the phenomena of divergence, control reversal, and flutter, where a build-up of static deflection or vibration is caused by the interactions of fluid forces and structural elasticity. These behaviors are controlled by the rigidity and inertia of the airframe. Provided the structure is reasonably stiff and the centers of mass in the aerodynamic surfaces are not too far aft, it is usually quite simple to avoid aeroelastic problems in small UAVs: issues mostly arise in larger high-speed aircraft, particularly those with swept wings. Nonetheless, such behavior can ultimately be catastrophic and has led to the failure of many engineering systems over the years, including perhaps most famously the loss of the first Tacoma Narrows suspension bridge.[13] We find that the claddings we apply to our foam aerodynamic surfaces increase their torsional stiffness so much that, combined with the use of CFRP spars and stiff control linkages, we do not experience aeroelastic problems in aircraft of the size we build. Even so, all aircraft designs should be checked for possible divergence, control reversal, and flutter problems during the preliminary design phase.

Static aeroelasticity problems, that is, those that are characterized by a steady increase in deflection rather than vibration, are broadly grouped into two main effects; divergence and – for surfaces containing control flaps – control reversal. Divergence arises when the action of the airflow over the wing gives rise to torsional loads that cause the wing to twist and so, by changing the angle of attack, increase the twisting moment until ultimately the wing fails. Control reversal arises when the hinge reactions induced by control flap deflection twist the wing sufficiently that its changed angle of attack more than counteracts the desired action of the flap deflection.[14] In both cases, the effects increase with the square of the air speed and are heavily dependent on the torsional rigidity of the structure. The designer needs to establish that the onset of these problems occurs at significantly higher speeds than the aircraft will experience in flight. A typical margin is that the divergence or control reversal speed should be more than 25% higher than the maximum dive speed in the Vn diagram. If this is not achieved, the torsional rigidity of the wing must be improved, either by the addition of material or by changes to the thickness-to-chord ratio of the outer surface.

Flutter is a vibration caused by the action of the aerodynamic forces, and depends both on the stiffness of the aerodynamic surface and also on its inertial properties. For surfaces containing control flaps, three types of flutter can be distinguished: torsional-flexural flutter, torsional control flap flutter, and flexural control flap flutter. Note that here we deal with what may be called classical flutter, where the flow remains attached and relatively smooth. Other forms exist that are not relevant to the kind of small UAVs being considered here, such as

- stall flutter, due to dynamic flow separation;
- transonic flutter, due to the interaction of moving shocks and resultant boundary layer separation;
- buffeting, caused by highly turbulent unsteady flow.

---

[13] Where wind blowing along Puget Sound excited the deck of the bridge in much the same way as flight speed can excite vibrations of wings, ultimately causing the bridge to collapse completely in November 1940.

[14] So, for example, if the starboard aileron is deflected upwards to decrease the lift of the starboard wing, yet the resulting torsional load causes the wing angle of attack to increase; the overall effect may be an increase in lift, that is, the reverse of what is required.

Torsional-flexural flutter is closely related to divergence but now the mass of the wing comes into play. If the center of mass of the wing lies behind the torsional axis, as a gust gives the wing an upwards impulse, it also tends to pitch the nose up, thus increasing the lift being generated. The rigidity of the wing counteracts this change, slowing the upwards motion before accelerating it back down again, now causing the nose to pitch downwards and thus causing an oscillatory behavior. If the frequency of this forcing motion nears the natural frequency of the structure, flutter will occur, which will tend to give rise to fatigue problems and ultimately structural failure. Again, an onset speed should be estimated for the airframe to establish that this lies sufficiently far above the maximum dive speed. To do this, one must estimate the natural frequency of the wing structure in both the primary flapping and twisting modes. This additionally requires any heavy masses in the aerodynamic surfaces to be considered in the analysis model; normally just the control surface servos need to be considered unless fuel or payload elements are carried by the wing spars.

Torsional control flap flutter arises when the wing twists around its torsional axis causing the control flap to be accelerated up and down. If the flap itself has a center of mass lying behind its hinge line, it will tend to deflect in the opposite direction to the main body of the wing. This inertial effect is similar to control reversal but is now oscillatory in nature. It is generally prevented by ensuring that either the control linkage on the flap is sufficiently stiff that it drives the flutter frequency higher than any likely aerodynamic excitation or the control surface center of mass lies forward of the hinge line, if necessary by the addition of added weights in the control surface leading edge. Flexural control flap flutter is also caused by the flap center of mass lying behind the hinge line, but now excited by flexural motions of the main wing rather than torsional ones. Again, this can be dealt with by a sufficiently stiff control linkage or by bringing the center of mass of the flap forward of the hinge line with added weights. Control flap flutter is sufficiently rare in small fixed-wing UAVs, and it consequences are sufficiently limited in terms of structural redesign, that in our view it need not be considered at the preliminary design stage of such UAVs.

14.4.1 Estimating Wing Divergence, Control Reversal, and Flutter Onset Speeds

It is possible to make a simplified estimate of the free-stream velocity that torsional divergence, control reversal, or flutter occurs at by assuming the wing to be unswept and of high aspect ratio, that is, by assuming a simple two-degree-of-freedom dynamic model of the wing. We start from the natural frequencies of the full wing in its primary flapping and twisting modes, as calculated using FEA (or indeed, as measured in the lab, see later), and use these with a simple linearized aerodynamic model as follows: First we define a two-degree-of-freedom model where the wing motion is characterized by its vertical heave h and torsional rotation α about its elastic axis (i.e., the axis about which rotation occurs and where applied forces cause only bending) being restrained by two springs of stiffness K_h and K_α, respectively, see Figure 14.23. The wing is taken to have mass M and polar moment of inertia about the elastic axis I (these are for just the port or starboard wing and not the total for the pair). The wing is also subject to an aerodynamic lift force L_{wing} acting at its center of lift x_{ac} in front of the elastic axis, while its center of gravity lies x_{cg} behind the elastic axis. The coupled equations of motion of the system are then approximately (noting the coupling caused by the fact that

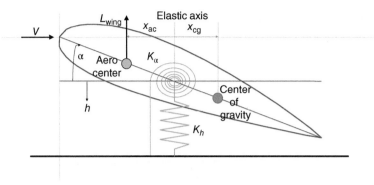

Figure 14.23 Two-degrees-of-freedom model of wing aeroelasticity.

the center of mass does not lie on the elastic axis) as follows:

$$-L_{\text{wing}} - K_h h = M\frac{d^2 h}{dt^2} + x_{\text{cg}} M\frac{d^2\alpha}{dt^2}$$

and

$$L_{\text{wing}} x_{\text{ac}} - K_\alpha \alpha = I\frac{d^2\alpha}{dt^2} + x_{\text{cg}} M\frac{d^2 h}{dt^2}.$$

We further assume that the two spring constants can be deduced from the natural frequencies in flapping and twisting as $K_h = \omega_{\text{flap}}^2 M$ and $K_\alpha = \omega_{\text{twist}}^2 I$. Assuming a steady-state thin airfoil behavior, which allows the effect of camber and fixed angle of attack to be superimposed later, the aerodynamic lift L_{wing} of one wing can be taken to be simply proportional to the twist as $L_{\text{wing}} = \alpha \frac{dC_L}{d\alpha} \frac{1}{2}\rho V^2 S/2$, where ρ, V, and S are the density, airspeed, and the wing area of both wings, respectively, as usual. This leads to

$$-\alpha \frac{dC_L}{d\alpha} \frac{1}{2}\rho V^2 S/2 - \omega_{\text{flap}}^2 Mh = M\frac{d^2 h}{dt^2} + x_{\text{cg}} M\frac{d^2\alpha}{dt^2}$$

and

$$x_{\text{ac}}\alpha \frac{dC_L}{d\alpha} \frac{1}{2}\rho V^2 S/2 - \omega_{\text{twist}}^2 I\alpha = I\frac{d^2\alpha}{dt^2} + x_{\text{cg}} M\frac{d^2 h}{dt^2}.$$

If we next assume a simple harmonic solution to these equations, of frequency ω and amplitude $\begin{Bmatrix} h_0 \\ \alpha_0 \end{Bmatrix}$, we get the following matrix equation:

$$\begin{bmatrix} M(\omega^2 - \omega_{\text{flap}}^2) & x_{\text{cg}} M\omega^2 - \dfrac{dC_L}{d\alpha} \dfrac{1}{2}\rho V^2 S/2 \\[2ex] x_{\text{cg}} M\omega^2 & I_\alpha(\omega^2 - \omega_{\text{twist}}^2) + x_{\text{ac}} \dfrac{dC_L}{d\alpha} \dfrac{1}{2}\rho V^2 S/2 \end{bmatrix} \begin{Bmatrix} h_0 \\ \alpha_0 \end{Bmatrix} = \begin{Bmatrix} 0 \\ 0 \end{Bmatrix}.$$

The nontrivial solutions to this equation arise when the determinant of the matrix is zero. This leads to a quadratic equation in ω^2 of the form $A\omega^4 + B\omega^2 + C = 0$, where

$$A = M(I - x_{cg}^2 M),$$

$$B = M\left((x_{cg} + x_{ac})\frac{dC_L}{d\alpha}\frac{1}{2}\rho V^2 S/2 - I(\omega_{twist}^2 + \omega_{flap}^2)\right), \text{ and}$$

$$C = \omega_{flap}^2 M(\omega_{twist}^2 I - x_{ac}\frac{dC_L}{d\alpha}\frac{1}{2}\rho V^2 S/2).$$

The point at which the roots of this equation become imaginary is the flutter speed, while the solution for $\omega = 0$ is the divergence speed. Note that for a classical thin airfoil section we can assume $\frac{dC_L}{d\alpha} = 2\pi$ per radian. (Note: for lower aspect ratio wings we could modify this using the span efficiency.) Consequently, the flutter speed can be found from $4AC = B^2$ or

$$4\omega_{flap}^2(I - x_{cg}^2 M)(\omega_{twist}^2 I - x_{ac}\pi\rho V^2 S/2) = ((x_{cg} + x_{ac})\pi\rho V^2 S/2 - I(\omega_{twist}^2 + \omega_{flap}^2))^2,$$

which is a quadratic equation in V^2 of the form $A'V^4 + B'V^2 + C' = 0$, where

$$A' = (x_{cg} + x_{ac})^2 \pi^2 \rho^2 (S/2)^2,$$

$$B' = \pi\rho S/2(4x_{ac}\omega_{flap}^2(I - x_{cg}^2 M) - 2I(\omega_{twist}^2 + \omega_{flap}^2)(x_{cg} + x_{ac})),$$

$$C' = 4x_{cg}^2 \omega_{twist}^2 \omega_{flap}^2 MI + (\omega_{flap}^2 - \omega_{twist}^2)^2 I^2.$$

Given the values of the natural frequencies, locations of the centers, and the inertial properties, it is then possible to solve for the flutter onset speed V_{flut}.

The divergence speed is found from $\omega_{twist}^2 I = x_{ac}\frac{dC_L}{d\alpha}\frac{1}{2}\rho V^2 S/2$, giving $V_{div} = \sqrt{\frac{\omega_{twist}^2 I}{x_{ac}\pi\rho S/2}}$. Notice that, if the aerodynamic center aligns with the elastic axis, x_{ac} is zero and divergence is not possible as would be expected, since the aerodynamic forces do not then cause the wing to twist; placing the main wing spar at the quarter chord point therefore helps eliminate divergence.

An essentially similar analysis can be carried out where a trailing-edge control surface is assumed to be present that alters the lift of the section in a way that is directly proportional to the control flap deflection δ; that is, we define the wing lift and moment coefficients as

$$C_L = C_{L,0} + \frac{dC_L}{d\alpha}\alpha + C_{L/\delta}\delta,$$

and

$$C_M = C_{M,0} + \frac{dC_M}{d\alpha}\alpha + C_{M/\delta}\delta,$$

where $C_{L/\delta} = \frac{dC_L}{d\delta}$ and $C_{M/\delta} = \frac{dC_M}{d\delta}$. Then it may be shown that the control reversal speed occurs when $V = \sqrt{\frac{-\omega_{twist}^2 I C_{L/\delta}}{\pi\rho c C_{M/\delta} S/2}} = V_{rev}$, where c is the mean aerodynamic chord used to nondi-mensionalize C_M, again assuming $\frac{dC_L}{d\alpha} = 2\pi$. Note that $C_{M/\delta}$ will be negative. The quantities $C_{L/\delta}$ and $C_{M/\delta}$ can be estimated for the control flap in question using a range of methods or taken from past practice. Thin airfoil theory gives $C_{L/\delta} = 2(\cos^{-1}(d) + \sqrt{(1 - d^2)})$ and

$C_{M/\delta} = 0.5(-aC_{L/\delta} + \cos^{-1}(d) - d\sqrt{(1 - d^2)})$, where d is the distance of the aileron hinge behind the mid-chord point as a fraction of the semi-chord (so for a 20% chord aileron $d = 0.6$), and a is the distance of the aerodynamic center in front of the mid-chord point as a fraction of the semi-chord (so, typically, 0.5 for foils where the center of lift is at the quarter chord point).

If a is taken as 0.5 and d as 0.6, then the ratio $C_{L/\delta}/C_{M/\delta} = 3.45/-0.64 = -5.39$. If we take this ratio and set x_{ac} to $c/4$, we can see that $V_{rev}/V_{div} = 0.5\sqrt{-C_{L/\delta}/C_{M/\delta}} = 1.16$, that is, wing divergence is likely to occur before control reversal. However, this would be applicable for an infinitely long wing with a full-length aileron; in practice $\frac{dC_L}{d\alpha}$ is reduced for finite-span wings, and the effectiveness of the control surface scales broadly with its extent along the wing, noting, however, that control surface effectiveness near the wing tip will be reduced by downwash effects.[15]

Hoerner [10] devotes an entire chapter to the behavior of control surface flaps and shows that the flap effectiveness ratio $\frac{d\alpha}{d\delta} = C_{L/\delta}/\frac{dC_L}{d\alpha}$ can be approximated as $4\pi\sqrt{(1 - d)/2}$, with typical values of 40% for surfaces extending over the rear 20% of the section, that is, $C_{L/\delta} \approx 2.5$ for $d = 0.6$. Experimental values of $C_{L/\delta}$ are shown to lie in the range 1.5–3.5/rad. Experimental values of $C_{M/\delta}$ are shown to be 75–80% of those predicted by thin airfoil theory and thus typically lie in the range -0.25 to -0.75 per radian, with a maximum around $d = 0.48$. Golland [27] shows the coefficients for an unswept wing with an aileron in the outboard half of the span as $C_{L/\delta}/C_{M/\delta} = 2.64/-0.72 = -3.67$. If we take this ratio, and again assume that x_{ac} is given by $c/4$, we can see that $V_{rev}/V_{div} = 0.96$, that is, control reversal is now likely to occur before divergence.

Given that an Abaqus model has already been built to assess the static stressing and stiffness of the structure, it is a simple matter to switch to a natural frequency calculation to ascertain ω_{flap} and ω_{twist} (noting that these are specified in units of per radian in the above analyses). All that is needed is to make sure that all the density properties of the material models are correctly specified and any lumped masses added before solving for the natural modes.

To solve for the modes, when creating the analysis step, select the "Linear perturbation" procedure type in the "Create step" window and then chose a "Frequency" analysis. This gives a form to complete that specifies the desired analysis. We choose to find the first 10 eigenvalues, leaving all other settings as default values. The natural frequencies and corresponding mode shapes can be found in the post-processing viewport by examining the displaced shapes. The natural frequency can be seen at the bottom of the viewport, and the corresponding mode shape occurring at this frequency is shown by the contour plot or by animation. Clicking the arrow icon at the top right of the viewport allows the various modes that were requested to be stepped through. The position of the elastic axis can be determined by examining the mode shape in twist; the axis lies along the nodal line in the mode shape. This can be revealed by examining a truncated contour plot on an undeformed plan view of the wing, see, for example, Figure 14.24. Note that this is not wholly along the spar line because the spar itself flexes, even in the twist mode.

The overall total mass of the FEA model and its various moments of inertia are printed to the Data File and should be used to check that correct totals have been generated with appropriate radii of gyration. Note that since only one wing is included in the preceding analysis when using the FEA analysis to generate inertia data and radii of gyration, only those parts that make

---

[15] Note also that the aspect ratio of the control surface itself plays a part when such aspect ratios reduce below 2; very short control surfaces should thus be avoided.

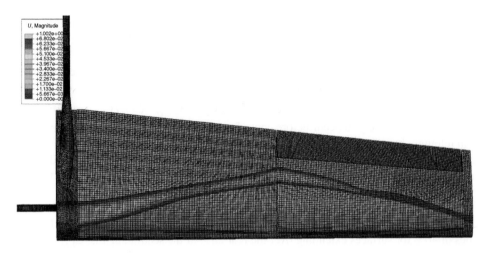

Figure 14.24 Truncated Abaqus contour plot of a first twist mode revealing the nodal line and hence the elastic axis.

up the wing under investigation should be included; while when establishing mode shapes and natural frequencies, the more complete the model the better. These results can then be combined with estimates for the aerodynamic coefficients to assess possible aeroelastic effects.

One key aspect of the natural frequency analysis is the constraint set imposed on the wings in the FEA model that is designed to represent their attachment to the fuselage. If the inboard edges of the SLS nylon parts that support the wings and tailbooms are heavily constrained, then it is likely that the flap and twist frequencies will be higher than those experienced in practice. If, alternatively, the SLS nylon parts are not restricted in motion with restraints just applied to the main spar, then the resulting frequencies can be too low. Of course, the best approach would be to model the fuselage itself and then use a contact model between the wings and fuselage, but this may involve more effort than can be justified for the aeroelastic checks at this stage of design. Our practice is to check the flap and twist modes with several differing constraint models and then use the resulting frequencies as a set to investigate possible aeroelastic problems. For Decode-1, the frequencies we found (along with those from subsequent experimental testing) are given in Table 14.3, while Figure 14.25 illustrates the first flap and twist mode shapes. Note that the twist frequency is relatively unaffected by these changes, and it is this frequency that dominates the aeroelastic behavior. Clearly, the unconstrained model best matches the experimental frequencies in this case.

For Decode-1, the Abaqus-calculated wing weight for just the port wing but including the entire main spar is 1.224 kg centered 0.134 m behind the wing leading edge (0.038 m or 12% of the mean chord behind the spar center), with moments of inertia about the origin of 0.867, 0.003, and 0.886 kg m^2 and thus a twist radius of gyration of 0.049 m or 15% chord.[16] Examination of the twist mode suggests that the elastic axis can be taken to approximately lie along the main spar, that is, at the quarter chord position. The ailerons are 25% of the chord and

---

[16] It is important to make sure that a reduced Abaqus model without the tail boom or any of the empennage is used when deriving the inertial properties of the wing since the simplified aeroelastic model used to derive the onset speeds does not include these, even though these features will probably be included when establishing the natural frequencies.

Table 14.3 Natural frequency results (Hz) using Abaqus modal analysis for the Decode-1 airframe.

Model	First flap mode	Second flap mode	First twist mode
Main spar clamped and inboard SLS part restrained in X, Y, and Z	9.2	53.4	66.7
Main spar clamped and inboard SLS part restrained in Y only	9.1	51.4	64.9
Main spar clamped and inboard SLS part restrained by torque peg	7.2	43.6	63.7
Main spar clamped only	6.3	40.1	63.1
Experimental (see Chapter 16)	5.8	31.7	59.3

Figure 14.25 Abaqus plots of first flap and twist modes for Decode-1 wing.

extend over 50% of the span. Plugging these numbers plus the lower set of calculated natural frequencies into the above formulae, and assuming thin wing theory, gives onset speeds of $V_{div} = 68.1$ m/s, $V_{rev} = 82.6$ m/s, and $V_{flut} = 63.6$ m/s, all comfortably above the maximum dive speed of 40 m/s. Reducing $\frac{dC_L}{d\alpha}$ to allow for a finite wing span would increase these speeds slightly.

It must be stressed, however, that all these calculations are greatly simplified from a full aeroelastic analysis. In particular, the assumption of steady-state aerodynamic equations is a significant approximation. It is possible to use an unsteady potential flow solution for the thin airfoil model and get better approximations, or to use a coupled aerodynamic and structural approach, using XFLR5, Fluent (in either steady or unsteady modes), Abaqus, and so on, but such methods lie beyond the scope of this book. And in any case, unless the initial estimates given in this section suggest that the various aeroelastic onset speeds are close to the maximum dive speed, they are almost certainly not warranted for the small fixed-wing UAVs considered here. We do, nonetheless, check the natural frequencies of the wings after assembly by simple experimental tests, as will be described shortly. We also routinely measure control surface characteristics in wind tunnel tests.

14.5   Summary of Preliminary Structural Analysis

To summarize, when building simple models to check the spars, either by hand or using FEA, stress levels should be well below the failure stress of the material to be used. In contrast, modest *local* overstressing of the nylon can usually be ignored, but not if it spreads significantly through the nylon or leads to overly large stresses being reported in the adjacent CFRP parts. When there is a concern about the design of the SLS parts, a more detailed geometrical model should be built and meshed using local seed control, and perhaps just of the junction under consideration, to see how typical flights loads will impact the design. The foam parts with outer cladding can also be analyzed for basic strength, though these results depend greatly on the thickness of the cladding model used and this is somewhat difficult to accurately assess. Provided local foam stresses do not greatly exceed those for the material in use, the design is probably acceptable as some local crushing of the foam will not be a problem. It is also important to understand that built-up structures that can be disassembled for transportation are intrinsically much more difficult to model than permanently bonded ones, since details of assembly and clearances around joints are extremely hard to model accurately. Aeroelastic checks should be carried out to ensure that divergence, control reversal, and flutter will not be a problem. Note, however, that the only fully satisfactory way of assessing the behavior of real built-up structures is by experimental testing, a subject we discuss in Chapter 16.

Once the CFD and FEA analysis processes are complete and the design has been reassessed, attention can be turned to any knock-on impacts on weight and center of gravity location.

15

Weight and Center of Gravity Control

Once the preliminary geometry has been defined in 3D CAD and a set of suitable analyses has been carried out, the total airframe weight and the position of the longitudinal center of gravity (LCoG) should be reassessed. These estimates will be made from CAD-based mass and centroid predictions plus the known weights and locations of the various bought-in components that will be used. It is standard practice at this stage to maintain these estimates in tabular form, usually as a spreadsheet. Control of the LCoG is, of course, vital to establish pitch stability of the aircraft. We always weigh our aircraft after building and before the first flight to establish the final maximum take-off weight (MTOW) and LCoG. We typically do this with sets of calibrated scales placed under the wheels. It is important when calculating LCoG values that the aircraft is horizontal and the contact points with the scales are in known locations.

15.1 Weight Control

Given that an adequate structural definition has been established, it should be possible to esti-
mate the weight of the aircraft with a good deal of precision; we typically work to the nearest gram. To do this, we try and avoid relying on manufacturer-stated weights for components; rather we prefer to weigh all the parts we intend to use in-house and add these to our weights build-up. If such weights are not available, some form of scaling will have to be used, see Tables 11.4–11.6. Table 15.1 shows a typical weight analysis for one of our aircraft, in this case the ducted wing unmanned air vehicle (UAV) already seen in Figure 4.22. Figure 15.1 shows this aircraft being weighed after final assembly. The component weights are all established by weighing the items to be fitted to the aircraft, while those of the selective laser sintering (SLS) and foam parts are taken from the CAD definition using a relative density of 0.95, which is based on weighing previously manufactured SLS nylon parts. Note that as the design progresses, further detailing of the CAD models for the SLS parts will rapidly intensify. This will, of course, modify the weights of these parts, but if a simple constant wall thickness, of say 2 mm, has been assumed for the initial structural model, the shift to internal stiffening of

Small Unmanned Fixed-wing Aircraft Design: A Practical Approach, First Edition.
Andrew J. Keane, András Sóbester and James P. Scanlan.
© 2017 John Wiley & Sons Ltd. Published 2017 by John Wiley & Sons Ltd.

Table 15.1 Typical weight and LCoG control table (LCoG is mm forward of the main spar).

Category	Part	No. off	Material	Weight each (g)	Total weight (g)	LCoG (mm)	Moment (g mm)
Spars	Main spar CG31.3/28.5 1 400 mm × 31 mm OD	2	CFRP	306	612	0	0
	Tail booms CG21.8/19.0 950 mm × 22 mm OD	2	CFRP	147	294	−536.67	−157 781
	Rudder posts and hinge pins CG10.0/08.0 490 mm × 10 mm OD	4	CFRP	26	104	−1 157.5	−120 380
	Aileron hinge pins 843 mm × 7.5 mm OD	2	CFRP	20	40	−196.88	−7 875
	Elevator hinge pin CG16.7/14.0 1 070 mm × 16 mm OD	1	CFRP	102	102	−1 070	−109 140
	Main threaded rods and nuts	3	steel	55	165	177.5	29 288
Foams	Main wings	2	Foam	261	574	−42.7	−24 518
	Ailerons	2	Foam	34	75	−217.04	−16 235
	Inner wings	2	Foam	20	44	−61.5	−2 706
	Rudder fins	2	Foam	100	220	−1 124.61	−247 414
	Rudder flaps	2	Foam	22	48	−1 124.61	−54 431
	Elevators	2	Foam	135	297	−1 125.67	−334 324
SLS	360 mm main fuselage with wing supports	1	Nylon	795	795	−83	−66 176
Nylon	250 mm fuselage with hatch	1	Nylon	454	454	496	225 397
	Conical rear fuselage	1	Nylon	235	235	−326	−76 598
	Front lower fuselage	1	Nylon	275	275	672	184 704

	Engine cowling	1	Nylon	34	34	737	25 045
	140 mm Fuselage section	2	Nylon	264	528	231	121 957
	Port duct	1	Nylon	1 020	1 020	−80.86	−82 477
	Port wing tip	1	Nylon	120	120	−95.94	−11 513
	Stbd duct	1	Nylon	1 020	1 020	−80.86	−82 477
	Stbd wing tip	1	Nylon	120	120	−95.94	−11 513
	Port tail connector	1	Nylon	160	160	−1 131.76	−181 082
	Port outer elevator end	1	Nylon	38	38	−1 123.24	−42 683
	Port inner elevator end	1	Nylon	40	40	−1 199.01	−47 960
	Port rudder cap	1	Nylon	38	38	−1 127.49	−42 845
	Stbd tail connector	1	Nylon	160	160	−1 131.76	−181 082
	Stbd outer elevator end	1	Nylon	38	38	−1 123.24	−42 683
	Stbd inner elevator end	1	Nylon	40	40	−1 199.01	−47 960
	Stbd rudder cap	1	Nylon	38	38	−1 127.49	−42 845
	Servo mounting plates (wing)	2	Nylon	14	28	−60	−1 680
	Servo mounting covers (wing)	2	Nylon	11	22	−60	−1 320
	Servo mounting plates (rudder)	2	Nylon	9	18	−1 125	−20 250
	Servo mounting covers (rudder)	2	Nylon	11	22	−1 125	−24 750
Servos	Ailerons Futaba S3470SV	2		58	116	−60	−6 960
	Rudders Futaba S3470SV	2		58	116	−1 125	−130 500
	Engine Futaba S3470SV	1		58	58	642.5	37 265
	Nose wheel Savox SC-1268	1		67	67	642.5	43 048
	Elevators MKS HBL380 X8	2		79	158	−1 125	−177 750
	Large control horns incl. screws	4		8	32	−592.5	−18 960

(continued)

Table 15.1 (*Continued*)

Category	Part	No. off	Material	Weight each (g)	Total weight (g)	LCoG (mm)	Moment (g mm)
	Large control horns support pads	4		1	4	−592.5	−2 370
Engine and motors	OS GF30 plus exhaust, ignition, propeller, spinner & fuel line	1		1 517	1 517	752.5	1 141 543
	DuBro 8 oz. fuel tank plus stopper, breather & filler lines and clunk	1		110	110	192.5	21 175
	Stainless SLS engine mount	1	Steel	130	130	697	90 667
	Hacker A50-12S V3 motors plus mounting nuts	2		345	690	−200.53	−138 366
	Two Jeti Advance 70 Pro SB speed controllers plus main harness	1		359	359	160	57 440
	Master Airscrew propellers E-MA1470T 14x7 three-bladed (tractor)	1		76	76	−240.53	−18 280
	Master Airscrew propellers E-MA1470TP 14x7 three-bladed (pusher)	1		76	76	−240.53	−18 280
Avionics	Futaba R6014 HS Receiver + ribon cable + two leads	1		35	35	282.5	9 888
	SkyCircuits SC2 autopilot with GPS and 2.4 GHz aerial and lead	1		414	414	407.5	168 705

	Material	Qty				
Pitot tube and connecting hose	brass	1	40	40	0	0
Overlander LiPo FP30 6S 22.2V 5 000 mAh 30C main motor battery		1	704	704	582.5	410 080
Spektrum LiFe 2S 6.6V 4 000 mAh avionics battery		1	243	243	524.5	127 454
Nano-Tech LiFe 30C 2 100 mAh 2S avionics battery		1	108	108	524.5	56 646
Double pole single throw 10 A switch plus local wiring harness		1	70	70	407.5	28 525
LED voltage indicator strips		2	4	8	500	4 000
2-6S LED balance voltage indicator		1	4	4	500	2 000
Baseboard (main)	Plywood	1	36	36	496	17 714
Baseboard (receiver)	Plywood	1	19	19	300	5 700
Baseboard (tank)	Plywood	1	19	19	160	3 040
Baseboard (speed control)	Plywood	2	19	38	0	0
Wiring in wings and tail booms		2	150	300	−400	−120 000
Servo linkages		6	7	42	0	0
Misc. cable ties and screws		1	50	50	0	0
Under-carriage Nose wheel and leg		1	79	79	642.5	50 758
Nose wheel upper steering column incl. collets, springs, and cap screws	steel	1	30	30	642.5	19 062
Steering arm bore 6 swg/5.0 mm plus springs		2	3	6	642.5	3 855
Main suspension, wheels, and axles		1	0		0	0
Totals				13 572	13	170 791

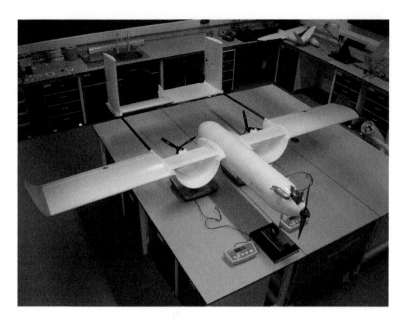

Figure 15.1   Channel wing aircraft being weighed after final assembly.

a thinner structure will reduce the weight of the SLS parts while leaving the centers of mass broadly unchanged. Then the impact of structural detailing will generally not adversely impact on either the overall aircraft weight or its LCoG position.

If at this stage the aircraft is significantly too heavy, some form of weight control exercise can be entered into (in our experience it is very rare for an aircraft ever to be too light). This can be very difficult to achieve, but typical measures could be as follows:

- Reduction in structural element sizes (typically by making elements thinner, especially where finite element analysis (FEA) reveals that stresses are low), although it is difficult to make substantial savings in this way without considerable design effort;
- Reducing battery sizes;
- Lightening wing covers, either by reducing foam thicknesses or by adopting less robust claddings;
- reducing the size of servos by accepting lower torque or by using a higher power-to-weight ratio and probably much more expensive items;
- fitting smaller diameter wheels.

Hopefully, the weight budget will not have been too greatly exceeded, but it is in the nature of all vehicle designs to increase in weight during design as the final build is approached, largely because extra items keep getting added to the build specification, either because they were simply not allowed for at the start or because higher specification items are selected or mission creep has set in. For this reason, it can be wise to add a design contingency at the outset of 5%, but this can, of course, become a self-fulfilling prophecy of weight growth.

15.2   Longitudinal Center of Gravity Control

Table 15.1 also shows the longitudinal center of gravity (LCoG) computation. To do this, the LCoG values of all the SLS nylon and foam parts are calculated by the CAD Program, while those for the bought-in components are established by their locations in the design drawings, assuming that the individual centers of gravity lie at the center of each component. The analysis shows that the estimated LCoG is slightly forward of the centerline of the main spar (which lies at the quarter chord point). This is, of course, for an aircraft without gasoline in the main tank, which itself lies forward of the quarter chord point, so the aircraft will have positive trim stability even when all fuel is used.

Should the LCoG not be as required at this stage, consideration must be given to changes to the overall geometry of the airframe or the positioning of heavy internal components like the batteries. When designing with SLS nylon fuselage elements, we find it a relatively simple matter to adjust the LCoG by simply changing the lengths of one or more fuselage elements; then provided there are some relatively heavy elements in the nose of the aircraft, the LCoG can be adjusted as required. In the case of the aircraft tabulated, the batteries and main gasoline engine all lie well forward, so rather small adjustments in fuselage length gave good control of the LCoG without the need for major redesign, though extending the fuselage does, of course, increase the weight.

16

Experimental Testing and Validation

Although computational methods have improved enormously over the last 20 years, it remains the case that the results obtained from computational fluid dynamics (CFD) and finite element analysis (FEA) cannot be relied upon to be completely accurate in a number of key areas. In particular, it is still very hard to predict the stall speed of an aircraft in landing configuration, to establish the ultimate failure loads for key aircraft structures, or to accurately compute the natural vibrational frequencies. These limitations mean that experimental testing retains an import role in the development of any aircraft design. Such tests can be used to validate computational results or to assess designs in the absence of computational modeling. Experimental testing is not, however, without its own problems. First and foremost is the cost involved, both in terms of the facilities needed and also in designing and building suitable aircraft parts for testing. Large facilities are expensive to build and maintain, and so it is relatively rare for prolonged experimental programs to be carried out for low-cost unmanned air vehicle (UAV) designs. Moreover, the parts used in testing must have sufficient structural integrity to survive in the wind tunnel or be realistic for structural load or vibration tests. This means that a good deal of design effort has to be expended in getting from the simplified geometries considered earlier to those that can be used in test part construction. We are fortunate in having routine access to the UK's largest academic wind tunnel which can house full-sized UAVs up to around 15 kg maximum take-off weight(MTOW) and entire built-up wings of larger aircraft. We also have a number of facilities to help carry out static and dynamic load tests on key structural elements. Therefore we usually do carry out quite detailed experiments on our UAVs. In all cases, the aims are to check that preliminary design calculations agree with reality and to update and adjust any empirical factors used during design, for both the current and subsequent aircraft projects.

The design of parts for experimental testing leads to a key choice in the design process. Should a short program of experimental validation be carried out on simplified parts to establish that preliminary design calculations have been adequately good predictors of performance before the costs of full detailing are entered into; or should a separate program of design

Small Unmanned Fixed-wing Aircraft Design: A Practical Approach, First Edition.
Andrew J. Keane, András Sóbester and James P. Scanlan.
© 2017 John Wiley & Sons Ltd. Published 2017 by John Wiley & Sons Ltd.

be set up to make fully detailed test parts? Of course, given suitable facilities, experiments can be carried out on the completed airframe before the first flight to finally establish airframe properties; but if any shortcomings are revealed at that stage, it can be very difficult and costly to correct them if hundreds of hours of detailed design effort has already been committed.

Our approach to this problem is based around the rapid manufacturing processes we adopt combined with a heavy focus on as much parametric CAD modeling during detail design as possible. We then carry out experimental tests on full-sized preproduction-quality parts that can lack some details, and iterate the detailing effort where necessary. The biggest weakness of this approach is that one tends to be somewhat cautious in terms of the selective laser sintered (SLS) nylon structures, which represents the largest commitment of manual detail design effort, and this can make such parts rather heavier than one would wish. Much effort can often be devoted to controlling weight and carrying out detailed stress calculations to ensure that such parts are as optimized as possible. In our experience, even quite poor design concepts can be made to perform adequately with good detail design, while poor detail design will wreck even the best concepts.

16.1   Wind Tunnels Tests

Wind tunnels have been used since the earliest days of flight: the Wright brothers built their own wind tunnel as part of the development of the world's first powered aircraft. Since that time, a great deal of practical experience has been gained and the most sophisticated tests now encompass models that are structurally flexible with active control surfaces and used in tunnels that permit real-time motions of the main body, allowing full aeroservoelasticity experiments to be carried out. We do not go that far in our wind tunnel programs, but we routinely test wings and full airframe configurations over a wide range of speeds and angles of attack (AoA), with variations in control surface deflections, both with and without propeller power to establish full lift, drag, and moment results. The primary result we seek is the stall speed of the aircraft, usually in the landing configuration since this plays such a critical role in the safe operation of small UAVs. If the estimation of landing speed has been overly optimistic, then flight operations can quickly lead to damaged airframes during difficult landings (hopefully the concept and preliminary design calculations will not have been so far out as to lead to designs that cannot generate enough lift given the available power to fly at all!). Secondly, we seek drag performance at the likely flying speeds so as to estimate glide angles and cruise performance. Finally, tunnel data can be used to calibrate previous computational runs and the initial spreadsheet calculations used at the start of the design process. This can be particularly important in improving the designs of subsequent aircraft. In all cases, we take great care to avoid any instances of flutter or divergence in the tunnel tests, first by having estimated likely onset speeds prior to test and second by careful control of the tunnel operating speed during measurements.

16.1.1   Mounting the Model

Given a part or complete airframe with the desired outer mold surface, the first problem faced in tunnel testing is mounting the model in the tunnel. Most tunnels use three-point mounting systems: a pair of main mounts that are side by side facing the airflow, and a smaller movable mount on the tunnel centerline either in front of or behind the main mounts, see again Figure 11.1. The movable mount can be used to change the AoA during testing. Often, the whole mounting system can be rotated about a vertical axis as well. Unless one is using dedicated tunnel test parts, some suitably strong locating points must be identified on the model for attachment. For whole-aircraft tests, it is normal to use the undercarriage mounting for this purpose, either through the wheel axle holes or where the undercarriage legs meet the main airframe. If the aircraft includes a catapult launch bar, this can also be used as a strong point for mounting. For tests on wings alone we use dedicated mounting rigs, which also incorporate a flat plate to provide the inboard boundary condition for the flow. We avoid using the tunnel wall for this because the tunnel boundary layer will be very different to that induced by the fuselage. The size of the boundary plate should be several times the root chord of the airfoil and also mounted away from the tunnel wall. In this case, we mount the wings vertically in the tunnel and change the AoA by rotating the mounting, see Figure 16.1.

Having attached the model to the tunnel mounts, great care must be taken to establish an appropriate datum AoA. For whole airframes, we tend to use the main fuselage as the datum since in the cruise condition this is ideally horizontal to minimize drag; for wings we adjust the datum so that the chord line used to define the root section is in line with the tunnel flow direction. Clearly, if tunnel results are to be compared to calculations, it is important that common AoAs are used.

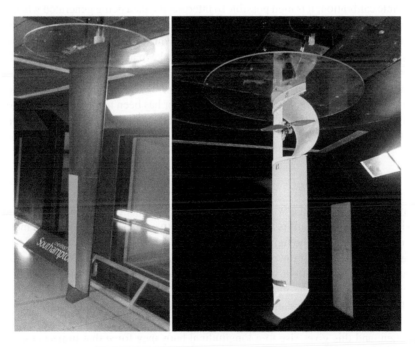

Figure 16.1 Decode-1 and channel wings on wind tunnel mounting rig. Note the circular boundary plate that stands in for the absent fuselage.

16.1.2   Calibrating the Test

Before any readings can be used from tunnel tests, it is vital that a careful and accurate calibration process is carried out. This needs to cover the following:

- The balances or force cells that record lift, drag, and moment (normally the tunnel support team will know these constants or they will be applied automatically and readings be supplied in known units);
- The relationship between tunnel mount vertical motion and changes in AoA (which will need measuring with an accurate inclinometer and will vary from model to model);
- The barometric conditions in the tunnel (which are highly temperature- and weather-dependent. Most tunnels have high-quality instruments in place to record these data. It is also not uncommon for these readings to change during a single session of testing, so such readings should be recorded throughout the test program);
- The wind speed in the tunnel (usually a Pitot-static tube is mounted on the center line of the tunnel working section ahead of the test area);
- The forces induced on the tunnel mountings in the absence of the test specimen (under all likely operating speeds, a series of runs should be made before mounting the test specimen, but including as much of the mounting system as possible);
- The gravitational forces acting on the mounts due to the presence of the test specimen with the wind tunnel stationary (at all likely orientations, a series of readings should be taken before turning on the tunnel fans).

Given suitable calibration, it is then possible to interpret the data being generated when the tunnel is in operation. Note, however, that the raw data from the tunnel sensors must be converted to lift and drag coefficients by suitable manipulation, which is usually best accomplished via an appropriately set up spreadsheet. It is always wise to keep all the raw data captured in case subsequent reanalysis is required. We also always have a series of estimated values at hand when capturing data so that any anomalous values coming from the tunnel instruments are immediately apparent. It is very frustrating if some error has been made in calibration or setup and this is not revealed until after the test runs have been completed and the model removed from the tunnel.

16.1.3   Blockage Effects

When a model is mounted in a wind tunnel, it partially blocks the flow of air going through the working section. The result is to increase the effective airspeed in way of the model over and above that recorded by the tunnel Pitot-static tube, which will lie in a region of the tunnel where there is no blockage. If the model is small compared to the tunnel and aerodynamically streamlined in shape, this effect will be small, but since aerodynamic forces are driven by the velocity squared, once the model starts to have a wing span of even 20% of the tunnel width or large AoAs with attendant stalled flows, the effects become important. Moreover, the variation in longitudinal velocity caused by blockage is not symmetrical forward and aft of the test specimen, and this gives rise to a longitudinal buoyancy force that impacts on the drag. Figure 16.2 illustrates the effect of the Decode-1 airframe at 6° AoA in our largest wind tunnel. The size of the blockage is readily apparent. The projected area of the aircraft is $0.233\,\text{m}^2$,

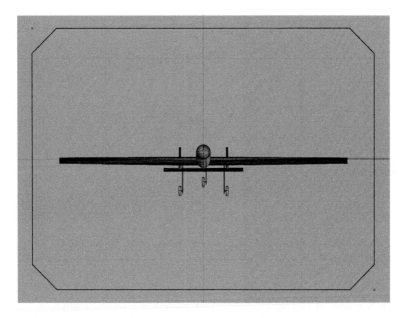

Figure 16.2   AirCONICS model of Decode-1 airframe in a representation of the R.J. Mitchell 11' × 8' wind tunnel working section at Southampton University, illustrating degree of blockage.

while that of the tunnel is $8.174\,\mathrm{m}^2$, that is, the area blockage is 2.85% without allowing for the wake.

Most tunnel operators will supply simple blockage correction factors for their tunnels based on frontal area of the specimen under test, tunnel speed and the type of flow regime being studied, and so on, which can be applied directly to lift and drag coefficients.[1] These corrections only form a starting point, however, since bluff bodies with appreciable wakes will lead to greater blockage than smooth airfoils at low AoAs, even for the same frontal area. Therefore, whole aircraft at modest AoAs but including undercarriage elements will lie somewhere between these extremes, meaning that deducing the correct blockage correction is far from straightforward. Even then, the corrections for lift and drag are generally not the same.

One way of attempting to calculate accurate blockage factors is via CFD models where the tunnel walls are able to be included or removed from the simulation. Direct comparisons between the lift and drag forces seen between the two models will then allow blockage correction factors to be deduced, although some care has to be taken in deciding how to model the boundary layer growth on the tunnel walls themselves (and, of course, building a high-quality boundary layer mesh on the tunnel walls can massively increase the cell count in the CFD simulation). If using CFD in this way, it is also important to calculate the blockage impact on lift well away from the zero-lift condition of course; we chose to run calculations at 6° AoA because this typically lies halfway between zero lift and full stall. Figure 16.3 shows

---

[1] See for example Road Vehicle Aerodynamic Design (p. 243) where the blockage correction for drag coefficients is given as $(1A/S)^{1.288}$, where A is the model frontal area, and S is the wind tunnel cross-sectional area, so for a model like Decode-1 with a frontal area of $0.233\,\mathrm{m}^2$ at 6° AoA, in our large tunnel of $8.174\,\mathrm{m}^2$ cross-sectional area, the correction factor is 0.963 or around 3.7% on the drag coefficient.

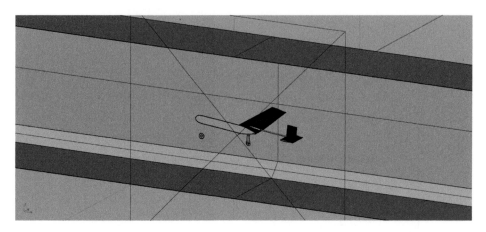

Figure 16.3   AirCONICS half-model of Decode-1 airframe in the R.J. Mitchell 11' × 8' wind tunnel prior to mesh preparation.

an AirCONICS model of Decode-1 at this AoA placed in a representation of our Mitchell wind tunnel ready for analysis, while Figure 16.4 shows the boundary layer and how close it is becoming to the disturbed flow around the aircraft wingtip along with a section through the mesh used. Comparison of this CFD model with one with widely spaced symmetry far-field

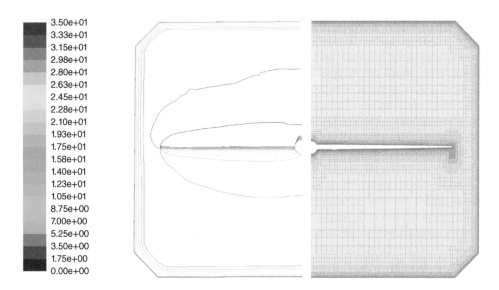

Figure 16.4   Section through Fluent velocity magnitude results and Harpoon mesh for Decode-1 airframe in the R.J. Mitchell 11' × 8' wind tunnel.

Note the extent of the boundary layer on the tunnel walls and the fine boundary layer mesh needed to resolve this, along with the refinement zone near the wing tip.

boundary conditions shows that the ratio of the lift coefficients at 6° AoA is 0.918 and that for the drag coefficients is 1.037 (greater than 1 because of the aforementioned buoyancy effects that cause the drag calculated in the model with the tunnel in this case to be slightly less than that without it, but note that, as ever, the drag forces are an order of magnitude less than the lift forces and so the likely errors in this correction are much larger). This lift correction is broadly in line with the various approximations found in the literature and makes clear the importance of correcting wind tunnel data for blockage affects before they are used. Figure 13.32 given earlier includes experimental data for the Decode-1 airframe adjusted for blockage corrections calculated in this way.

16.1.4   Typical Results

To illustrate the type of results that can be captured during a wind tunnel campaign, we next give a series of results for the Decode-1 airframe as measured in the Mitchell tunnel. In all cases, averaging has been applied to the raw results where possible but some slight variations from the expected trends are noted at some points in the plots.

Basic Results

Figure 16.5 shows the effect of flight speed (Reynold's number (Re)) on the aerodynamic coefficients when the control surfaces are in the neutral position. Several comments can be made:

- The lift force is not affected by Re, only the stall, as would be expected, while stall occurs at higher AoA for higher Re, (some data for higher speeds are not available, because the loads and vibrations were too high to carry on the experiments safely).
- The side force is generally low, and arises only because of slight manufacturing asymmetry, except during the onset of stall which occurs on one wing slightly before the other.
- The pitch moment decreases with the AoA, because the lift on the elevator increases, but this effect seems to be evident mainly at low Re, when the wing wake has less affect on the tail.
- The roll moment is generally almost zero, except during the onset of stall as for the side force.
- The slight aircraft asymmetry, particularly at low Re, generates small yaw coefficients.

Elevator Effectiveness

Figure 16.6 shows the elevator effect on the drag, lift, and pitch when varying the control angle and the wind speed. The pitching moment clearly increases when the deflection angle increases, as expected. This is clear for all the speeds, but the effect with respect to the AoA changes with the speed. At lower speeds, the trend is generally decreasing with AoA. This is reasonable because the lift at the elevator decreases the pitch. However, at higher speeds the pitch increases for negative AoA and decreases for positive AoA. The total lift generally decreases with the elevator angle, because the elevator lift decreases. The drag trend is not

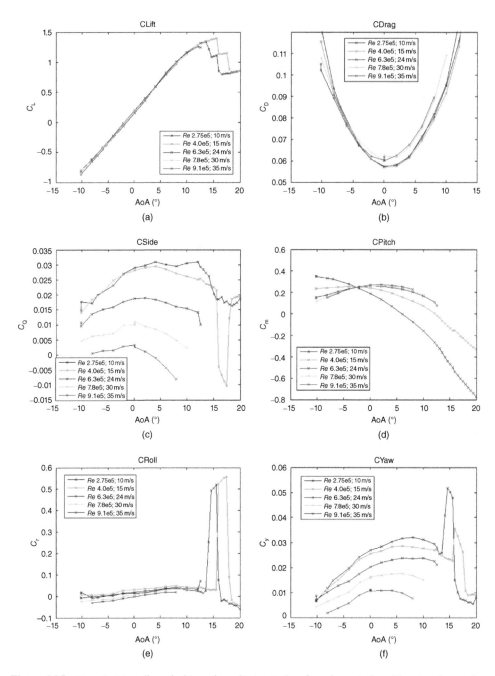

Figure 16.5 Decode-1 baseline wind tunnel results (control surfaces in neutral positions) under varying wind speed. (a) Lift coefficient. (b) Drag coefficient. (c) Side coefficient. (d) Pitch coefficient. (e) Roll coefficient. (f) Yaw coefficient.

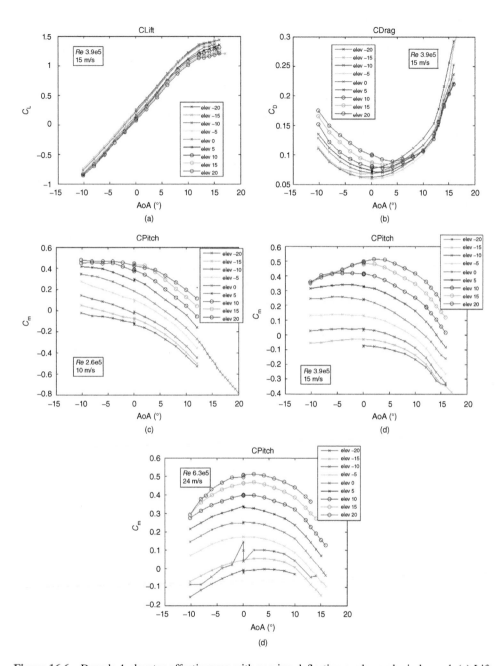

Figure 16.6 Decode-1 elevator effectiveness with varying deflection angles and wind speed. (a) Lift coefficient at 15 m/s. (b) Drag coefficient at 15 m/s. (c) Pitch coefficient at 10 m/s. (d) Pitch coefficient at 15 m/s. (e) Pitch coefficient at 24 m/s.

so clear. Here, the general effect seems to be to shift the minimum drag point to higher AoA and higher values. There are two main effects involved: The elastic effect on the tail due to the higher load at higher AoA is to change the actual AoA. Furthermore, the wake coming from the wing and the fuselage changes the flow impacting the tail. This could be due to the combination of the elevator and wing lift, because the wing lift increases with the AoA and this effect is opposite to the elevator one, being more effective at higher speeds. Furthermore, in these results, drag plays a role too, because the moment axis is not the wing centerline, and therefore the behavior is more complex.

Rudder Effectiveness

Figure 16.7 shows the rudder effect when varying the control angle. Varying the wind speed shows no effect on the general trend. The lift is not affected by the rudder, while the drag increases with the rudder deflection, as expected. The pitching moment is slightly affected because of the rudder drag effect. More important are the effects on the side force, yaw, and roll. The side force and yaw moment are AoA-independent, but vary with the rudder control angle as expected. However, note that a rudder setting of $0°$ does not correspond exactly to zero side or yaw results because of a slight misalignment of the rudder with respect to the control settings. At high AoA, the rudder effectiveness is reduced by the wake from the fuselage and wing. The side force coefficient range is about ±0.1, which translates into a force equal to ±34 N. Roll is affected by the rudder because the lift application point on the fin is above the wing centerline. This is more evident at lower AoA when the fin is not affected by the wing and fuselage wake. At very high AoA, the stall changes the trend because of the asymmetry of the stall behavior.

16.2   Airframe Load Tests

Although structural analysis methods are now highly sophisticated, it remains the case that deducing the behavior of a built-up airframe that contains multiple material types with a plethora of joints and attachments and subject to a wide range of forces is still far from simple. It is therefore useful to validate any structural analysis with a short program of experimental testing. Generally we limit these to static tests, measuring deflections and stresses, and vibration tests to establish wing/control surface structural dynamic behavior, and to see how engine and other sources of vibration may be transmitted to sensitive on-board equipment: it is particularly difficult to accurately predict the vibrational behavior using structural analysis methods. Where we can afford it, we also carry out destructive testing of key components to establish failure modes; we also do this to test aircraft after they have completed all the flying we wish to carry out so as to gain further data to support subsequent design activities.

16.2.1   Structural Test Instruments

The most basic structural test that engineers are familiar with is the simple longitudinal tensile test used to establish fundamental material properties. Loading machines can vary from simple hand-powered rigs to large servo hydraulic systems capable of exerting many

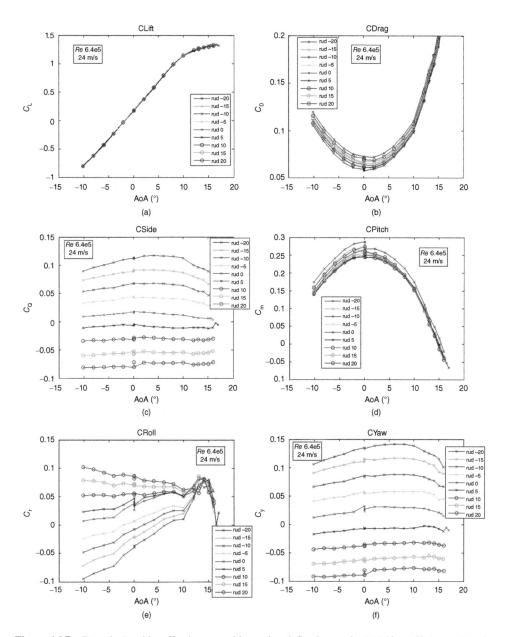

Figure 16.7 Decode-1 rudder effectiveness with varying deflection angle. (a) Lift coefficient at 24 m/s. (b) Drag coefficient at 24 m/s. (c) Side coefficient at 24 m/s. (d) Pitch coefficient at 24 m/s. (e) Roll coefficient at 24 m/s. (f) Yaw coefficient at 24 m/s.

tons of force. We rarely use such methods in our work, as we tend to rely on external suppliers to provide information on material properties, although from time to time we have assessed the strength and stiffness of 3D printed specimens just to confirm the data we have received. Rather, we seek to test the parts and airframes we have designed and built directly. The primary instrument in such tests is the displacement dial gauge (sometimes referred to as a dial test indicator or DTI). Such gauges can accurately measure deflections to an accuracy of 0.025 mm and are relatively cheap and easy to deploy, see Figure 16.8. The are usually based around a spring-loaded plunger that keeps a sensor tip in light contact with the part under test as it deflects. If we have particular concerns about the stress levels being reached in a component, we sometimes also deploy strain gauges on those areas of concern. Strain gauges are formed from fine filaments of wire embedded in a plastic foil system that can be glued to the surface of the part being investigated. We use the type with windings in three axes at 45° intervals so that full stress results can be obtained and the directions of the principal stresses deduced. If we are carrying out laboratory vibration tests, we use piezoelectric accelerometers and force transducers coupled to electromagnetic shakers, see Figure 16.9. To gather vibration data in flight, we use a lightweight piezoelectric accelerometer coupled to a battery-powered data capture system that writes to an on-board micro-SD card, see Figure 16.10.

Figure 16.8   Dial gauge in use to measure aiframe deflection during static test in the lab.

Figure 16.9 Lab-quality force transducer, piezoelectric accelerometers, and electromagnetic shakers.

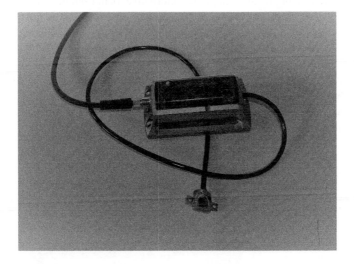

Figure 16.10 Flight-capable piezoelectric accelerometer and data-capture system.

16.2.2 Structural Mounting and Loading

To carry out a load test on an airframe or airframe part, it is important that a safe and secure mounting system is devised that not only holds the parts under test but also prevents any damage being caused to the airframe by the actions of the mount itself. It is very easy to crush fragile aircraft fittings by the clumsy application of clamps or by resting parts over the sharp edges of hard work surfaces where stress concentrations can occur. Equally, however, it is important that any flexibility in the mounting system is accounted for when taking measurements of part deflections. So, for example, when testing a wing or wing spar, a mounting has to be provided that stands in for the fitting to the main aircraft fuselage. We typically use wooden supports that surround the Carbon-fiber-reinforced plastic (CFRP) spar tube, which can then be clamped to the spar and a hard point in the laboratory, sometimes with the extra precaution of a binding of tape between the wood and the spar, see Figure 16.11, or using a specially made

Figure 16.11   Mounting system for wing and main spar assembly under sandbag load test.

clamping block with foam liner, see Figure 16.12. We would never apply clamps directly to the spar itself for fear of causing local damage. An alternative, better but more expensive, solution is to make a subsection of the fuselage mounting in the same material as the final design will use and treat this part as being sacrificial: this more accurately simulates the actual support the spar would see, of course. We also ensure when carrying out tests with dead weights that should anything fail during test, the weighted items cannot fall very far, that they will land on suitable soft surfaces, and that staff will not be injured should this happen. Although we do not always deliberately cause parts to fail under structural test, one must always allow for this to happen and take appropriate safety precautions. Figure 16.13 shows a wing assembly being tested with sandbag loading.

16.2.3   Static Structural Testing

Perhaps the most important reason for carrying out structural testing is to validate the previous sets of calculations used when designing the structure. Ideally, one would replicate the load

Figure 16.12   Clamping system for main spar.

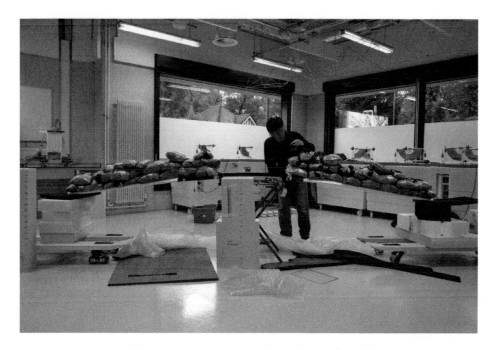

Figure 16.13   Wing assembly under sandbag load test.

case and support conditions used during analysis in the structural experiment. This is rarely possible for full flight conditions without access to a wind tunnel, as generating detailed pressure variations across wing surfaces is very difficult using weights or force actuators. If a wind tunnel is available, attempts can be made to set up displacement measurement indicators at wing tips and then record deflections at various air speeds and AoAs. If this approach is to be followed, care needs to be taken to ensure that the measurement system does not significantly impact the flow field. It is also difficult to generate significant airframe loads in a tunnel test without exciting noticeable vibrations in the structure. Consequently, we normally restrict our wing structural tests to sandbag loadings and match these against calculations set up to match what we can achieve in the laboratory. Then, if good agreement is gained, one can be confident that results for other load cases will also be acceptably accurate.

Another key purpose of structural tests is to establish critically loaded parts and failure mechanisms by increasing loads until the point of failure is reached. In general, airframe structures do not fail completely catastrophically without there first being significant signs of distress. While these may not be readily visible during flight, they should be observable during a controlled lab test. Such tests can reveal those areas of the structure that need further design effort. They will also show which parts of the structure are relatively unaffected by extreme loading and are thus good candidates for weight-saving exercises. Figure 16.14 shows the onset of failure in an selective laser sintered (SLS) nylon wing part in the area between the main wing spar penetration and the torque reaction peg. Figure 16.15 shows an undercarriage leg and SLS mounting structure under load test.

Figure 16.14 Partial failure of SLS nylon structural component during sandbag load test. Note the significant cracks and large deformations.

Figure 16.15 Load testing of an undercarriage leg and associated SLS nylon mounting structure. Note the dummy carbon-fiber tubes present to allow the SLS structure to be correctly set up.

16.2.4   Dynamic Structural Testing

Exciting a wing structure with oscillatory inputs is known as ground vibration testing (GVT) and is an essential preliminary ground test normally conducted prior to the beginning of flight testing. The main objective is to obtain the primary natural frequencies, mode shapes, and damping. To do this, the wing is excited near the root and aft of the main spar at a suitable strong point on the structure (such as at the torque reaction lug). This allows the shaker to excite both flapping and twisting modes of the wing at the same time. Accelerometers are fixed centrally near the tip and across the chord from front to back closer to the root so as to be most sensitive to the dominant flap and twist modes that need to be assessed. The system is then typically first driven with pseudo-random white noise, and a fast Fourier transform system is used to display the frequency response functions recorded by the accelerometers, suitably normalized by the excitation force transducer reading. This will reveal the various natural frequencies, and the widths of the response peaks show the damping levels.

To establish the mode shapes, the vibration signal is then changed to a pure tone at the frequency of interest, making sure that excessive motions are not stimulated by accident and the structure is illuminated with a powerful strobe. Stroboscopic illumination allows the tester to positively identify the associated mode shape for each peak in the frequency response spectrum. It is important to distinguish between the higher flap modes and the first twist mode: the first flap mode is usually obvious without stroboscopic lighting. This series of tests should confirm the values used in the previous aeroelastic analysis and thus remove any concerns over divergence, control reversal, or flutter. If the frequencies predicted by structural analysis are significantly different from those revealed by experiment, the analysis should be revisited to try and establish why. The most likely causes will be differing boundary conditions or the failure to include all the items contributing to the overall mass of the wing. These generally lead to the calculated frequencies being higher than those found from experiment. Errors of 10–20% are quite common and should be allowed for when predicting flutter, divergence, and control reversal onset speeds by adopting suitable margins of safety.

Figures 16.16 and 16.17 show Decode-1 on-ground vibration test, while Figures 16.18 and 16.19 show the frequency response plots for two accelerometer positions with the cursors set for the two modes of interest. The relevant natural frequencies are seen to be 5.78 and 59.25–60.00 Hz. These values are slightly lower than those predicted by the FEA model

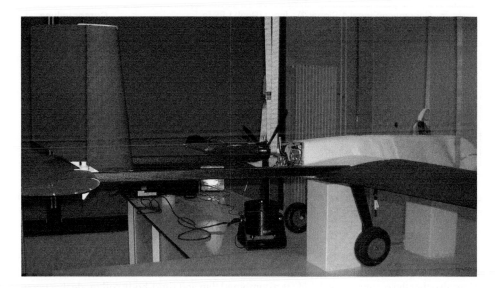

Figure 16.16   Ground vibration test of a Decode-1 wing showing support and mounting arrangements.

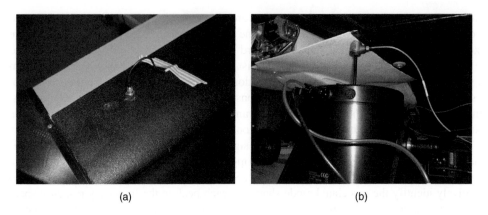

(a)   (b)

Figure 16.17 Ground vibration test of a Decode-1 wing ((a) accelerometer on starboard wing tip: (b) shaker and force transducer near wing root).

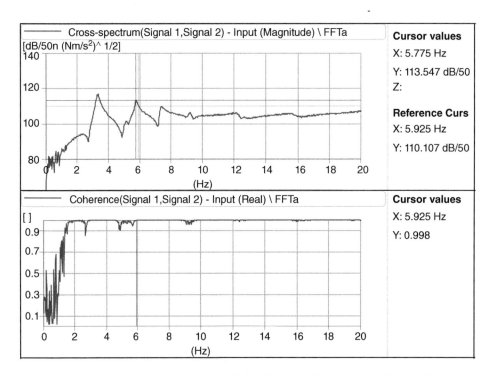

Figure 16.18 Frequency response from ground vibration test of a Decode-1 wing: accelerometer on port wing tip and cursors on first flap mode.

described earlier. This is generally found to be the case as noted above: here, the errors are between 6% and 8%.

Note also that the resonance at 3.1 Hz is a combined flap and rigid-body roll mode that occurs because of the way the model was mounted. This is easily distinguished because the wing tips

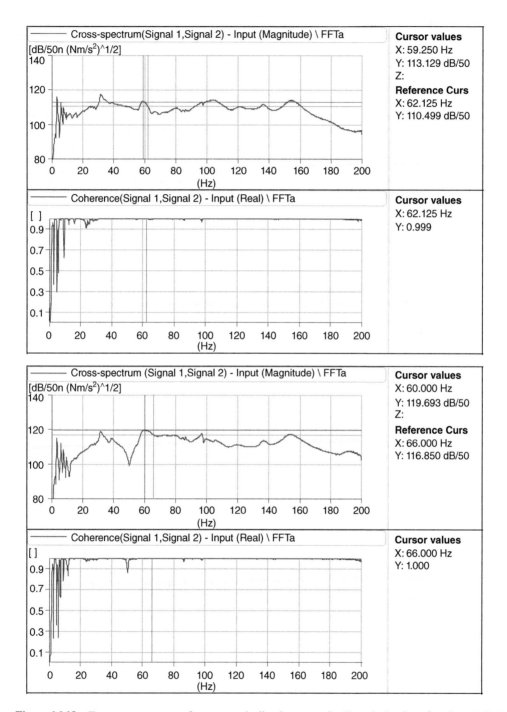

Figure 16.19   Frequency response from ground vibration test of a Decode-1 wing: flapping mode accelerometer placement (upper) and twisting mode placement (lower), cursors on first twist mode.

move in anti-phase with each other. The mode at 7.3 Hz is caused by flexure of the entire tail in opposition to motion of the wing and is again easily identified by its mode shape: this mode is also seen in the FEA model. That at 33 Hz is the second flap mode that occurs before the first twist mode, as was predicted by FEA and which can be identified by stroboscopic illumination. The low-frequency rigid-body modes can also be eliminated from consideration by observing that they shift significantly if the airframe supports are changed and tests repeated: for example, by testing with the aircraft sitting on its own suspension and then by supporting it on mounting blocks or from a soft hanging suspension. The coherence values in the response plots confirm that nonlinear and signal-threshold effects are not significant at the resonant frequencies, being at 0.998 or better at the frequencies of interest. Tests should also be carried out on both wings to check that natural frequencies do not differ appreciably; if they do and the measured wing masses are similar, this can point to deficiencies in the manufacture that should be addressed before flight testing.[2]

In Figure 16.19(upper), the accelerometer is placed at the wing tip directly above the main spar, while in Figure 16.19(lower) the accelerometer is placed at half span and on the trailing edge. Thus the upper set of results accentuate flapping mode responses, while the lower focus on the twisting mode. The magnitudes of the two responses can be seen to change by around 6 dB between the two plots in the way expected, given the accelerometer placements.

The first flap and twist modes have damping ratios of 0.026 and 0.05–0.1, respectively.[3] The damping stems mainly from the foam-plus-cladding nature of the structure, which has quite good intrinsic damping characteristics compared to pure CFRP or glass-fiber reinforced plastic (GRP) structures.

16.3 Avionics Testing

Avionics systems also need to be extensively tested before committing the airframe to flight. If the avionics malfunctions, it is highly likely that the aircraft will crash. In many jurisdictions, there is a mandatory set of ground tests the avionics have to be put through before every flight; we assume that a competent and qualified pilot will be in charge and thus such tests will routinely be carried out and no flight is commenced unless all is satisfactory. Even so, it is not good practice to wait until reaching the airfield before testing the on-board systems. Indeed, we would not commit to final avionics build without first carrying some preliminary experimental testing of the intended configurations in the lab. To do this, we construct a full-scale plan view drawing of the airframe and attach this to a rigid plywood baseboard: this is termed the "iron-bird", see Figure 16.20. To this, we attach all the avionics we propose to fly and wire it up with appropriate length cables laid out as we expect the final harnesses to lie. This system can then be extensively soak-tested for reliability and any interference problems. To do this, we energize any ignition systems to make sure that spark generation does not cause any problems. If there are generator systems to be included in the design, we include these and power them by dedicated electric motors. Sensitive items of equipment will be subjected to vibration test, also while powered up.

---

[2] We once lost a student aircraft in flight because of a defect in a main spar, which would have been revealed by such tests had they been carried out.

[3] The width is measured as that required for the height of the peak to be reduced to 0.7071 of its peak value, or by 3 dB, and the damping ratio is then given by $\zeta = 0.5\Delta\omega/\omega$: here, the width of the flapping and twist modes are 0.3 Hz and 6–12 Hz – the twist mode resonance being somewhat difficult to exactly define.

Figure 16.20 SPOTTER iron-bird being used to test a complete avionics build-up: note motors to spin generators in a realistic manner.

When carrying out vibration tests on components, it is important that any mounting does not significantly change the natural frequencies of the parts being tested as compared to their behavior in flight conditions. For wings, clamping at the spar mounting point, as in static structural load tests, is normal practice. However, for small components and on-board instrumentation, achieving realistic mounting can be very difficult in practice, and often one has to accept simple free-free mounting conditions as simulated by supporting components in soft springs or elastic bands, see Figure 16.21.

Figure 16.21 Avionics board under vibration test. Note the free-free mounting simulated by elastic band supports. In this case, a force transducer has been placed between the shaker and the long connecting rod that stimulates the board. The in-built accelerometer in the flight controller is used to register motions.

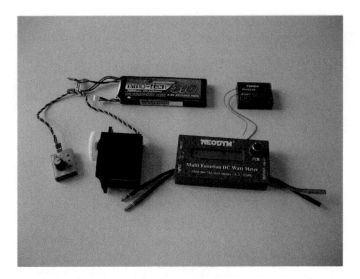

Figure 16.22 Typical Servo test equipment: (front left to right) simple low-cost tester, large servo, motor speed tester with in-built power meter, and servo control output; (rear) avionics battery and standard primary receiver.

Having tested all the avionics on the "iron-bird", components can next be tested in the airframe during construction. So, as each part is installed, it should be checked for correct functionality before proceeding. Many elements of the wiring are tedious to remove from a completed airframe and so should be thoroughly checked before being "buried" by further assembly operations. Careful attention should be paid to servos during this stage; they should not be coupled up without first being tested for adequate movement. Thus the linkage to the control surface or other item being moved should be disconnected and the servo powered up with an avionics battery and servo tester or receiver/transmitter combination (see, e.g., Figure 16.22). Servo testers are simpler to use in the lab, but ultimately all servo testing must, of course, be carried out using the intended primary transmitter and receiver system. Mechanical adjustments can then be made to linkages during build to ensure that full movement of the servo does not result in the device being "stalled" by a limited range in the mechanical movement of the components being driven.

17

Detail Design: Constructing Explicit Design Geometry

Next we turn to the direct creation of "explicit" geometry (as opposed to the "implicit" geometry we have mostly used so far, and which is generally used only during the early stages of design[1] – implicit geometry is essentially a compact representation of a number of less abstract parameters). Ultimately the goal will be to declare an unambiguous, fully defined artifact that is referred to as detail geometry. This artifact is made up of a series of geometric parameters that explicitly define the shape. In this chapter, examples are illustrated using the Solidworks computer aided design (CAD) software package,[2] which provides many useful capabilities including mass property analysis, cost modeling, and others.

17.1 The Generation of Geometry

A detailed description of the overall design stage logic is given in Appendix A. The detail design process is shown in Figure 17.1. It will be noted that, within the overall logic, generation of detailed geometry only starts after much preliminary activity/thinking/decision making has taken place. A premature start to the creation of detail geometry can be very inefficient, result in much wasted effort, and is likely to result in considerable unnecessary iteration and perhaps substantially suboptimal solutions.

In a typical low-cost or student design project, it is likely that the concept and preliminary design phases will make much use of spreadsheets of the sort described earlier, which hold a number of implicit design parameters. During the preliminary design phase, the geometry becomes sufficiently complex to justify the start of explicit geometry modeling (embodiment) for five reasons:

1. visualization (does the geometry look right; has there been a silly mistake – this is why our spreadsheets usually include very basic sketch views);

---

[1] An example of implicit geometry is "wing area," which is a key attribute for early hand or spreadsheet calculations and constraint analysis.
[2] http://www.solidworks.com/.

Small Unmanned Fixed-wing Aircraft Design: A Practical Approach, First Edition.
Andrew J. Keane, András Sóbester and James P. Scanlan.
© 2017 John Wiley & Sons Ltd. Published 2017 by John Wiley & Sons Ltd.

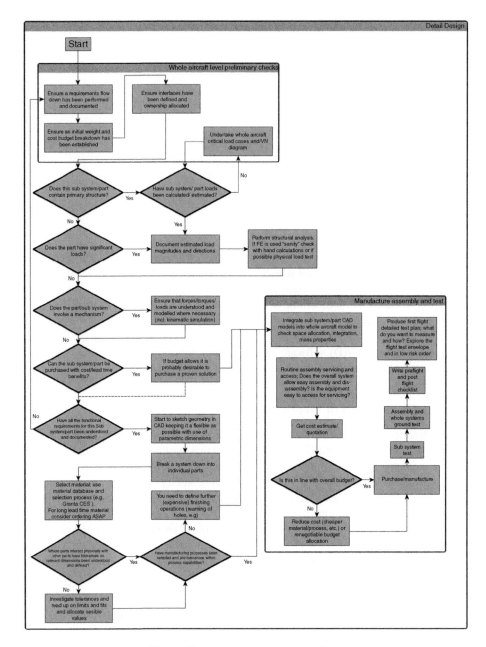

Figure 17.1 Detail design process flow.

2. space allocation (do all the proposed systems fit within the fuselage/wing, etc.);
3. support for physics-based analysis of aerodynamic and structural performance calculations;
4. accurate calculation of mass properties;
5. start of detailing.

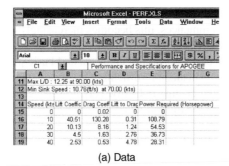

(a) Data

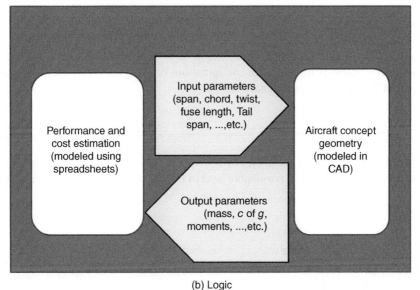

(b) Logic

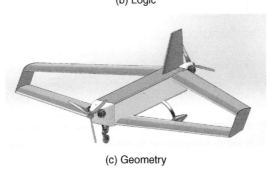

(c) Geometry

Figure 17.2   The structure of well-partitioned concept design models.

During the early design stages, flexibility needs to be preserved, as design parameters remain fluid. Therefore the goal is to construct geometry (assemblies) that can cope with large changes in parameter values. The "Holy Grail" is to achieve parametric geometry that can easily be modified externally in order to facilitate quick and easy search of a design space, as illustrated in Figure 17.2, see also Gudmundsson [15]. As the definition becomes more complex and explicit geometry creation commences, this becomes increasingly difficult.

In what follows, it is assumed that certain initial design decisions have been made such as configuration, engine selection, constraint analysis, and basic payload/ range calculations. Until these decisions have been made, it is difficult to justify significant effort in explicit geometry modeling because it could change radically. Having selected a configuration, one has to make initial decisions concerning the following:

- engine location (tractor, pusher)
- main wing location
- undercarriage configuration
- control surfaces.

An example configuration study for a twin-engine UAV is shown in Figure 17.3. The rapid generation of reasonably realistic outline configurations is one of the great benefits of adopting the AirCONICS suite during preliminary design.

17.2 Fuselage

During preliminary design, our use of the AirCONICS approach has also allowed us to develop the lifting surface geometries to a reasonable level of fidelity suitable for first-pass analysis and

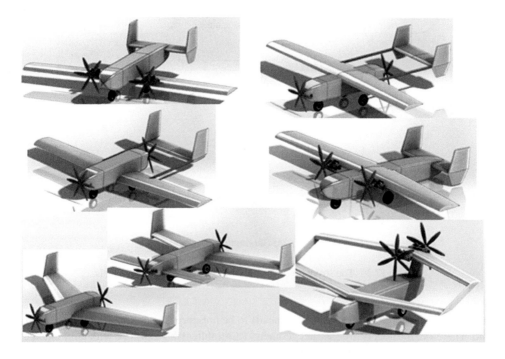

Figure 17.3 Example configuration studies.

further detailing. Thus far in the design process, however, relatively little thought has been given to the precise shape of any fuselage elements. Typically, the two main functions of a conventional fuselage are to

- provide a low-drag "package" for payload (cameras, cargo, other sensors, etc.) and systems (engines, avionics, batteries, fuel tanks, etc.), and
- structurally connect masses and sources of load.

In order to construct the "package," it is useful to have both a structural construction approach in mind and dimensionally correct geometry of the main items that have to be fitted into the fuselage. To lower costs, the construction method may well impose some restrictions on the range of possible fuselage shapes. One way of providing the package geometry is to construct an accurate 3D model of each element to be housed; examples for engines are shown in Figure 17.4. Clearly, such models require a significant amount of effort to construct.[3]

An alternative that requires less effort is to create a simple solid object that is an approximate representation of the object onto which the CAD tool can drape "decals," which follow the contours of the solid. This can generate very realistic geometry from photographs without a great deal of effort. Such objects can also be given accurate mass properties to aid center of gravity (CoG) calculations. An example is given in Figure 17.5, which is a camera model that we constructed for a UAV project.[4]

At the concept/preliminary design stage, the level of effort required for such high-fidelity component models is possibly not appropriate (if they are not already available), and hence a third alternative is to directly use sketches or photographs in the fuselage model. Furthermore, photo-realistic or complex 3D models at the concept stage can slow down rebuild times potentially, making the model unwieldy and frustrating to use. Figure 17.6 shows 2D drawings of the 3D models given in Figure 17.4. These are a very lightweight representations of the geometry but are good enough to undertake the construction of fuselage geometry concepts.

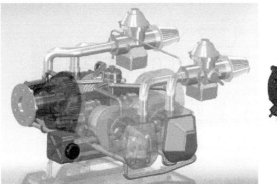 

Figure 17.4   Example 3D models of Rotax aircraft engine and RCV UAV engine. Courtesy of Chris Bill and RCV Engines Ltd.

---

[3] For some brought-in components, the manufacturers will make available suitable CAD models to customers.

[4] Of course, this can be achieved only if there are orthogonal photographs available. Instructions on how to create realistic looking approximate models are given under "Decals" in the Solidworks help files.

Figure 17.5 Example of images used to create realistic looking 3D Solidworks geometry model.

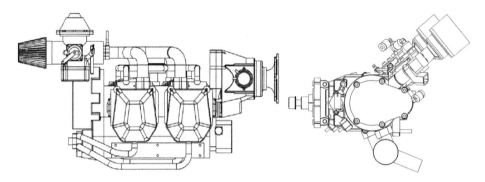

Figure 17.6 2D side elevations of Rotax aircraft engine and RCV UAV engine. Courtesy of Chris Bill and RCV Engines Ltd.

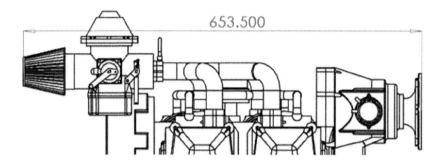

Figure 17.7 Scaling dimension added to drawing (mm). Courtesy of Chris Bill.

If one does have a 3D model of the object that one wants to put in a fuselage, it is very easy to create a 2D drawing from this. However, as we will be planning to put a 2D representation directly into our Solidworks fuselage 3D model, it is important to add in an accurate reference dimension so that the image can be scaled to match the fuselage, see Figure 17.7. This drawing needs to be saved as a picture format (we use *.jpg). It is also important to crop the image as close to the geometry as possible to prevent the empty borders obscuring other parts of the fuselage model.

Figure 17.8   "Spaceframe" aircraft structure.

The designer can now start to think about how to "package" these objects into a low-drag fuselage shape. Although we cannot calculate the drag at this stage, we can use some good aerodynamic design principles to try and minimize drag by minimizing the frontal and wetted areas as well as having a generally "smooth" profile.

At this stage, the *importance of having parametric geometry* becomes all too apparent. We are essentially "hand sketching" a shape around the main fuselage objects. In order to develop a successful low-drag aircraft, we will need to use good engineering practice of analysis and iteration to find the best solution. Detailed CFD studies of the shape will ultimately allow it to be "fine-tuned"; hence the need for a flexible parametric geometry at this stage. In order to do this, we need to sketch a series of closed "bulkhead" profiles that encompass our fuselage objects. At this stage we also need to understand what fundamental materials and processes we are going to use to construct our fuselage. This has a constraining effect on the sorts of shapes that we can produce. A space frame structure (Figure 17.8) cannot be used to create a smooth double-curvature fuselage shape. Therefore we cannot use circular or elliptical "bulkhead" profiles if we intend to use such a structural approach.

For the low-cost UAVs associated with student projects, double- curvature fuselages are difficult to achieve. Double curvature generally implies complex, costly manufacturing processes such as composite molding (for nonmetallic) or stretch-forming/press-forming (for metallic materials) or even large nylon SLS 3D-printed parts that are relatively expensive.

17.3   An Example UAV Assembly

An example student UAV is illustrated in Figure 17.9. The structural philosophy adopted is one of a composite tube/plate load- carrying skeleton with hollow-foam single-curvature bodies

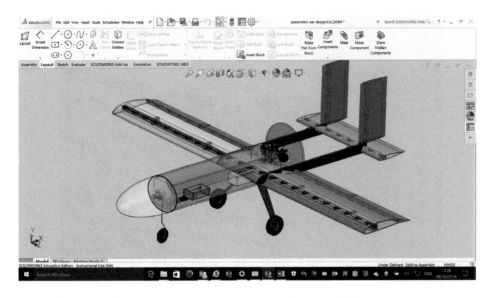

Figure 17.9   Illustrative student UAV assembly.

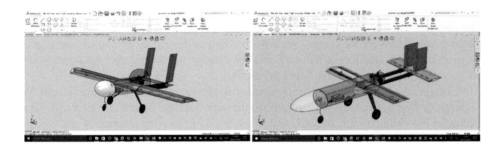

Figure 17.10   UAV assembly model can be modified by changing design table parameters.

for flying surfaces and fuselage. The foam can be covered in a variety of reinforcing materials including Mylar (of various thicknesses), glass, and carbon-fiber cloths. Finally, for nose cones and double-curvature surfaces, vacuum forming is a quick, cheap, and low-cost process and this is used for the nose cone in the example.

The Solidworks assembly shown is parametric and can be modified to respond to externally driven parameters as demonstrated in the somewhat exaggerated examples in Figure 17.10. All of the significant masses are included in order to help accurately predict overall mass properties such as maximum take-off weight (MTOW) and CoG location.

It is very helpful if the guidelines on systems engineering given in Section 8.4 have been followed when starting to generate explicit geometry. This is particularly important if many people are producing geometry models. Essentially, the construction of geometry by a team demands clarity over who is doing what and very precise interface definitions and ownership thereof.

In order to manage what becomes quite a complex geometry model, the following procedure is recommended:

17.3.1 Hand Sketches

At the concept design stage, many people find hand-sketching the quickest way to illustrate and communicate ideas. Once a particular concept has been shortlisted for embodiment, a scale-sketch can be produced as in Figure 17.11. Most of the information can be captured in a side elevation and plan view as shown. However, to be rigorous, a front elevation is also required. For example, the two sketches shown fail to unambiguously capture the geometry of the main undercarriage legs.

If an AirCONICS model is not going to be used to seed the detail design process, such hand sketches can be used to start to create the CAD geometry and therefore need to be reasonably accurate scale sketches with a dimensioned feature in each sketch. Because we are going to later create parametric variables, these do not to need to be exact. These sketches can be imported into a CAD tool to provide an initial starting point for generating exact parametric geometry. The two hand sketches have been positioned orthogonally in Solidworks and scaled appropriately as shown in Figure 17.12. Each sketch has been made partially transparent to allow emerging geometry to be visible through the sketch.

17.3.2 Master Sketches

Based on these hand sketches, exact "master sketches" can now be created. First, all significant dimensions need to be captured in a set of of 2D fully dimensioned CAD sketches: typically a side elevation, plan view, and front elevation. These sketches need to be constructed at the top level of the assembly hierarchy and not associated with any particular part.

Illustrated in Figure 17.13 is the fully dimensioned exact CAD plan view sketch being created on the imported hand sketch, which is used as an initial template.

The two master sketches for the UAV assembly are shown in Figure 17.14. Note that all the features within these exact sketches are fully dimensioned and therefore fully defined. In

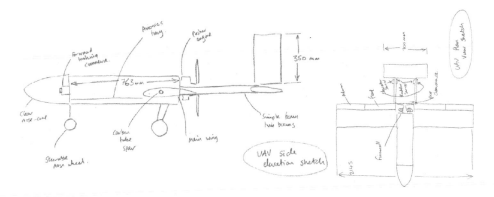

Figure 17.11 Plan and side view hand sketches.

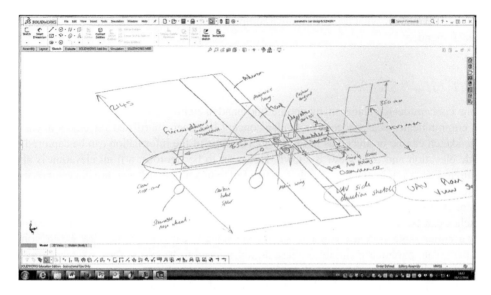

Figure 17.12   Hand sketch scaled and positioned orthogonally in Solidworks.

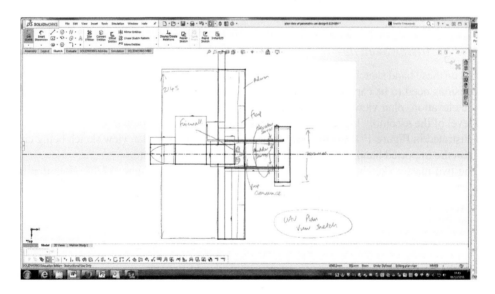

Figure 17.13   Exact, dimensioned sketch being created on hand-sketch outline.

Solidworks, the sketch changes from blue to black to signify that it is fully defined: in other words, unambiguous and fixed in space.

Within these sketches, it is important that meaningful names are given to each dimension to make the assembly and associated design table more comprehensible. As parts are defined or brought into the assembly, the key dimensions need to be connected with the master sketch to "synchronize" the whole assembly and conform to the "Systems Engineering" interface definitions.

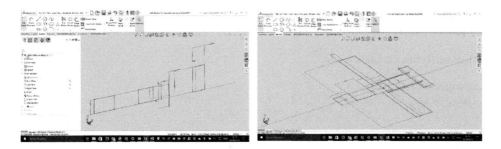

Figure 17.14   The "master" driving sketches in the assembly.

	wing span@plan view	tialplan span@plan view	tailplane chord@plan view	fuselage width@plan view	nose length@plan view	length of engine@plan view	wing chord@plan view	front bulkhead dist@plan view	wing box bulkhead dist@plan view	eng blkhd dist@plan view	tail spar dist@plan view	prop diamter@plan view	boom prop clearance@plan view	boom width@plan view	fin height@side elevation	engine height@side elevation
Default	1145	700	400	250	430	125	250	399.3	240	130	700	115.6	5	15	251.4	30
example 1	2145	700	200	250	330	125	350	399.3	240	130	700	315.6	5	15	351.4	30
example 2	1645	400	400	150	230	125	450	399.3	240	130	700	315.6	5	15	351.4	30

Figure 17.15   Design table for example UAV.

Having identified all the key master geometry parameters in high-level sketches belonging to the assembly, it is very easy to create a design table (spreadsheet), which can then be linked to other spreadsheets that contain design calculations. An extract of the design table for the example UAV is shown in Figure 17.15.

17.4   3D Printed Parts

The following section addresses the detailed geometry associated with 3D printed parts. While 3D printing is useful for complex geometry, it should not be used for simple geometry that can be more cost effectively produced with laser cutters or subtractive operations such as turning/milling.

17.4.1   Decode-1: The Development of a Parametric Geometry for the SLS Nylon Wing Spar/Boom "Scaffold Clamp"

Functionally, this complex part

- connects two foam wing panels;

- makes a structural connection between the wing spar and one of the tail booms; and
- allows the structural connection to be disassembled easily and repeatedly.

The key performance goals are to

- minimize weight, and
- minimize drag.

17.4.2   Approach

Multifunctional parts such as this can get very complex, and so it is desirable to partition the part into a number of subparts. This makes the development/management of the geometry more convenient. It also removes the need for a single, complicated, and often fragile "history tree" on which all geometry relies. By partitioning the geometry into functional/geometrical units, the creation of flexible parametric models is facilitated.

17.4.3   Inputs

The input is some reference geometry based on an imported file in the form of a Standard ACIS Text (.sat) file which is an ASCII text file, here generated from our AirCONICS study. This geometry, derived from a higher level conceptual model, is used to drive all the dependent geometry and features. This initial input geometry is shown in Figure 17.16. It has been saved as a Solidworks part and added to an assembly. Two reference geometry features (spar tube axis and boom tube axis) have been added, and the part has been lined up with the three default assembly reference planes.

Sketches have been added to the top-level assembly, and all the key dimensions and curves have been extracted from the .sat file and added to these sketches. This makes all

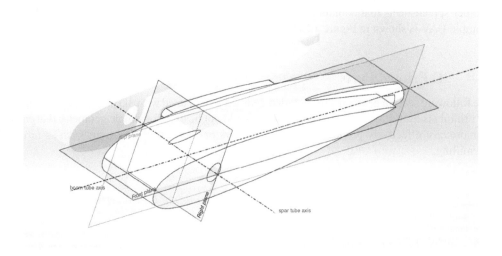

Figure 17.16   Input reference geometry.

these dimensions explicitly available to the subparts. This .sat file geometry is not sufficient, however, to drive the entire detail of the part, and further dimensions and decisions are required to fully define the final part.

17.4.4   Breakdown of Part

The initial part has been broken down into subparts as follows, see Figures 17.17 and 17.18:

1. wing spar tube
2. boom tube
3. skin
4. outer rib

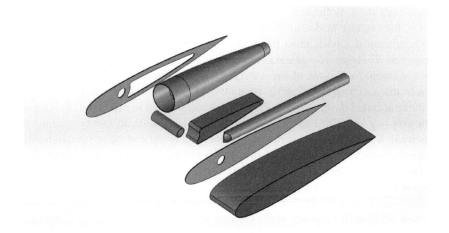

Figure 17.17   The input geometry modeled as partitioned parts.

Figure 17.18   The assembly generated from reference geometry.

5. inner rib
6. foam insert spigot
7. boom tube fairing
8. skin stringer stiffening.

17.4.5 Parametric Capability

Ideally, the detailed geometry should automatically reconfigure/scale to any changes to the input reference geometry and other explicitly defined input variables/dimensions. In practice, it is very hard to achieve this from manually constructed models. An alternative way of achieving this is to capture all the logic and associated relationships in software code and use this to fully generate the detail model. This requires a substantial amount of effort, and for component topologies/functions that are repeatedly modified and reused, this can be worth the effort. However, this is beyond the scope of this example, and in this model manual construction has nevertheless been used. With care, manual models can, within limits, be made parametric. In other words, they can be made sufficiently robust to allow a limited range of changes to the inputs. Where manual models most frequently fail is when changes to inputs cause features to be eliminated. This then causes errors in the relationships between parts and other features.

To test the initial robustness of this emerging detailed model, a scaling feature has been inserted into the initial input reference geometry. This simulates the changes that might be made to the conceptual design. The scaling feature can be used to make global changes to the input geometry (e.g., making the entire geometry 20% larger). It can also be used to "stretch" the geometry in less than three dimensions. Hence, you could choose to stretch the length of the part in only one direction. The problem with this is that it can cause the downstream model to fail because circular features become ellipses, for example. For this reason, testing of this model was restricted to global 3D scaling tests.

By scaling the input reference geometry, the detailed geometry model can be "debugged" to ensure that it faithfully responds to the changes without generating errors. This is shown in Figure 17.19, which shows that the input reference geometry has downscaled by 30% (inner)

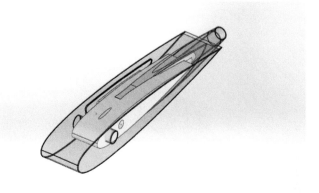

Figure 17.19   "Debugging" the detailed model.

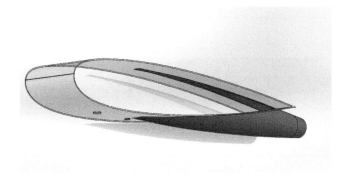

Figure 17.20   Trimming of boom tube fairing.

and the previously generated detailed geometry (outer) has not yet been rebuilt. On pressing the rebuild button, successful scaling of the detailed model can be checked and any errors understood and fixed.

A number of steps now need to be carried out, including the following:

- Modification of internal geometry (see, e.g., trimming of the boom tube fairing, Figure 17.20);
- Ensuring that the part is capable of being manufactured; in the case of the nylon SLS process, closed voids need to be eliminated;
- Editing of part geometry to meet functional goals consistent with the process capabilities of the selected manufacturing method – in other words, the geometry needs to allow parts to fit together;
- Addition of further detail that does not exist in the conceptual input geometry.

All parts can now be edited as independent objects. The advantages of this compared with creating the detail model as one monolithic or multibody part are twofold. First, it allows very large and complex geometries to be manipulated more easily. Very large, complex geometries can create very large file sizes. It can be tedious having to load a very large geometry when the user only wants to modify a small feature. Second, it allows a model to be decluttered quickly and easily. Just having to open the item you are interested in allows you to focus only on that part of the geometry. A disadvantage of using this approach is where relationships between many "parts" are required. This may require the whole assembly to be loaded and only the relevant "parts" made visible.

17.4.6   More Detailed Model

The more detailed model is illustrated in Figure 17.21. This shows the added detail of carbon-fiber tube clamps, access holes, and example stringers for stiffening the skin. This entire geometry has been produced in the context of the input geometry from the higher level conceptual geometry. If this conceptual geometry changes, then the detailed model should change accordingly. However, there are limits to changes that can be made. If, for example,

Figure 17.21   Final detailed model.

the topology of the part changes, then the detailed geometry is likely to fail because of a broken link to something that no longer exists.

17.4.7 Manufacture

This assembly can now be saved as a single body ".stl" file, which is the format used as the input for most additive manufacture machines. At this stage, certain other detailed features might be put into the geometry. For example, fillets might be defined between flat features. Unfortunately, fillets are particularly fragile features in parametric CAD models. It is therefore better to leave fillet features out of the parametric model to prevent errors on rescaling. Only when the higher level geometry has been "frozen" should the part be fully detailed by undertaking the (unfortunately tedious) task of filleting and the introduction of other minor details prior to manufacture.

17.5 Wings

This section encompasses not only the main lift-generating parts of a conventional aircraft but also the empennage surfaces such as horizontal and vertical tail surfaces. In other words, all these surfaces can essentially be classified as "wings" for the purposes of this section and the parameterization techniques outlined applied to all these parts of the aircraft.

Careful selection of wing geometry is of particular importance in ensuring that a good overall aircraft design is achieved. It is important that wing geometry is well parameterized in order to be able to make adequate modification to improve such things as cruise drag, stall characteristics, and so on.

For the purposes of this section, it is assumed that a wing section profile has been selected. This is often an early consideration for aircraft design taking into account the type of aircraft and mission being addressed. While it is desirable to develop parametric geometry models that allow section profiles to be modified, this takes considerably more effort in tools such as Solidworks. Therefore, here we assume that the level of parameterization for the wing surface is confined to

- span
- twist
- taper (root chord, tip chord) and
- sweep.

Further architectural sophistication can easily be achieved by having "multipanel" wings, where for each panel, all of the above variables can be defined separately. Indeed, for many light aircraft, the wing is a distinct multipanel configuration. Figure 17.22 is a photograph of the wing of a PA-28 aircraft showing the leading edge kink at the junction of the two essentially straight wing panels.

The wing parameterization technique outlined here cannot produce elliptical wing surfaces such as those employed on the Spitfire aircraft, but since such shapes are essentially used to control induced drag and this can be dealt with by suitable changes to camber or twist, this is of little concern – we tend to avoid elliptical wing shapes when using foam-cored wings.[5]

Figure 17.22 Multipanel wing of PA-28. Photo courtesy Bob Adams https://creativecommons.org/licenses/by-sa/2.0/ – no copyright is asserted by the inclusion of this image.

---

[5] We did use an elliptical planform on the SULSA aircraft but mainly for aesthetic reasons rather than aerodynamic ones; also, on that aircraft the entire wing was an SLS nylon printed structure.

17.5.1   Wing Section Profile

Having selected a suitable wing section profile, a set of coordinate points needs to be generated. There are many spreadsheets that can be found on the Web that can generate these coordinates. Figure 17.23 gives an example of an NACA four-digit coordinate calculation, in this case producing the NACA 2412 section that is widely used on light aircraft.[6]

In whatever way the coordinates are produced, it is important that one generates enough points to generate an accurate, smooth curve, which is then going to be imported into Solidworks. The data needs to be formatted into columns of x, y, and z coordinates (even though the z coordinates will all be zero) and saved as a text file. From within Solidworks, this can now be imported as a .txt file to generate a curve using the "Insert/Curve/Curve through XYZ points" command (Figure 17.24).

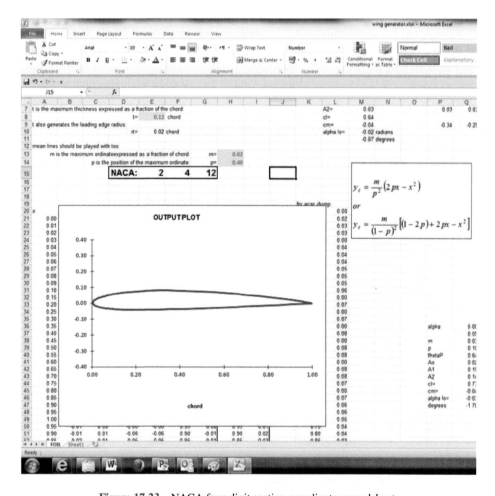

Figure 17.23   NACA four-digit section coordinate spreadsheet.

---

[6] A particularly good source of coordinate data can be found at http://airfoiltools.com/plotter/index.

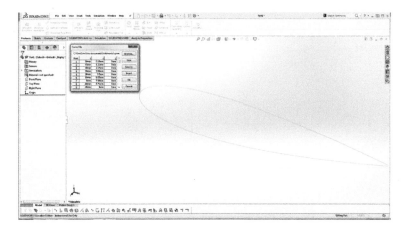

Figure 17.24   Curve importing in Solidworks.

In this case, 60 points are used to create an adequately smooth curve. Notice that by zooming enough into the rear edge of the section, it is apparent that it is an open shape. This will be important later on.

This initial curve is reference geometry and cannot be directly used to construct solids and surfaces. The next stage is to create a sketch using this reference curve. To do this, we open a sketch on the *XY*-plane that the curve sits on.

We then need to carry out the following:

- Use the "convert entities" command to convert the reference curve to a sketch curve – the curve turns black when this is carried out to indicate that it is fully defined, Figure 17.25.
- At this stage, it is useful to close the shape (if open), see Figure 17.26. This can simply be accomplished by drawing a line between the two ends of the curve (if it is small; which it should be if we have at least 60 points in the input *.txt file), Figure 17.26.

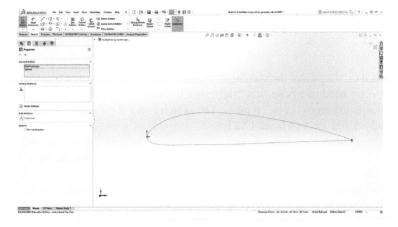

Figure 17.25   Use of "convert entities" in Solidworks.

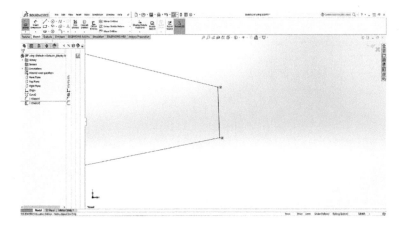

Figure 17.26   Closing the 2D aerofoil shape.

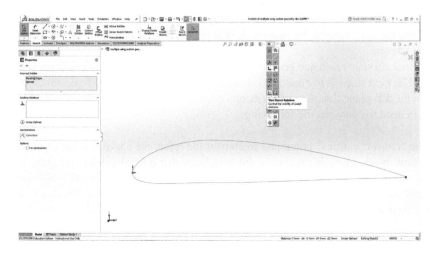

Figure 17.27   Deleting sketch relationship with reference geometry.

- Next we need to "decouple" the sketched curve from the reference curve so that we can scale it later on – we need to ensure that "view sketch relations" is selected in the view palette and then delete the green "cube" symbol. The sketch curve should turn blue to indicate that it is not fully defined now, Figure 17.27.
- Then we must construct some reference geometry to constrain the section and allow it to be scaled. Touching vertical and horizontal reference lines therefore need to be defined, Figure 17.28.
- A number of constraints now need to be defined to tie the section geometry to this reference "scaffold"; the curve needs to pierce the junction point of the lines, and the rear end of the section needs to be tied to the end of the horizontal line. A dimension can now added to the horizontal line to scale the wing section, Figure 17.29.

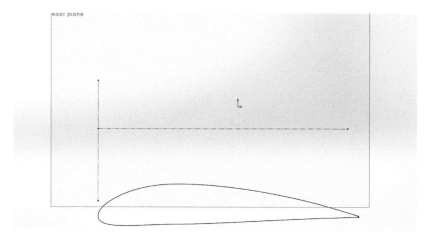

Figure 17.28   Reference geometry.

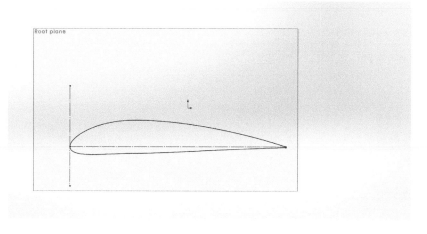

Figure 17.29   Constraining curve to reference "scaffold" geometry.

17.5.2   Three-dimensional Wing

We now have an accurate closed wing section profile that can be arbitrarily scaled. We next need to create some 3D reference geometry so that we can place two independently scaled wing sections in relative positions in 3D space. We can then loft between these two. Figure 17.30 shows the resulting placement of the two sections in relation to each other. A loft between these two will generate a wing surface whose sweep, taper, root chord, and twist can be controlled (all these steps are automated in the AirCONICS suite of course).

Figure 17.31 shows the finished result. Note also

- the naming of reference planes to provide clarity;

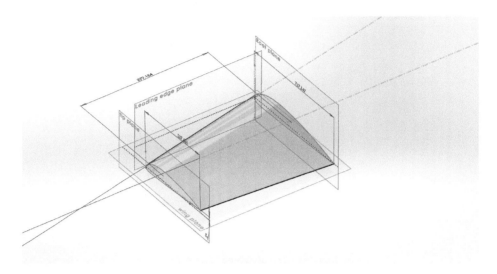

Figure 17.30 "3D" scaffold to define the relative positions in space of two independently scalable wing sections.

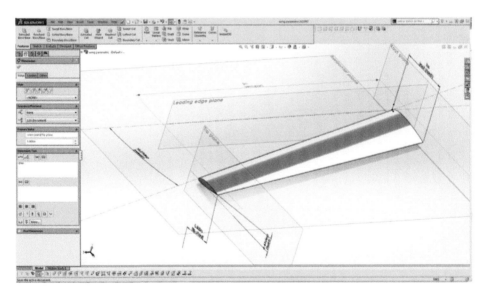

Figure 17.31 Wing surface with span, twist, taper, and sweep variables.

- the naming of dimensions with meaningful text – this becomes very important when complex geometry, design tables, and external files are used. It is easy to link the wrong dimension when it is merely labeled "D15 @sketch23."

Figure 17.32 shows how a PA-32 aircraft-type wing can be modeled using two panels. Of course, the number of variables needed to define this wing has now risen to 9 ($4n + 1$, where n is number of panels).

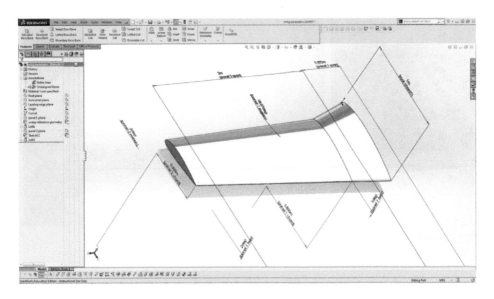

Figure 17.32   Multipanel wing.

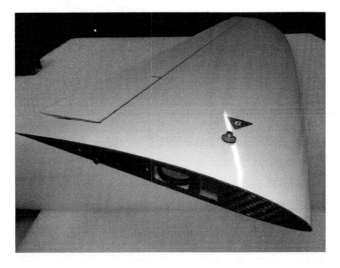

Figure 17.33   Example of a double-curvature composite wing.

Having now created the surface geometry, an approximate structural model needs to be derived from this surface in order for a good mass estimate to be made.

As with the fuselage, a structural philosophy and manufacturing method needs to be decided before creating further geometry. Stressed skin composite structures give the aerodynamicist the most geometrical freedom including double-curvature surfaces that would result if an elliptical planform was chosen (Figure 17.33). Composite structures are, however, notoriously expensive compared to other methods of construction. In light and microlight aircraft, it is

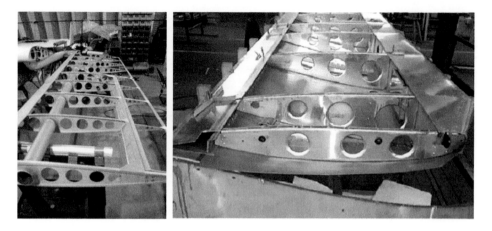

Figure 17.34   Fabricated wing structures.

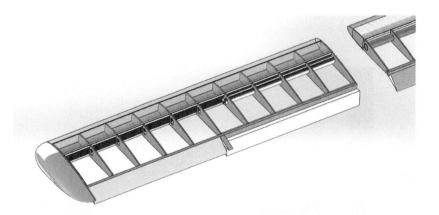

Figure 17.35   Simple wooden rib and alloy spar structure.

generally cheaper to manufacture a fabricated wing using a range of materials such as aluminum spars, wooden ribs, or polyester fabric covering, see Figure 17.34. As already seen, in most of our UAVs we adopt carbon-fiber spars supporting foam cores with fiber or Mylar coverings.

Having made decisions regarding wing structures and materials, a more realistic wing can now be constructed based on the aerodynamic surface. For example, Figure 17.35 shows a very simple wing structure popular in lightweight microlight aircraft designs.

A more sophisticated structure is shown in Figure 17.36. This assumes hollow glass-fiber leading and trailing edges with glass-fiber ribs (a material type has been assigned to the structure). A table has been created in this model that gives an accurate mass prediction for this geometry.

The geometry created in Figure 17.36 is based on the parametric wing surface given in Figure 17.31. Hence the model can still have the flexibility of the original parametric

Figure 17.36   Parametric wing structure.

model. A useful way of creating the ribs is to use a Boolean operation to create a multibody part.

Appendix C shows a worked example of this approach to generating a detailed parametric CAD model for a manned aircraft.

Part IV

Manufacture and Flight

18

Manufacture

The main structures of most small fixed-wing unmanned air vehicles (UAVs) are essentially built using fiber-reinforced plastics of some kind with the addition of foam, plywood, and metallic parts where needed. To these are assembled a whole raft of externally sourced components, which are then connected via suitable wiring harnesses. As has already been made clear in Parts II and III of this book, we have adopted the use of 3D printing (rapid prototyping) to deal with a great deal of the more complicated geometries needed in our UAVs and have limited the use of fiber-reinforced material to stock shapes that can be bought in and used with little further effort, though they still usually form the main load-bearing backbones of most of our airframes (see Table 18.1 for typical properties). We thus avoid the use of molds with glass-/carbon-fiber-reinforced plastic (GFRP/CFRP) lay-up processes completely; molds are expensive and add a layer of tooling that we no longer believe necessary, though many commercial UAVs are still produced in this way and we often manually clad our foam parts in fiber materials.

18.1 Externally Sourced Components

To hold costs down and to make the best use of economies of scale, we buy in as much of our UAVs in part form as possible. The cost savings achieved by using commercial off-the-shelf (COTS) parts can be prodigious. A good-quality aeromodeler-grade petrol engine will cost a few hundred dollars. A specialist engine from a dedicated UAV manufacturer will cost thousands, while a military-grade engine developed specifically for a particular UAV project may well cost hundreds of thousands by the time development costs have been allowed for. It is thus much more cost effective to limit one's design choices to make use of what is available off the shelf and can be simply bought in, than to try and customize everything for the aircraft in hand. And by using 3D printing we turn a whole series of custom-designed and aircraft-specific airframe parts into being externally sourced as well: we simply send our part designs as CAD files to external manufacturers with the prerequisite 3D printing facilities and they then ship back the parts we need in the quantities required, see Figure 18.1. This approach to manufacture means that on a typical UAV it is only parts made of hot-wire-cut foam and the prototype wiring looms that we make in-house; all the rest of our airframe manufacture then simply

Small Unmanned Fixed-wing Aircraft Design: A Practical Approach, First Edition.
Andrew J. Keane, András Sóbester and James P. Scanlan.
© 2017 John Wiley & Sons Ltd. Published 2017 by John Wiley & Sons Ltd.

Table 18.1 Typical properties of carbon-fiber-reinforced plastic (CFRP) tubes.

Property	Units	CF fabric	CF unidirectional
Density	g/cm^3	1.60	1.60
Young's modulus 0°	GPa	70	135
Young's modulus 90°	GPa	70	10
Ultimate tensile strength 0°	MPa	600	1500
Ultimate compression strength 0°	MPa	570	1200
Ultimate tensile Sstrength 90°	MPa	600	50
Ultimate compression strength 90°	MPa	570	250
Poisson's ratio		0.3	0.3

Figure 18.1 3D SLS nylon parts as supplied from the manufacturer.

becomes an assembly operation with standard hand tools. Even the foam is produced directly from our CAD files using fully digital approaches.

18.2 Three-Dimensional Printing

We use 3D printing extensively in the manufacture of bespoke parts for our UAVs. We use selective laser sintering of nylon (for airframe parts) and of stainless steel (for highly loaded parts such as engine bearers) and fused deposition modeling of ABS for nonstructural parts where complex shaping is needed (such as for wing tips). This renders the production of parts into essentially a CAD-based design task followed by outsourced manufacture direct from the CAD files. The companies that operate such machines have a fast turn-around providing high-quality parts with very good repeatability. Figure 18.2 shows a stainless steel selective laser sintering (SLS) printed engine bearer for a gasoline engine.

18.2.1 Selective Laser Sintering (SLS)

SLS works by fusing fine-grained powder with a laser that scans a bed of the working material. This creates a thin laminar structure. A wiper then covers the fused part with a further thin

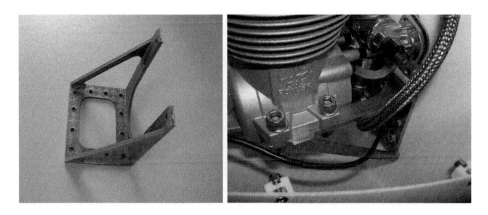

Figure 18.2 3D SLS stainless steel gasoline engine bearer after printing and in situ.

layer of powder, which is then fused to the first, thus growing a three-dimensional object layer by layer, with the parts being supported by the lower layers as they are "grown." At the end of the print, a large "cake" of powder is left, within which lies the fused part. By removing the unfused powder (typically by suction or blowing with compressed air), the desired parts are exposed. These then need cleaning internally to remove any unwanted powder before use, see Figure 18.3. With nylon, this is all that is required. With metal SLS, it is normal to start the process on a metal base plate that must be cut off from the finished part (typically by using a wire cutting machine). In either case, the finished parts have a slightly rough surface finish

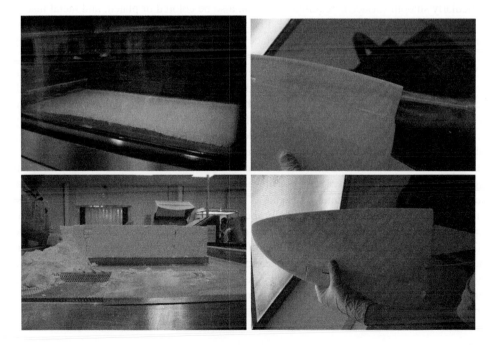

Figure 18.3 3D SLS nylon manufacturing and depowdering.

Table 18.2 Typical properties of SLS nylon 12.

Property	Units	Value
Color		white
Density of laser sintered part	g/cm^3	0.9–0.95
Young's modulus	MPa	1700± 150
Tensile strength	MPa	45± 3
Elongation at break	%	20± 5
Bulk modulus	MPa	1240± 130
Melting point	$^\circ$C	172–180
Vicat softening temperature B/50	$^\circ$C	163
Vicat softening temperature A/50	$^\circ$C	181
Coefficient of thermal expansion	K^{-1}	1.09×10^{-4}
Poisson's ratio		0.39

For isotropic materials, the shear modulus G, bulk modulus K, Poisson's ratio v, and Young's modulus E are related as $2G(1 + v) = E = 3K(1 - 2v)$; for the tabulated values, SLS nylon is not isotropic.

that has the grain size of the raw powder. This can be removed by polishing or filling if a particularly smooth surface is required. Parts can also be colored or plated, and metal inserts can be added into the nylon. Generally we do not bother with further surface treatment for our parts. Moreover, the slightly roughened surface we find highly suitable for gluing with epoxy resins if required, either to attach parts or to carry out repairs.

When using SLS, the only limitations on design freedom are, first, it must be possible to remove any unwanted powder so fully enclosed cavities cannot be made in this way; second, there is a minimum wall thickness that can be achieved (around 1 mm); and third, the maximum size of part is constrained by the dimensions of the build chamber in the machine (currently for the machines our suppliers use to sinter nylon this is 700 mm × 380 mm × 580 mm). See Table 18.2 for typical SLS nylon properties. In metal the build chambers are generally smaller (our supplier's machines have a maximum chamber size of 250 mm × 250 mm × 325 mm) but slightly thinner wall thicknesses can be achieved (down to around 0.5 mm). In either process, the build chamber in which sintering occurs is kept at a high temperature to aid the process. This means that allowance must be made for the shrinkage that occurs on cooling; our suppliers deal with this by scaling our designs before printing so that we do not have to consider the effect. It is also the case that the orientation of the part during construction has slight influences on the finished part. Curved surfaces, if not subsequently polished, show the lines where layers of powder end, for example, and there are slight variations in material properties. In general though, very high quality parts with good structural properties result, allowing highly functional components to be made.

Figure 18.4 Small office-based FDM printer. Parts as they appear on the platten and after removal of support material.

18.2.2 Fused Deposition Modeling (FDM)

Since the machines required for SLS manufacture tend to be very expensive and the depowdering process can be quite messy, it can be useful to turn to alternative 3D printing technologies for nonstructural parts. We have for many years operated our own, small FDM machine for making parts up to 330 mm × 100 mm × 100 mm in dimension, see Figure 18.4. In FDM, a plastic filament (generally ABS) is squeezed through a heated nozzle so that it emerges in a semiliquid form. This is then laid down onto a plastic platten in a chamber held just below the melting point of the plastic. The emerging plastic thus fuses to that already produced, and layers are built up at about 0.7 mm at a time. The major restrictions with this process is that overhanging structures can be made only if the resulting degree of overhang is limited (in practice a 45° angle can be achieved). If a greater overhang is required in the final part (such as in an arch or circular hole), a scaffold of sacrificial material has to be inserted to support the ABS filament and then this must be removed after printing. This is a tedious and not always successful, so when using FDM we typically design the part such that a build orientation can be achieved with no significant overhangs. On the more positive side, FDM does allow fully enclosed voids to be produced provided the overhangs are controlled. We have printed double-skinned parts in this way, which are linked by internal baffles at the appropriate angles. A selection of FDM-printed parts is shown in Figure 18.5. In all cases, the parts that result are highly orientation-dependent in terms of mechanical properties, being very weak in the layer-to-layer direction. This can be overcome by using such parts in a prestressed form by including tension rods to link and compress the FDM components. The fuselage of the aircraft in Figure 18.6 is made from ABS using FDM in this way.

18.2.3 Sealing Components

It is possible to manufacture fuel tanks using 3D SLS printing, although this makes sense only when they also form an integral part of the main structure. If nylon is used for this

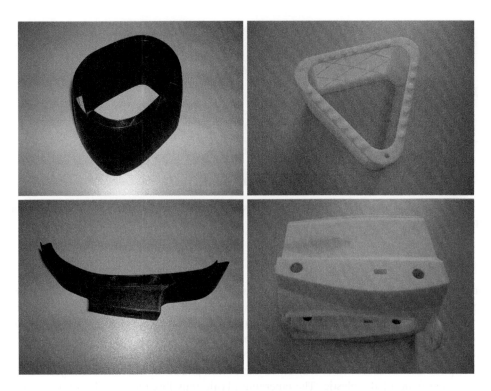

Figure 18.5   FDM-printed ABS fuselage parts.

Figure 18.6   Aircraft with FDM-printed fuselage and wing tips.

purpose, we find that, as manufactured, it is very slightly porous, so the finished tank must be internally sealed with an appropriate fuel-safe compound (we use Kreem Fuel Tank Liner[1]). Before applying the sealer, it is important that a thorough mechanical cleaning is carried out to remove any residual powder from the manufacturing process. We load the tanks with bead blasting balls and shake them well before emptying the balls out and rinsing with a degreaser.

18.3 Hot-wire Foam Cutting

Except for rather small UAVs, we make use of lightweight closed-cell plastic foam to manufacture most of the aerodynamic surfaces of our aircraft. This foam is cheap to buy and available in a wide range of sizes, densities, and colors. We mainly use foam blocks designed for floor insulation in the UK. This has anisotropic properties, see Table 18.3. We cut the foam using hot-wire cutting machines, one of which was manufactured in-house, see Figure 18.7, and a much larger one that was commercially sourced, see Figure 18.8. These machines work by dragging a heated stainless steel wire held in tension through the foam, melting a path just a little wider than the wire as it travels. By using suitable stepper motors to control the end points of the wires, complex aerodynamic shapes can be cut.

When using a hot-wire cutter to make wing parts, it is advisable to design the airfoils to have blunt trailing edges or trailing edges with a definite radius. Sharp trailing edges are difficult to cut, as the hot wire tends to melt fine edges. Holes can be cut for ribs and spars, but they

Table 18.3 Typical properties of closed-cell polyurethane floor insulation foam.

Property	Units	Value
Young's modulus	MPa	3.64
Tensile strength	kPa	485
Shear modulus	MPa	10.1
Shear strength	kPa	391
Compressive modulus in rise direction	MPa	21.3
Compressive strength in rise direction	kPa	434
Compressive modulus in in-plane direction	MPa	10.4
Compressive strength in in-plane direction	kPa	241
Poisson's ratio		≈ 0.2
Mean bending strength – crushing failure (as per BS EN 12089:2013)	kPa	789.4
Mean bending strength – shearing failure (as per BS EN 12089:2013)	kPa	502.5
Density	g/cm^3	0.025
Poisson's ratio		0.4

Direction refers to the orientation during block manufacture.

---

[1] http://www.kreem.com/fueltankliner.html.

Figure 18.7 In-house manufactured hot-wire foam cutting machine. This cuts blocks of foam up to 1400 mm × 590 mm × 320 mm.

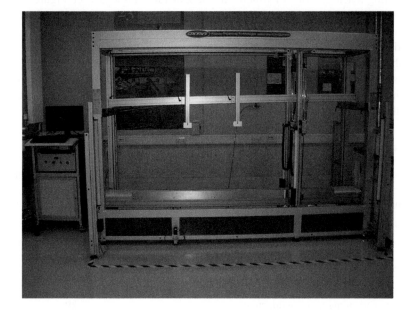

Figure 18.8 Large hot-wire foam cutting machine.

must be defined so that the wire will cut in from an outer surface. Moreover, creating hollow airfoil sections can be difficult for very thin wings, as the walls of the resulting airfoil sections can be too thin, causing the wire to burn through the section. Foam cutters are also able to cut tapered geometry using different end profiles, but care has to be taken to ensure that the sections at either end line up correctly and are traversed at the correct speeds to make sure that the generated conics are as desired. There are also limitations on the maximum taper that can be cut using the machines. High levels of taper result in the wire being almost stationary at one end during manufacture, which tends to burn holes in the foam. By cutting the blocks in two orthogonal directions before removing the finished parts, all sorts of shapes can be produced. Figure 18.9 shows some typical foam wing parts and the parent blocks.

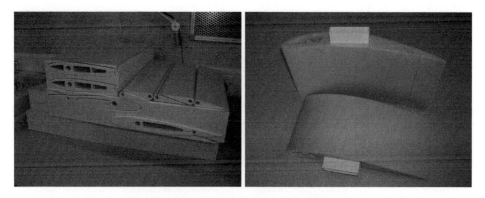

Figure 18.9 Hot-wire-cut foam wing parts: (Left) The original material blocks with and without cores removed; (right) with FDM-manufactured ABS joining parts.

18.3.1 Fiber and Mylar Foam Cladding

Although hot-wire-cut foam can be used to make very light and cost-effective aerodynamic parts, the approach suffers from some drawbacks. First, the finished surface is not perfectly smooth; second, any local loading must be fed into the foam via load spreaders to avoid local crushing damage; and third, the foam is usually not resistant to fuels of any kind. To deal with all these issues, and to provide robustness against ground handling damage, we almost always clad our hot-wire-cut foam parts after manufacture. This also significantly strengthens the foam in bending and, because it closes off the cut lines needed to make interior lightening, access, and spar holes, significantly increases torsional rigidity. We do this either by applying very fine glass-fiber or carbon-fiber tissue using thin resins or by covering the foam with sheets of heat-shrinkable aero-modeler wing film or thin Mylar attached with spray-on contact adhesives. This is one of the few remaining manufacturing tasks in our UAV builds that relies on skilled workers to produce the best results, although with a little practice and some care, cladding parts is not that difficult. If fiber tissue has been used, the resulting surface, though tough and fuel proof, is not completely smooth. This can be overcome via judicious use of fillers, but this adds weight, noticeably so on large wings. Figure 18.10 shows a selection of aircraft wings with differing forms of cladding. Table 18.4 gives typical bulk properties of glass-fiber-reinforced materials; but note that our claddings ultimately fail by delamination of the bond between the foam and the cladding and not by failure of the cladding itself. Also note that, when applying very thin glass fiber layers, the ratio of fiber to resin is lower than for thicker parts, and consequently the stiffness and strength properties achieved in our glass-fiber claddings are typically less than half the values given in the table.

18.4 Laser Cutting

Although we now almost exclusively use hot-wire-cut foam to make our wings, we have built wings using assemblies of laser-cut plywood components glued together with epoxy resins. These are then covered with thin aero-modeler-grade films, see Figure 18.11. Such wings can be extremely light and stiff, but they are rather time consuming to assemble and require some skill to achieve the best results. The film coverings can also be susceptible to ground-handling

Figure 18.10 Foam wings after cladding: glass fiber, Mylar, and filled glass fiber.

Table 18.4 Typical properties of glass-fiber-reinforced plastics.

Material	Specific gravity	Tensile strength (MPa)	Compressive strength (MPa)
Polyester resin (Not reinforced)	1.28	55	140
Polyester and chopped strand mat laminate 30% E-glass	1.4	100	150
Polyester and woven rovings laminate 45% E-glass	1.6	250	150
Polyester and satin weave cloth laminate 55% E-glass	1.7	300	250
Polyester and continuous rovings laminate 70% E-glass	1.9	800	350
E-Glass epoxy composite	1.99	1770	—
S-Glass epoxy composite	1.95	2358	—

The composites have Young's modulus of 72–85GPa and Poisson's ratio 0.25.

damage. Laser cutting does, however, guarantee very accurate component sets to start with. Laser cutting can also be easily applied to sheets of acrylic to form lightweight reinforcement patches. Currently we try to restrict the use of such laser-cut parts to avionics base boards, wing ribs at the ends of hot-wire-cut foam parts, servo mountings, and horn reinforcement patches, see Figures 18.12 and 18.13.

Figure 18.11 Aircraft with wings fabricated from laser-cut plywood covered with aero-modeler film.

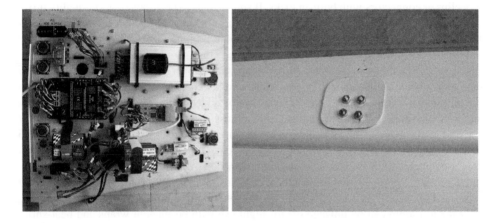

Figure 18.12 Avionics base board and servo horn reinforcement made from laser cut plywood.

Figure 18.13 Foam reinforcement ribs made from laser-cut plywood.

18.5   Wiring Looms

As has already been noted in previous chapters, the automated manufacture of customized wiring looms is generally not affordable for low-cost, low-volume UAV manufacture. Even if a fully geometrically and functionally detailed CAD representation is prepared, assembly still requires lengths of wire to be cut and soldered to connectors, plug, sockets, and so on, by hand. This can be aided in a number of ways, however. First, a fully detailed logical wiring diagram should be prepared, which notes the color and thickness/grade of each wire in the loom. This should also show terminations and plug breaks, See, for example, Figure 18.14. Then a full-scale printed plan view drawing of the aircraft showing all parts to be connected should be printed and glued to a plywood baseboard: the so-called iron bird, see Figure 18.15. At this point, wires can be cut and compared for length to the iron bird before connectors are soldered or crimped in place. Each wire should be labeled or color-coded during manufacture.

It is good practice to use only connectors that have an auxiliary locking mechanism to prevent disconnections being caused by accident or vibration. This can take the form of extra safety clips or connectors with built-in locks. If soldering is to be adopted, good-quality fluxed solder should be used, and the technician carrying this out should be provided with a dedicated soldering station with suitable cable holding clamps, see Figure 18.16. If crimped connectors are to be adopted, proper professional-grade crimping pliers should be used to form the crimps. In all cases, terminations should have shrink-wraps placed over exposed conductors and strain reliefs added where bulkheads will be penetrated or other chaffing or stress raisers might occur. Once terminated, each wire or group of wires can be added to the iron bird and the prototype loom gradually built up. When complete, the harness should be functionally tested by connection to all the avionics items to be fitted to the aircraft. We carry out interference and soak tests with spring-loaded servos and motor-driven generators and ignition systems on the iron bird, for example. If all is well, the loom can then be covered in cable wrap and test-fitted in the aircraft to see if any cable lengths need adjusting. At this point, either the test loom can be used to specify wire lengths and bundles for subsequent professional loom manufacture or the test loom can be finished off for flight use with permanent wraps and heat-shrink termination.

18.6   Assembly Mechanisms

By this stage, parts manufacture should be complete, and all that is required to produce the flight-ready aircraft is assembly with hand tools. Such assembly typically occurs in three phases. First, some components are assembled into the airframe structure with the assumption that they will never need removing. Second, some parts that are subsequently removed only for major service or replacement after failure, such as servos, receivers, motors, and so on, are assembled. Finally, some major components of the aircraft are usually designed to be separable for storage and transport. Thus on most UAVs it is rare to leave the wings or tail attached to the fuselage when flying is complete. These three categories of assembly tend to lead to different approaches when designing assembly mechanisms:

- Parts in the first category can be glued into place with epoxy resins or bolted in using cap-screws that are then completely inaccessible (we always apply thread-locking fluid to such screws).

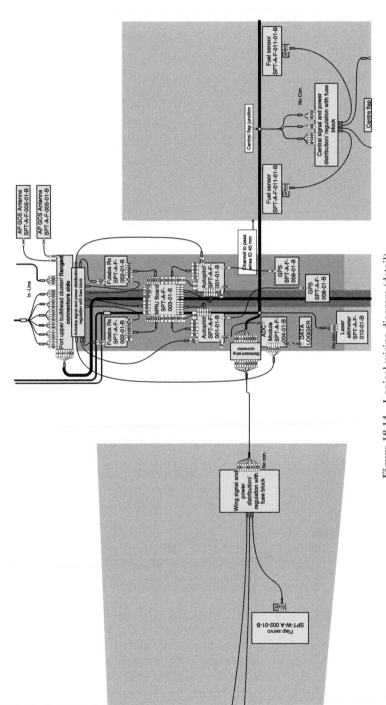

Figure 18.14 Logical wiring diagram (detail).

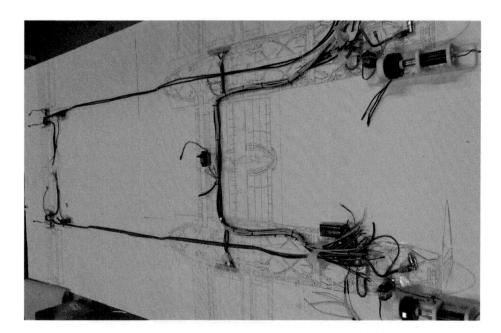

Figure 18.15   Iron bird for building wiring looms.

Figure 18.16   Soldering station (note the clamps, heat-resistant mat, and good illumination).

- Parts that may need removal for replacement or service must be fixed with screws, threaded rods, or 3D printed clips that facilitate removal. When designing the fixings for such parts, care must be taken to allow for access with suitable hand tools such as hex-drivers for cap-screws, and so on, and to get hands in to attach cables or pipes. Sometimes it is useful to make small access holes in the fuselage parts to facilitate this; alternatively, suitable hatches must be designed in.

- Parts that must be disassembled every time the aircraft is stored or transported should ideally require very few screws to be removed. We often use tool-free locking pins and clips and bayonet systems to facilitate assembly of the main components of the airframe.

18.6.1   Bayonets and Locking Pins

Since we now routinely build our fuselages from SLS nylon, and because the build chambers in the machines available to us are of limited size, we often find it necessary to build up our fuselages from multiple SLS parts. This has naturally led us to develop bayonet joining systems that allow parts to be effectively and quickly linked. We have then made best use of this need to provide access to the interior of our airframes. The bayonets we use require a 60° twist to engage, and we then prevent them from undoing by the insertion of quick-release pins, see Figure 18.17.

We often also use quick-release locking pins to secure the wings on our aircraft to the fuse-lage, see Figure 18.18. Here the main boom slides into a mating hole in the fuselage, and the

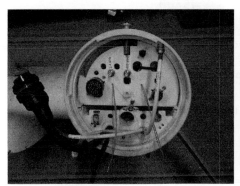

Figure 18.17   Female and male bayonet produced in SLS nylon with quick-release locking pin.

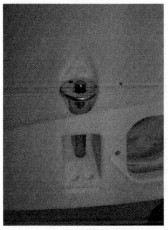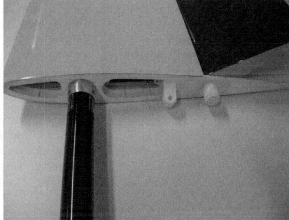

Figure 18.18   Quick-release pin fitting used to retain a wing to a fuselage (note lug on wing rib).

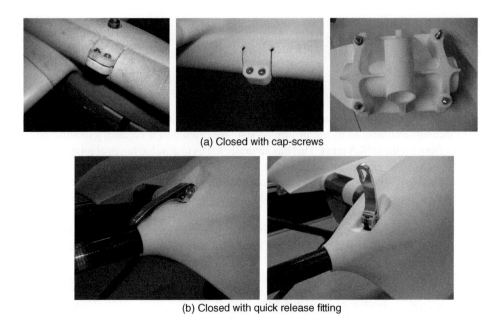

(a) Closed with cap-screws

(b) Closed with quick release fitting

Figure 18.19   SLS nylon clamping mechanisms.

two nylon lugs then locate in matching holes in the fuselage with one of the lugs being shaped to accept the retaining pin.

18.6.2   Clamps

To attach the main structural carbon fiber reinforced plastic (CFRP) booms and spars to our aircraft, we use clamping arrangements to avoid the need for drilling retaining holes in the parts. We find CFRP is not easily or cleanly drilled, so we avoid this where possible. Where the CFRP part is not meant for routine removal, we design such clamps to be closed with cap-screws, see Figure 18.19a. For clamps that retain wings or tailplanes, where these need to be repeatedly removed for storage and transport, we use quick-release systems so that the clamps can be undone without tools, see Figure 18.19b.

18.6.3   Conventional Bolts and Screws

To hold in parts that are not routinely being removed we make extensive use of stainless steel cap-screws and washers with nyloc nuts or plain nuts and thread-locking fluid. We often find it convenient to place the nuts in appropriately molded hexagonal holes in the SLS nylon and be retained there with cyanoacrylate glue so that only a cap-screw driver is needed to fix the bolt; it is often difficult to gain access to both ends of such screws to apply both driver and spanner, see Figure 18.20.

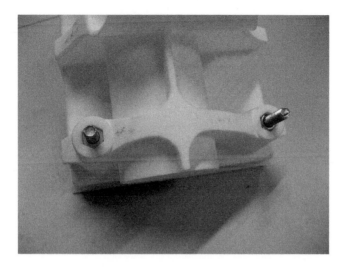

Figure 18.20 Cap-screws and embedded retained nuts, here on an undercarriage fixing point.

Figure 18.21 Transport and storage cases.

18.7 Storage and Transport Cases

Having assembled a new UAV that may represent a considerable investment in manufacturing cost and effort, we think it sensible to provide storage and transport cases, custom-sized to suit the airframe. Such cases are not very expensive and a number of companies around the world produce suitable items to supplied dimensions very rapidly. Ideally they should have armored corners and edges, locks, foam-lined interiors to match the airframe, and wheels and handles to facilitate handling. It is also convenient to have hinged lids with stays. Figure 18.21 shows two of our cases.

19

Regulatory Approval and Documentation

Having designed and built a new airframe, preferably backed up by calculations, wind tunnel tests, and simple experiments, the next stage in airframe development is a flight trials program. The aim of the trials is to establish that the whole aircraft system performs as expected and to establish limits on flying such as take-off performance, stall speed, and maximum maneuver loads. However, before any flying can take place, the trials team must ensure that they have appropriate regulatory approval for flights. Here we will describe a typical process based on our current experience. The process varies from one jurisdiction to another but all generally encompass the same main aspects. The end point of the process is a fully documented aircraft backed up by trials data and with appropriate approvals for operational flight. In general we do not seek full certification for our aircraft since this usually involves certification of the design, build, and operational organizations as well as the aircraft itself. Rather, we seek "exemptions" – that is to say, specific approvals from the regulatory authority that our flights can permitted by, or exempted from, the relevant air navigation laws in their jurisdiction.

19.1 Aviation Authority Requirements

Aviation authorities often classify aircraft in terms of their take-off weights and adjust the regulatory framework accordingly. In the UK, this is currently based on the dry maximum take-off weight (i.e., the maximum weight without fuel). Following such an approach, small aircraft of the type this book discusses are typically categorized into groups: for example, aircraft weighing less than 7 kg, those from 7 to 20 kg, and those from 20 to 150 kg. Generally for aircraft in the heavier range, the operators must supply significant additional material to the regulator to demonstrate the airworthiness of the airframe before any permission to fly will be granted. In this section we will therefore discuss aircraft in a notional 20–50 kg category since dealing with lighter airframes normally just involves a subset of these requirements.

Small Unmanned Fixed-wing Aircraft Design: A Practical Approach, First Edition.
Andrew J. Keane, András Sóbester and James P. Scanlan.
© 2017 John Wiley & Sons Ltd. Published 2017 by John Wiley & Sons Ltd.

Aircraft certification is a complex and, in the case of unmanned air systems (UASs),[1] a rapidly evolving field. The Aviation Authority always need to establish that the aircraft to be flown is itself airworthy and also that the operators can safely fly it, while ensuring that such flights do not infringe the relevant laws, particularly with respect to where the aircraft flies. Often, the Authority has separate divisions concerned with airworthiness and flight operations, and when planning to fly heavier aircraft, both divisions must be satisfied.[2] Generally, initial airworthiness is established by the submission of suitable documentation on the airframe, often followed by an inspection by suitably qualified staff – if the Authority is satisfied that the design and build team are appropriately experienced, they may, at their discretion, waive the need for airframe inspections. The operations division will be concerned with where flights are to be carried out and for what purposes. They will be particularly concerned with the skills of the pilot, the precise flight paths, and their likely duration. Sufficient details will need to be documented and supplied to convince the Authority that the intended flights are safe and legal. Major concerns will be the likelihood of any forced landings or excursions outside the planned flight airspace. To help operators produce satisfactory documentation, many authorities publish an "Operations Manual Template", which will be divided into major parts and numerous subject sections, see, for example, Tables 19.1–19.4. To meet these various requirements, our practice is to work on four separate but linked documents:

1. *A system description.* This describes the aircraft and the ground-based control systems and includes a full parts list.
2. *An operations manual.* This sets out how to set up, check, fuel, fly, recover, and maintain the aircraft and its GCSs.
3. *A safety case.* This details all the possible failure modes, what their likelihoods are, and any mitigations in place to reduce the scope for harm.
4. *A flight planning manual.* This describes the flight location, the operational team, the missions to be flown, and the data to be recorded.

In what follows we set out how we construct these documents, but it is stressed that this is not the only way of presenting information to the certifying authority – this is what works for us. By using a strict version control and release approval approach to these documents, it also helps us keep a record of the state of the aircraft at any given time (all our documents include tabulated version control lists, indicating what changes each new version has included and where these have been made in the manual). It is inevitable when developing a new airframe design that early flights will reveal changes in the design that will be beneficial in some way or another. The great temptation is to simply make these changes in an ad hoc and undocumented manner. Not only is this a poor approach to professional design but it also means that the airframe can stray away from the design approved by the regulators: clearly any significant changes must be agreed with the approving body *before* the aircraft is flown. Equally, the regulator will not wish every trivial change on the airfield to result in a renewed application for permission to fly. A degree of judgment will be called for, as will a good working relationship with the team at the regulating authority.

---

[1] The UAV and its ground control system or GCS.
[2] Note that all aircraft operators should consult their local aviation authority to establish current air regulations before flying.

Table 19.1 Typical small UAS operations manual template part Ai.

Section	Subject	Comment
Part A	Introduction	
1	Contents	Brief list of the OM contents
2	Introductory statement including outline of operations	Include statement of compliance with any permission and the requirement that operational instructions contained within the manual are to be adhered to by all personnel involved in the operation
3	Definitions	Include any common acronyms if necessary
4	Document control and amendment process	To ensure that the OM remains up to date and that different versions are not being used. Amendments should be sent to the Authority. Suggest including a version number
	Organization	
5	Structure of organization and management lines	Organogram and brief description
6	Nominated personnel	As appropriate, for example, Operations Manager, Technical Manager, Chief Pilot, Other Pilots
7	Responsibility and duties of the Person in Charge of the UAS	Articles 86, 87, and 166 of the UK Air Navigation Order, 2009, may provide some useful text for this section as determined by the operator (despite relevance to manned aircraft in the case of 86/87)
8	Responsibility and duties of support personnel in the operation of the UAS	Operators may use an assistant to help with the operation of the aircraft. Give a brief description of this person's responsibilities and duties
9	Brief technical description of UAS and roles	Full technical description can be in technical manuals or added as an appendix
10	Area of operation	Geographic scope, and so on. Likely operating areas – for example, building sites, open countryside, roads, and so on
11	Operating limitations and conditions	Minimum and maximum operating conditions in compliance with the regulations and conditions of any Authority permission

19.2 System Description

At its most basic level, our System Description sets out the leading particulars of the aircraft such as weight, wing span, powerplant, maximum flight speed, endurance, and so on. This is the bare minimum the regulator will wish to know; see, for example Table 19.5. In addition, we find it useful to work up from a complete parts list to describe the design logic, the whole

Table 19.2   Typical small UAS operations manual template part Aii.

Section	Subject	Comment
Part A	Operational Control	
12	Supervision of UAS operations	A description of any system to supervise the operations of the operator
13	Accident prevention and flight safety program	Include any reporting requirements (see, e.g., CAP 722)
14	Flight team composition	Make-up of the flight team depending on type of operation, complexity, type of aircraft, and so on
15	Operation of multiple types of UAS	Any limitations considered appropriate to the numbers and types of UAS that a pilot may operate if appropriate
16	Qualification requirements	Details of any qualifications, experience, or training necessary for the pilot or support crew for the types of UAS and the roles employed by the operator
17	Crew health	A statement and any guidance to ensure that the "crew" are appropriately fit before conducting any operations
18	Logs and records	Requirements for logs and records of flights for the UAS and by the pilots

airframe, and its GCS. We find it helpful to maintain this parts list in a dedicated database with a whole range of entries for each part from maintenance schedule, through suppliers and costs to possible failure modes and any failure mode mitigations: then each table in the documentation stack can simply link to the relevant entries in the database. This simplifies the upkeep of the manuals considerably. The System Description also includes the design Vn diagram (to indicate the design loads, see Chapter 14), the predicted performance and, as they become available, the actual performance results from the acceptance flight tests. When all acceptance tests are complete, the set of results from these tests, as recorded in the Systems Description, are a particularly important set of data for use by subsequent operators.

19.2.1   Airframe

When describing the airframe, we first set out the general configuration and summarize the manufacturing methods used (and as already described in earlier parts of this book). We then step through each major part of the airframe with a simple description, photographs, exploded views, cut-away drawings, and any particular notes on aspects that may not be obvious. For one of our large aircraft, this would include sections on the following:

1. *Fuselage*. Broken down into subsections such as nacelles, main fuselage, integral fuel tank, and so on;
2. *Wings*. Showing attachment points and the size and location of control surfaces;
3. *Tail*. Showing attachment points and control surfaces;

Table 19.3 Typical small UAS operations manual template part Bi.

Section	Subject	Comment
Part B	Operating procedures	
1	Flight planning/preparation	
1.1	Determination of the intended tasks and feasibility	
1.2	Operating site location and assessment	(a) the type of airspace and specific provisions (e.g., Controlled Airspace) (b) other aircraft operations (local aerodromes or operating sites) (c) hazards associated with industrial sites or such activities as live firing, gas venting, high-intensity radio transmissions, and so on (d) local byelaws (e) obstructions (wires, masts, buildings, etc.) (f) extraordinary restrictions such as segregated airspace around prisons, nuclear establishments, and so on (suitable permission may be needed) (g) habitation and recreational activities (h) public access (i) permission from landowner (j) likely operating site and alternative sites (k) weather conditions for the planned event using available information from aeronautical charts, the UK Aeronautical Information Service (www .ais.org.uk), digital imagery (Google Earth/Maps, etc.), Ordnance Survey maps, and so on
1.3	Risk management	Identification of the hazards, risk assessment, mitigating procedures
1.4	Communications	Contact numbers for other local aircraft operations
1.5	Pre-notification	If the flight is to be performed within an Aerodrome Traffic Zone, or near to any aerodrome or aircraft operating site, then their contact details should be obtained and notification of the intended operation should be provided prior to takeoff. It may be necessary to inform the local police of the intended operation to avoid interruption or concerns from the public
1.6	Site permission	Reference to document confirming land owners permission
1.7	Weather	Methods of obtaining weather forecasts. Consideration of UAS limitations
1.8	Preparation and serviceability of equipment and UAS	Pre-use checks and maintenance

4. *Undercarriage*. Illustrating attachment strong points and any suspension and steering components;

5. *Engines*. Including details such as those shown in Table 19.6;

6. *Payload*. Often describing a separate pod or bay on the aircraft and indicating the variations in payload that can be accommodated and how such items are fixed.

Table 19.4 Typical small UAS operations manual template parts Bii, C, and D.

Section	Subject	Comment
Part B	Operating procedures	
2	On-site procedures and Preflight checks	
2.1	Site survey	Visual check of operating area and identification of hazards
2.2	Selection of operating area and alternate	Size, shape, surrounds, surface, slope. Landing zone for an automatic "home" return should be identified and kept clear
2.3	Crew briefing	To cover the task, responsibilities, duties, emergencies, and so on
2.4	Cordon procedure	Adherence of separation criteria
2.5	Communications	Local and with adjacent air operations if appropriate
2.6	Weather checks	Limitations and operating considerations
2.7	Refueling	Or changing/charging of batteries
2.8	Loading of equipment	Security
2.9	Preparation and correct assembly of the UAS	In accordance with the manufacturers instructions
2.10	Pre-flight checks on UAS and equipment	May be covered in other technical manuals
3	Flight procedures	These procedures may be contained in the "operators manual" or equivalent but should cover all necessary matters including safety
3.1	Start	
3.2	Takeoff	
3.3	In flight	
3.4	Landing	
3.5	Shutdown	
4	Emergency procedures	
4.1	Appropriate to the UAS and control system	Should consider all those events that might cause the flight of the UAS to fail or be terminated
4.2	Fire	Risk and preventative measures should be considered relevant to the type of UAS power sources and fuel
4.3	Accidents	Considerations, responses, and so on
Part C	Training	
1	Details of the operator training programme	Training and checking requirements for pilots and support crew as determined by the operator to cover initial, refresher, and conversion syllabi
Part D	Appendices	
1	Copy of Authority Permission	This will provide immediate reference to the conditions under which the operations are to be conducted when applicable
2	Other documents	As considered necessary

Table 19.5 Typical summary airframe description.

Item	Characteristic	Units
Length	2.18	m
Wing span	3.92	m
Wing area	1.46	m^2
Aspect ratio	10.5	
Empty weight	23.7	kg
Maximum fuel capacity	8.4	l
Maximum fuel mass	6	kg
Maximum endurance (estimated)	5	h
Maximum payload mass	5	kg
Powerplant	2× OS GF40 with generator	

Table 19.6 Typical engine characteristics.

Item	Characteristic
Type	Four stroke
Power	3.75 hp
Displacement	40 cc
Bore	40 mm
Stroke	31.8 mm
Cylinders	1
System weight	1.2 kg
Engine-only weight	1.17 kg
Prop shaft thread	5/16 UNF
Propeller	18× 8 to 20× 10 (two blade)
RPM range	1800–9000 rpm
Fuel	1:50 Mix synthetic oil to unleaded petrol
Carb. type	Walbro WT1070

19.2.2 Performance

The performance section of the System Description will set out the specifications and performance of the aircraft in various modes of flight, but pay particular attention to takeoff and stall. It will also contain kinetic energy calculations for an unpremeditated descent (i.e., crash) or

Table 19.7   Typical aircraft performance summary in still air.

Item	Characteristic	Units
Maximum speed	41	m/s
Cruise speed	30	m/s
Stall speed clean	17	m/s
Stall speed take-off flaps	15	m/s
Stall speed landing flaps	12	m/s
Maximum flap extended speed	20	m/s
Take-off distance	40	m
Distance to clear 10 m obstacle	62	m

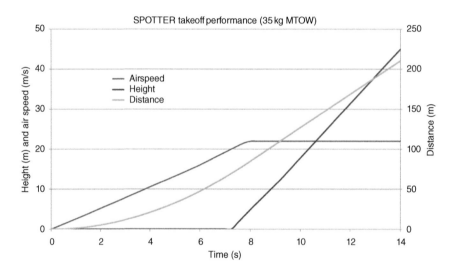

Figure 19.1   Typical take-off performance.

other loss of control to demonstrate that the total kinetic energy in such scenarios complies with the relevant laws (such as UK CAP722). Table 19.7 summarizes typical performance information, while Figure 19.1 illustrates take-off performance.

19.2.3   Avionics and Ground Control System

In our experience, provided the design team is reasonably competent and the aircraft under consideration is of a fairly conventional configuration, the regulators tend to be much more focused on the avionics and GCS of the UAS than the basic airframe of the UAV, since it is these aspects that ensure the aircraft avoids collisions, does not stray away from the designated flight area, and can be satisfactorily flown. If any form of autonomous operation is envisaged, the focus on the control system will be particularly intense. The regulator will wish to be satisfied that the pilot in command maintains full situational awareness and the ability to intervene if any mishap looks likely to happen. Our approach to this is based on a combination of using redundancy wherever we can along with as many mitigations as possible against causing harm. So, for example, we now choose to adopt twin-engine designs with duplicated control systems whenever possible on our larger UAVs.

When documenting the avionics, we always include a full set of wiring schematics and information on the radio links being used, see Figure 19.2 and Table 19.8. When using autonomous autopilots, we typically have separate radio channels for these in addition to those used for

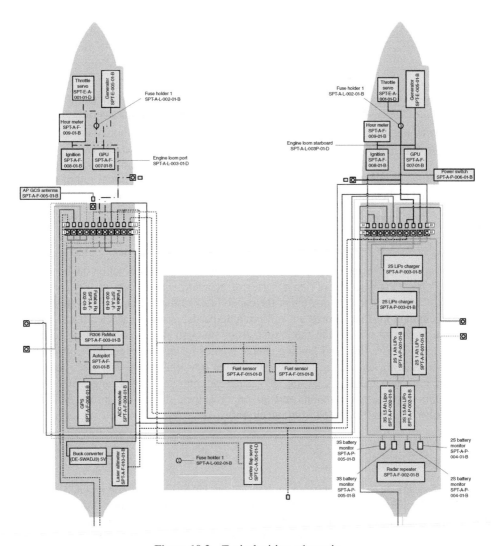

Figure 19.2 Typical wiring schematic.

Table 19.8 Radio control channel assignments.

Channel	Function	Type	Normal position	Failsafe position
1	Roll	Stick	As demanded	$2°–3°$ port
2	Pitch	Stick	As demanded	Neutral
3	Throttle 1	Stick	As demanded	Idle
4	Yaw	Stick	As demanded	Neutral
5	Flaps	Switch (flaps)	Off	Off
6	Throttle 2	Mixed to throttle 1	As demanded	Idle
7	AP On	Switch (AUX1)	As demanded	VLOS: off/EVLOS: on
8	RRS	Failsafe	Signal <1.5 ms	Signal >1.5 ms

the normal manual radio control systems all our aircraft carry, typically both being duplicated on the airframe. We generally do not duplicate the pilot's radio control transmitter since we consider switching transmitters during an emergency procedure to be more risky than having multiple receivers bound to a single transmitter. We generally do operate with two GCSs active at all times; however, since these are much more complex, we opt for a single radio link bound to the autopilots so that this can rapidly be switched between GCSs. Clearly, all of this must be documented in the manual and operational sequences set out in the OM.

19.2.4   Acceptance Flight Data

At the start of the flight trials program this section is blank, but as information on the airframe is recorded from flight tests, we add a series of standard sets of data to this part of the System Description since they document the capabilities of the airframe. To do this, we use an instrumentation kit that records air and ground speeds, altitudes, accelerations, attitude, and so on. Typically we include the results from standard radio control range tests plus take-off, climb, cruise, stall, and landing characteristics for the airframe, fuel consumption, idle, throttle change, and engine failure tests for the powerplant plus any tests carried out on the autonomous capabilities of the UAV such as communications range, auto-takeoff, auto-landing, way-point navigation, and so on. Together, these help inform subsequent operators of what can be expected from the aircraft.

19.3   Operations Manual

While the System Description sets out to describe the UAS (the UAV and its GCS), the OM details how the UAS should be operated in general terms. In our practice it does not detail actual flight plans and locations or individual personnel but does set out requirements and roles along with the procedures to be followed. We use a separate document to set out the details of particular flight plans, locations, and teams of operators: this allows for new flight plans and operator teams to be filed without the need to rewrite the OM.[3] We essentially follow the template of Tables 19.1–19.4 in setting out the manual.

19.3.1   Organization, Team Roles, and Communications

The roles and skill levels of the various people needed to conduct safe flights must be set out in the OM. This will include at least the pilot and the individual in charge of operations (in our practice, these are separate roles since the pilot must be free to focus on the aircraft and not be distracted by other aspects of operation). We would normally also add a system maintenance engineer and, if an autonomous GCS system is to be used, a suitably qualified operator in addition to the pilot. Sometimes it is useful to have assistants to enforce a cordon around the operational area. The manual should also set out how these individuals communicate with each other and those not involved in operations, such as spectators or general members of the public not involved in the flights. As noted in Table 19.2, consideration must be given to training, qualifications, crew health, and logs.

19.3.2   Brief Technical Description

We find it helpful to repeat a brief technical description in the OM since this document is typically the first one that is referred to on the airfield.

19.3.3   Operating Limits, Conditions, and Control

We next set out the basic types of flying the aircraft can be used for, such as the following:

1. Operation within visual line of sight (VLOS). In the UK, the aircraft must remain below 400 ft above local ground level and within a 500 m radius of the pilot, who must be able to maintain unaided visual contact.
2. Operation within VLOS of *two* pilots where handover is carried out between pilots who are not separated by more than 800 m. In the UK, the aircraft must remain below 400 ft above ground level and within a 500 m radius of one or other pilot, who must be able to maintain unaided visual contact while they are in charge of the aircraft. Suitable handover arrangements must be in place and agreed with the regulators.
3. Operation within extended visual line of sight (EVLOS). For example, the aircraft must remain below 400 ft above ground level and within a 1000 m radius of the pilot, who must be able to maintain visual contact to ensure that no other aerial vehicle is in the area and representing a possible collision hazard. In this mode, the pilot relies on autopilot operation of the aircraft for flight control since it will not be possible to have sight of control surface actions or small motions of the aircraft.
4. Operation beyond line of sight (BLOS). The aircraft can fly to arbitrary locations, but an acceptable process for dealing with collision avoidance and guarding against flying away from the planned flight path must be in place. Typically, this might involve the use of a

---

[3] Many regulators require updated details if operators, locations, or activities change significantly for the larger classes of UAVs.

designated danger area combined with external (radar) monitoring of the aircraft's location, or alternatively a network of observers who can maintain visual contact with the aircraft and have reliable radio links to the pilot.

In each case, we detail how the team will operate the aircraft in terms of the staff involved and the control systems fitted to the aircraft, typically ranging from a simple, hand-held model aircraft transmitter to a fully configured flight operations vehicle with multiple staff and a range of communications systems and channels.

19.3.4 Operational Area and Flight Plans

This section sets out the typical requirements for the airfield and the nature of any flight plans that must be defined prior to flying. We leave specific details of actual airfields and flight plans to the Test Flight (or other Flight Planning) Manual: this decouples the basic aircraft documentation from what is needed as each new flight scenario is considered (since we typically do not achieve full aircraft certification, rather relying on individual exemptions for each set of flights, this proves an effective way to manage the documents[4]). As noted in Table 19.3, consideration must be given to a wide range of issues associated with the site, risks, communications, permissions to fly (other than from the main regulator), and any weather limitations.

19.3.5 Operational and Emergency Procedures

The bulk of the manual is then devoted to detailed instructions on flying the aircraft, see Table 19.4. This starts with the agreed crewing, flight plan, and location and works through airfield and weather checks, airframe assembly/pre-flight checks, fuelling and engine starting, takeoff, flight, landing, shutdown, and any post-flight checks or disassembly procedures. We make extensive use of checklists, which are also copied to laminated waterproof sheets that the flight crew can mark up as the items concerned are dealt with on the flying field.

The manual will also include a series of well-established procedures for dealing with all foreseeable emergency conditions (it is, of course, vitally important that in the event of an emergency the whole operational team know what their roles are and have to hand any contact information and means of communication to the emergency authorities). We always include emergency sections for unplanned or forced landing/ditching of the aircraft, excursions above the allowed operational ceiling/beyond the allowed operational radius, or other "fly-away," fire, and injury to anyone involved.

19.3.6 Maintenance Schedule

The parts list in the System Description will include a full list of all components used in the airframe down to the last nut and bolt, along with a complete maintenance schedule. For some

---

[4] To achieve a fully certified airframe, regulators generally require the certification of the design, manufacture, and maintenance organizations involved, and this is typically beyond the scope of small UAS teams.

parts, this will be a simple visual check that no visible damage has occurred, while for others, such as the engines, a service cycle based on running hours will be specified. The OM will refer to this and note how and by whom maintenance is to be performed and what logs are to be kept. Engine maintenance logs, battery charging logs, and control system update logs are particularly important. We do not generally repeat the full parts list in the OM.

19.4   Safety Case

The purpose of the Safety Case is to set out a risk assessment process, detail all the perceived failure mechanisms, list the operational hazards such failures may give rise to, any mitigations that have been put in place, and then summarize all this in an accident sequence table. This table starts with each hazard, notes the consequences and how these might arise, what the risk is, and then by referring to the appropriate mitigations in place (if any) establishes the residual risk in each case. Every effort must be made to step through all possible scenarios to ensure as complete a risk assessment as possible is in place. Tables 19.9 and 19.10 set out the probability and severity definitions we use, while Tables 19.11 and 19.12 define our risk classification matrix and risk classes.

Table 19.9   Risk probability definitions (figures refer to flight hours).

Probability	Occurrence		
Frequent	Likely to occur many times	Several times a month	Probability 10^{-2}
Likely	Probable; likely to occur often	Monthly	Probability 10^{-3}
Infrequent	Likely to occur several times	Annually	Probability 10^{-4}
Seldom	Likely to occur at some time	Once in several years	Probability 10^{-5}
Unlikely	Improbable; may exceptionally occur	Unlikely	Probability 10^{-6} or less

Table 19.10   Accident severity definitions.

Severity	Definition
Catastrophic	Aircraft/vehicle destroyed; multiple deaths
Major	Major damage/serious incident; loss of life or multiple severe injuries/occupational illnesses
Moderate	Moderate damage/significant incident; several severe injuries/occupational illnesses
Minor	Minor damage/minor incident; a single severe injury/occupational illness, and/or minor injuries/occupational illnesses
Negligible	Little consequence; a single minor injury/occupational illness

Table 19.11   Risk classification matrix.

Severity	Probability				
	Frequent	Likely	Infrequent	Seldom	Unlikely
Catastrophic	A	A	A	B	C
Major	A	A	B	C	C
Moderate	A	B	B	C	C
Minor	B	B	C	C	D
Negligible	B	C	C	D	D

Table 19.12   Risk class definitions.

Risk class	Occurrence	
A	Intolerable/extremely high	Requires urgent action/risk reduction
B	Undesirable/high	Requires action, and should only be accepted when risk reduction is impracticable
C	Tolerable/medium	Acceptable
D	Acceptable/low	Acceptable with no unnecessary risks

19.4.1   Risk Assessment Process

The risk assessment process starts with a typical failure effect, such as a loss of engine power, notes a cause, for example, an engine stalling (specifying which aircraft parts are involved and any warnings or automatic procedures in place), and then defines how probable this is (such as "likely," see Table 19.9). Each failure effect is then mapped to an operational hazard, for example, the aircraft might start to loose altitude, so that the probability of each hazard occurring can be assessed, with each hazard then in turn being mapped on to possible accidents. The resulting accident list then records which hazard leads to which accidents and their severity in terms of the outcomes in Table 19.10. Finally, given this information, the accident sequence and mitigation table can be constructed to establish the residual risks associated with each hazard using Tables 19.11 and 19.12. The following subsections illustrate this process for typical effects, hazards, accidents, and mitigations.

19.4.2   Failure Modes and Effects

A failure modes and effects analysis (FMEA) is normally conducted at a component level and subsequently at a system level for each new design. This is used to establish a failure effects list. For each part in the parts list, as detailed in the System Description, this list contains

Table 19.13 Typical failure effects list (partial).

Part no.	Description	Possible failure mode	Mitigation	Failure probability given mitigation, inspection, and service	Risk
SPT-A-F-006-01-B	Autopilot GPS antenna	Loss of signal from satellite	Autopilot initialization sequence indicates lack of signal before takeoff. Autopilot can maintain control if GPS signal is lost in flight. Manual override is an option	Unlikely	C
SPT-A-F-007-01-B	Generator power unit (GPU)	Damage due to overvoltage, overheating, or impact	Overvoltage highly unlikely if correct ground power supply battery is used. Heat sink and fan help prevent overheating. Second GPU provides redundancy	Unlikely	C
SPT-A-F-008-01-B	IG-04 Ignition unit	Overvoltage or disconnection from/damage to loom.	Use high-quality connectors, do not overstress loom. Telemetry warns GCS operator if engine fails in flight. Second engine allows aircraft to return to base	Unlikely	C

the following entries: part no., description, possible failure mode, mitigation, failure probability given mitigation, inspection and service intervals, and risk. If all the UAS manuals refer to a common parts database, errors and duplication are avoided when constructing this list. Table 19.13 shows a short section from such a list.

19.4.3 Operational Hazards

A list of generic operational hazards is then produced based on the Failure Effects list, System Description documentation, analysis of previous operations, and risk assessment meetings, see, for example, Table 19.14. These should include ground operations and flight areas beyond the anticipated operational flight envelope. It is also normally assumed that all operations will be conducted at a dedicated airfield with perimeter fence, controlled access, and away from main roads, centers of population, and limited or controlled air movements. If it is likely that there will be large crowds, major roads, or conurbations within

Table 19.14 Typical hazard list (partial).

HZ-ID	Hazard	Relevant failure effects	Summary	Probability	Possible accidents
HZ-01	Aircraft fails to achieve flight and ditches	FE-01, FE-02, FE-03, FE-04, FE-05, FE-06, FE-10	Aircraft fails to reach a safe height during takeoff and ditches on the airfield	Infrequent	AC-01, AC-05
HZ-02	Recovery to unprepared location	FE-01, FE-02, FE-09, FE-11, FE-12.1, FE-12.2, FE-12.4, FE-12.3a, FE-12.3b	The aircraft lands at a location that is unprepared, that is, the desired landing area is obstructed	Infrequent	AC-02
HZ-03	Aircraft moving on ground	n/a	If the aircraft is not secured then it will travel if under power, potentially exposing people to risk. This hazard applies during all takeoffs and landings	Likely	AC-02

the operational area, then the severity of a crash landing might become "catastrophic" and additional mitigation will be needed before the regulatory authority is likely to approve flights. If there are likely to be other local air movements, then impact with other air vehicles is an increased risk and the pilot will need to obtain permission for flight from the local airfield control.

19.4.4 Accident List

The accident list should contain a complete list of accidents that could potentially occur as a result of the hazards identified, see, for example, Table 19.15.

19.4.5 Mitigation List

The mitigation list should detail the steps taken to reduce the probability of a hazard resulting in an accident or reduce the severity of an accident; see, for example, Table 19.16. We normally subdivide these into five aspects with an appropriate naming convention for each as follows:

Table 19.15 Typical accident list (partial).

AC-ID	Accident	Relevant HZ-ID	Summary	Severity
AC-01	Injury or death during takeoff	HZ-01	Death during the take-off phase, that is, from point when takeoff is commanded to point when aircraft is above head height. Exposure time approximately 10 s for each takeoff	Major
AC-02	Injury or death during landing	HZ-02	Caused by collision of aircraft with people on the ground while the aircraft is landing at a chosen location. Exposure time approximately 10 s for each landing	Major
AC-03	Injury or death while aircraft is on the ground	HZ-04	Caused by collision of aircraft with people on the ground while the aircraft is not in flight	Major
AC-04	Death due to mid-air collision	HZ-09, HZ-10, HZ-06, HZ-05	This requires collision with a manned aircraft which results in the death of one or more of its occupants. Otherwise the accident is covered by AC-05 due to debris from the aircraft	Catastrophic
…				
…				
AC-10	Personnel accident due to exposure	HZ-18	This describes the situation where personnel suffer from an occupational illness due to wind chill and exposure	Minor
AC-11	Damage to local surroundings	HZ-06, HZ-07, HZ-08, HZ-09	Damage to man-made surroundings (structures, vehicles, or vessels) or natural environment due to collision, fire, release of hazardous materials from aircraft/ancillary equipment, or debris from aircraft crash	Moderate

1. *MT-D-##*. Mitigations implemented during the design of the aircraft;
2. *MT-P-##*. Mitigations implemented during the planning of flight operations;
3. *MT-O-##*. Mitigations implemented while operating the system;
4. *MT-E-##*. Mitigations implemented through care of equipment and
5. *MT-H-##*. Mitigations to ensure health and safety of personnel.

Table 19.16   Typical mitigation list (partial).

MT-ID	Mitigation	Reference or information
MT-D-01	Dual redundancy of avionics and batteries	System description
MT-D-02	Dual redundancy of propulsion system	System description
MT-D-03	Multiple servos and control surfaces per axis	System description
...		
MT-P-01	Select site that allows a choice of take-off/landing directions to suit wind direction	Operations manual
MT-P-02	Safe climb and approach route within designated area	Operations manual
...		
MT-O-01	Thorough pre-flight checks	Operations manual
MT-O-02	Takeoff and flight only in favorable meteorological conditions	Operations manual
MT-O-03	Limited fuel load: do not carry fuel far in excess of that required for the planned flight plus reserve	Operations manual
...		
MT-E-01	Aircraft stored and transported in dedicated flight cases	Operations manual
MT-E-02	Transmitter and GCS protected from damage (impact, moisture, sand, etc.) during transport/storage and operation	Operations manual
MT-E-03	Post-operation checks of equipment	Operations manual
...		
MT-H-01	Continual assessment by HSE manager	Operations manual
MT-H-02	Warm and wind-proof clothing	Operations manual
MT-H-03	Ear protection to be worn if in close proximity to ground-running aircraft	Operations manual

19.4.6   Accident Sequences and Mitigation

Having constructed the failure effects, hazard, accident, and mitigation lists, it is then possible to bring all this information together to detail the accident sequences and mitigation list; see, for example, Table 19.17. Again, the use of suitable databases will help manage this process: in our work we typically have around 20 hazards, over 10 accident types, and more than 50 mitigations in place and so bringing these all together in the final sequences and mitigation list is a nontrivial task. Note that each hazard may lead to several accidents and that each hazard/accident combination may be subject to multiple mitigations when arriving at the Residual Risk. Also, if any of the Residual Risks is above a category C, further mitigation should be put in place unless this is completely impractical.

Table 19.17 Typical accident sequences and mitigation list (partial).

HZ-ID	Hazard	AC-ID	Accs.	Risk	Accident sequence	MT-ID	Mits.	Res. risk
HZ-03	Aircraft moving on ground	AC-02	Injury or death during landing	A	Aircraft overruns during takeoff or is not arrested during landing and continues on to impact personnel. For this accident sequence to occur, personnel must be in proximity to the moving aircraft in the over-run area	MT-P-06	Select site with limited public access	C
	Likely		Major		It is considered highly unlikely that personnel would be present on the over-run area during landing. The probability of the accident sequence, given the probability of the hazard, is assessed as Seldom.	MT-P-07	Safe area designated for personnel not directly involved in aircraft operation	
HZ-04	Aircraft unsafe on ground	AC-03	Injury or death while aircraft is on the ground	B	The aircraft or propeller strikes personnel while not in flight. For this accident sequence to occur, personnel must be in proximity to the moving aircraft or spinning propeller. Their awareness of its location and condition (training, procedures) is a factor	MT-P-06	Select site with limited public access	C
	Infrequent		Major		The nature of the injury will be also affected by a number of factors including the force of impact. The probability of the accident sequence, given the probability of the hazard, is assessed as unlikely	MT-P-07	Safe area designated for personnel not directly involved in aircraft operation	

19.5   Flight Planning Manual

The final manual in our documentation set is the Flight Planning Manual. During initial development and test flying of a new airframe and GCS, this sets out the full sequence of acceptance tests we intend to carry out along with the test location and team who will be doing the flying. Then as each test in the sequence is successfully completed, we migrate the results to the System Description so that, on completion of all acceptance tests, all the results are completely moved into the System Description. For subsequent operational flying, a completely new document is then produced for each mission, which sets out purpose, activities, location, flight crew, and so on. This is generally produced on a case-by-case basis for submission to the regulators and can be quite brief for simple flight missions.

In setting out flight manuals, we often duplicate a short description of the system at the beginning of the manual for convenience. Next we include details of the flight location, specifying full address, GPS coordinates, and local map references (in the UK we use the Ordnance Survey grid reference system). We also include an aerial view of the location showing the normal boundaries for flight (for VLOS flying in the UK, this is currently a 500 m radius circle around the pilot's station) and any specific out-of-bounds areas (the area where spectators may be standing, nearby buildings or other runways, etc.). We also set out site-specific emergency procedures, including all relevant contact telephone numbers for the local emergency services and repeat a summary of the no-go conditions for which flying is prohibited (such as weather or failure to complete flight checks). Next, a short set of general procedures are laid down, referencing the more extensive operations manual:

1. *Aims and general procedures.* Why the flight is being conducted, under what authority, maximum flight times, and so on;
2. *FCS conditions of operation.* Which modes of control will be used (fully manual flights, autopilot-assisted flights, fully automated flights, etc.);
3. *Communication between UAV pilot and GCS operator.* A summary of the protocols being used between pilot and ground station operator;
4. *Conflicting air traffic.* Actions to be taken if any crew member sees a potential air traffic conflict;
5. *Flight data.* The data to be recorded on each flight together with the means for capturing it;
6. *Flight control system software management.* The processes to be used in setting either the manual transmitter or GCS control logics (since one may have multiple aircraft, it is imperative for the team to ensure that the correct logics are being used for each aircraft).

Then the actual flight crew are detailed, and since this document may be used to seek regulatory approval, a short CV of each crew member is specified. Next, the individual test or mission details are provided. For acceptance purposes we typically include the tests set out in Chapter 20. We use the on-board autopilot to measure a great many aircraft characteristics, but if this is not possible, some other means of measuring speeds, accelerations, angles, climb rates, and so on, will be needed. For each flight, standard records are kept of the participants, location, date, time of start, duration, forecast and measured wind speeds and directions, fuel at start and finish, and payload on board, together with any notes or comments arising during the flight. All flights should begin with standard pre-flight and center of gravity (CoG) checks. All flight test records should be countersigned by the relevant design authority before being accepted for use in the system description.

20

Test Flights and Maintenance

20.1 Test Flight Planning

When developing a series of test flights, we first consider what we are attempting to measure and the safest order in which to carry out these flights. By definition, test flying is bound to be a higher risk activity than simple operation once all the system behaviors have been established. We always carry out initial tests at a properly controlled test airfield and with an experienced test pilot. We have for many years worked with Paul Heckles of the Paul Heckles Flight Centre.[1] Paul is an enormously experienced pilot who has on many occasions managed to cope with unexpected events during flight testing and bring our aircraft safely back to the runway. He has also spent many hours helping our students learn the importance of a disciplined approach to flight testing while still making the experience a great deal of fun. Here we set out our approach to planning test flights.

20.1.1 Exploration of Flight Envelope

The most fundamental aim of test flying is to establish the safe flight envelope of the aircraft. This should, of course, be very similar to that planned for during the design process, but it is only by carrying out flight tests that this can be finally assured. There are a number of fundamental flight behaviors that must be established for all aircraft:

1. confirmation of center of gravity location
2. control response
3. overall performance measurement
4. stall angle
5. stall behavior
6. measuring lift/drag by glide testing
7. propulsion system behavior and endurance
8. calibration of pitot static system.

[1] http://www.paulhecklesrc.co.uk/

Some of these tests will be carried out on the ground, either with the aircraft suitably restrained or by fast taxiing tests. Once these are completed flights can begin.

20.1.2   Ranking of Flight Tests by Risk

It is of course vital that tests start with the simplest action and using only the most conservative setting of control surface deflections. Then, once the pilot and team start to gain confidence, more demanding test sequences can be attempted. We certainly would not begin to carry out any autopilot-controlled flying until an exhaustive set of manual test flights in close proximity to the pilot and in benign weather conditions had been completed.

20.1.3   Instrumentation and Recording of Flight Test Data

In all operations of unmanned air systems (UASs), it is important that good-quality logs and records are kept, see Figure 20.1 for example. This matters as much for routine operation as it does for tests during system development. Clearly, details need to be kept of airframe and engine hours to enable the specified maintenance schedule to be followed, as do details of charging cycles on any batteries that are repeatedly charged and discharged. In addition to normal pilot logs, during test flying we typically log all autopilot data to onboard secure digital (SD) cards and/or to the ground station computer for subsequent analysis. This should allow for positional and acceleration data to be logged as well as control surface inputs. We have also found it useful to attach small, self-contained video cameras to a wing tip and rudder

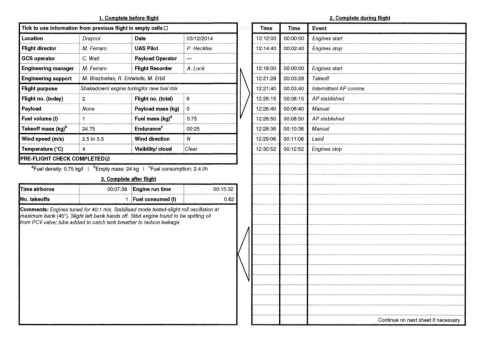

Figure 20.1   Typical flight log.

tip so that footage of the aircraft's behavior is readily available and can be correlated against the autopilot data. Should any failure in-flight or during taxi tests then occur, it will be much simpler to understand the nature of what has happened. If the aircraft is being flown under a regulatory authority permission, this will also assist in filing any incident reports that may be needed. To coordinate all this, it is useful to designate an airframe engineering role that is distinct from the other flight team roles to ensure that suitable data is properly curated.

20.1.4   Pre-flight Inspection and Checklists

Before commencing any flight trials, it is important to ensure that the airfield and aircraft are all as they should be. It is now standard practice in the aviation industry to do this by way of a series of well-established and routine checklists. These should cover the following:

1. *Departure*. Checks at base carried out before even leaving for the airfield.
2. *Post-assembly*. Carried out on the airframe once it has been put together having been removed from its transport cases or hangar.
3. *Pre-flight*. The standard checks on the aircraft before each flight; see, for example, Figure 20.2.
4. *Flight procedures*. An overall list of checks that pertain to the type of test flight to be carried out; see, for example, Figure 20.3.
5. *Site*. To ensure that the airfield in use is fully understood and meets all the requirements for safe operation.

20.1.5   Atmospheric Conditions

Atmospheric conditions play an important role in all flight operations. Therefore, before flying, suitable measurements of prevailing conditions for the flight test should be made and recorded. At the most basic level, this will be the wind speed and direction: at the very least a small hand-held anemometer should be available to the team. Pressure and humidity can also be useful in understanding engine behavior and limits on glide angle (lift and drag are of course affected by air density). The team should also have defined the worst case weather under which the test may go ahead. This should be done before reaching the flying field so that nobody is tempted to test-fly in marginal conditions.

20.1.6   Incident and Crash Contingency Planning, Post Crash Safety, Recording, and Management of Crash Site

Although every step should be taken to ensure safe and uneventful tests, it remains the case that all forms of flight can result in unplanned aircraft behavior. This can vary from brief losses in control due to radio link failures right through to crashes and the total destruction of the aircraft. It is therefore good practice to have suitable contingency plans in place along with procedures for dealing with crashes should they occur. If a sensible series of test flights have been mapped out and good safety procedures are in place, even when a major crash happens it should not lead to any injuries to people on the ground or damage to third-party

SPOTTER pre-flight checklist version 1.04

A General
- [ ] 1 Post-assembly check completed
- [ ] 2 Centre of gravity within limits

B Fuel
- [ ] 1 Sufficient fuel for planned flight
- [ ] 2 Fuel cap in place and tightened

C Manual RC check
- [ ] 1 RC Transmitter ON
- [ ] 2 Transmitter battery voltage sufficient (above 6 V)
- [ ] 3 Correct model memory selected (SPOTTER)
- [ ] 5 Aircraft power switches ON
- [ ] 6 Check control throws and directions are correct (including throttle)
- [ ] 7 If secondary transmitter is used- repeat steps 1–8
- [ ] 8 If secondary transmitter is used- test handover functionality

D Autopilot check
- [ ] 1 Ground station battery or power supply sufficient for planned operations
- [ ] 2 Ground station and autopilot comms ON
- [ ] 3 Set waypoints in SkyCircuits Plan or Toolkit software and upload to autopilot
- [ ] 4 Check and confirm that positions of waypoints match GPS detected positions
- [ ] 5 Check SkyCircuits FLIGHT software is connected to the autopilot
- [ ] 6 Check for normal autopilot startup sequence (full and free control surface movements)
- [ ] 7 Set airspeed to zero by pointing aircraft into wind and partially enclosing both Pitot tubes

E Payload check (if a new one is fitted)
- [ ] 1 Power connected as necessary
- [ ] 2 Front and rear retaining bolts in place
- [ ] 3 Rear pylon fairing in place and secure
- [ ] 4 Check payload functionality

Figure 20.2   Typical pre-flight checklist.

structures or equipment. The whole purpose of the first few flight tests is to flush out any unexpected behavior or weakness in the aircraft in a setting that will not lead to a major incident.

If, however, a crash does occur, a known crash procedure should be invoked. Provided a thorough test program is being followed, it is most likely that such an event will lead to the aircraft impacting the ground within the perimeter of the test site and well away from those conducting the test. It is the essential that precautions are taken to deal with any hot or hazardous materials at the crash site. These will most likely be engine-, fuel-, and battery-related. Suitable fire extinguishers should be readily available and the test team knowledgeable in their use. Provided the team members are sure that there is only a very low

SPOTTER flight procedures checklist version 1.00

A Prestart

☐
1 Pre-flight check successfully completed
2 RC transmitter(s) ON
3 Ground station ON
4 Aircraft power switches ON
5 Check full and free control surface movements with correct directions

B Engine start

☐
1 Aircraft moved clear of safe area
2 Aircraft restraint in place
3 Team member in place behind port nacelle to operate choke and support aircraft
4 Team member in place in front of port engine with electric starter
5 Call "clear prop" and ensure people are out of prop plane
6 Use starter to turn port engine. Apply choke and throttle as necessary
7 Pilot to set throttle to idle
8 Team members to re-position ready to start starboard engine
9 Use starter to turn starboard engine. Apply choke and throttle as necessary
10 Check both engines stable at idle
11 Check both engines stable at full throttle
12 If necessary, tune carburettor and re-check idle and full throttle

C Power check

☐
1 Face aircraft in to wind and restrain
2 Ensure surrounding area is clear of persons, objects and debris
3 Check both engines stable at idle
4 Check both engines stable at full throttle
5 Check full and free movement of controls

D(M) Takeoff (manual)

☐
1 RC transmitter set to autopilot OVERRIDE
2 GCS connected and operator ready
3 All GCS set controls to REMOTE
4 Check battery voltages via autopilot telemetry
5 Set QFE
6 Autopilot airspeed set and indicating correctly
7 Check takeoff area and local airspace are clear and unobstructed. All personnel should be at least 20 m away from aircraft (except pilot)
8 Check wind speed and direction, confirm suitability of takeoff direction
9 Set flaps as required
10 Pilot to announce and perform takeoff when ready and satisfied

Figure 20.3 Typical flight procedures checklist.

risk of fire, attempts should be made to disconnect and isolate such items from the rest of the crash site. Photographic records should also be taken. When the team members are sure that there are no remaining hazards and all has been recorded, the debris should be gathered up and the area thoroughly cleaned. Lastly, all incidents should be written up and the relevant aviation authority contacted if the aircraft is being flown under an authority-issued permit or exemption.

D(A) Takeoff (auto)

☐ 1 GCS connected and operator ready
☐ 2 Set RC transmitter to AUTO
☐ 3 Set all GCS controls to AUTO
☐ 4 Check battery voltages via autopilot telemetry
☐ 5 Set QFE
☐ 6 Autopilot airspeed set and indicating correctly
☐ 7 Check takeoff area and local airspace are clear and unobstructed. All personnel
   should be at least 20 m away from aircraft (except pilot)
☐ 8 Check wind speed and direction, confirm suitability of takeoff direction
☐ 9 Pilot to announce takeoff when ready and satisfied
☐ 10 Ground station operator to invoke auto takeoff

E In flight

☐ The pilot and operator should follow the agreed flight plan. Deviation should only
occur if a significant factor changes (such as unexpected weather, equipment
malfunction, intrusion by other aircraft, etc.). If a problem is suspected, it is prudent
to abort the mission, land and diagnose. Flight recorder to note timings and
significant events

F(M) Landing (manual)

☐ 1 Set RC transmitter to autopilot OVERRIDE
☐ 2 Set all GCS controls to REMOTE
☐ 3 Set flaps as required
☐ 4 Check wind speed and direction, confirm suitability of landing direction
☐ 5 Pilot to call "landing" to alert ground crew
☐ 6 Check landing area is clear and unobstructed. All personnel should be at least 20 m
   away from aircraft (except pilot)
☐ 7 Pilot to perform landing and bring aircraft to a stop

F(A) Landing (auto)

☐ 1 Set RC transmitter to AUTO
☐ 2 Set all GCS controls to AUTO
☐ 3 Set flaps as required
☐ 4 Check wind speed and direction, confirm suitability of landing direction
☐ 5 Pilot to call "landing" to alert ground crew
☐ 6 Check landing area is clear and unobstructed. All personnel should be at
   least 20 m away from aircraft (except pilot)
☐ 7 GCS operator to invoke auto landing
☐ 8 Pilot to manually bring aircraft to a stop if nececessary

G Shutdown

☐ 1 Pilot to taxi aircraft to a safe part of the site
☐ 2 Pilot to STOP both engines
☐ 3 Check with GCS operator that the aircraft can be powered down
☐ 4 Aircraft power switches OFF
☐ 5 RC Transmitter OFF
☐ 6 Ground station OFF

Figure 20.3 *(Continued)*

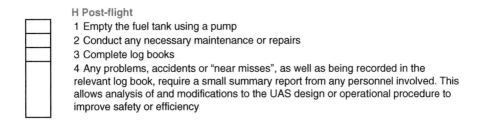

H Post-flight

1 Empty the fuel tank using a pump

2 Conduct any necessary maintenance or repairs

3 Complete log books

4 Any problems, accidents or "near misses", as well as being recorded in the relevant log book, require a small summary report from any personnel involved. This allows analysis of and modifications to the UAS design or operational procedure to improve safety or efficiency

Figure 20.3 (*Continued*)

20.2   Test Flight Examples

We next list a typical set of test flights for a 25 kg UAV and ground control station (GCS) capable of full autonomous takeoff, mission operation, and landing. The sequence follows the logic just set out and will provide all the information needed to complete the system description that will be used with the completed UAS.

20.2.1   *UAS Performance Flight Test (MANUAL Mode)*

The very first set of self-powered tests seek to assure the design team that the airframe has basic airworthiness.

1. Ground handling. (a) Aircraft speed and direction controllable on ground, (b) UAV can idle and remain stationary on a grass surface suitable for takeoff without intervention, (c) control can be maintained in a fast taxi situation, (d) undercarriage is serviceable, and (e) no loss of aircraft control signal.
2. Aircraft handling and performance at takeoff, low-speed flight, cruise, high-speed flight, and landing to a satisfactory level for the UAV pilot to continue with flight testing.
3. With aircraft in chosen take-off configuration with flaps at specified angle, measure the take-off run. Keep flaps in take-off (mid) position. (a) Take-off run measured, (b) take-off run to clear 10.7 m obstacle, (c) autopilot-measured average climb rate recorded within the first 5 s after takeoff, and (d) aircraft climb average airspeed measured by autopilot.
4. Aircraft to clean configuration. Climb to height over airfield. (a) Max. clean climb rate into wind measured by autopilot.
5. Straight and level clean cruise speed measured. (a) full control inputs on ailerons, rudder, and pitch, both directions. Confirm response is adequate and equal in both directions, (b) fly UAV within approach to clean stall; check full control authority, and (c) autopilot-measured air speed.
6. Aircraft clean max. cruise speed measured by autopilot (no flutter seen). (a) Apply suitable control inputs to confirm handling suitable.
7. Gliding. Clean setup over the airfield at maximum suitable height. (a) Maintain straight and level into wind, (b) autopilot-measured average descent rate, and (c) autopilot-measured approximate airspeed.
8. Gliding. Take-off flap set up over the airfield at a suitable height. (a) Maintain straight and level into wind, (b) autopilot-measured average descent rate, and (c) autopilot-measured approximate airspeed.

9. Gliding. Landing flap setup over the airfield at suitable height. (a) Maintain straight and level into wind, (b) autopilot-measured average descent rate, and (c) autopilot-measured approximate airspeed.

10. Steep turns. Clean setup. At safe height and suitable maneuvering speed. (a) Autopilot airspeed measured, (b) maximum angle of bank achievable at constant height level to the left at approximate autopilot airspeed and approximate autopilot measured throttle position, and (c) maximum angle of bank achievable at constant height level to the right at approximate autopilot airspeed and approximate autopilot measured throttle position.

11. Steep turns. Take-off flaps. At safe height and suitable maneuvering speed. (a) Autopilot airspeed measured, (b) maximum angle of bank achievable at constant height level to the left at approximate autopilot airspeed and approximate autopilot measured throttle position, and (c) maximum angle of bank achievable at constant height level to the right at approximate autopilot airspeed and approximate autopilot measured throttle position.

12. Steep turns. Landing flaps. At safe height and suitable maneuvering speed. (a) Autopilot airspeed measured, (b) maximum angle of bank achievable at constant height level to the left at approximate autopilot airspeed and approximate autopilot measured throttle position, and (c) maximum angle of bank achievable at constant height level to the right at approximate autopilot airspeed and approximate autopilot measured throttle position.

13. Stalling clean. Climb to height then maintain straight and level. (a) Autopilot height measured and (b) idle power stall into wind – stall speed and height loss.

14. Stalling clean climb. Climb to height. (a) Full power stall into wind – stall speed and height loss.

15. Stalling take-off flaps. Climb to height then maintain straight and level. (a) Autopilot height measured and (b) idle power stall into wind – stall speed and height loss.

16. Stalling landing flaps. Climb to height then maintain straight and level. (a) Autopilot height measured and (b) idle power stall into wind – stall speed and height loss.

17. Max. level speed. Climb to height away from personnel. Increase airspeed incrementally. Maximum speed measured by autopilot.

18. With aircraft in take-off configuration with flaps at specified angle, measure take-off run. Flaps to zero immediately after takeoff. (a) Take-off run measured, (b) autopilot average climb rate recorded within the first 5 s after take-off, and (c) aircraft climb average airspeed measured by autopilot.

19. If possible, take off with zero flap. Measure take-off run. (a) Take-off run measured to be, (b) autopilot average climb rate recorded within the first 5 s after takeoff, and (c) aircraft climb average airspeed measured by autopilot.

20. Landing with full flap setting. (a) Flap position at specified angle, (b) approach speed autopilot approximate average, and (c) measure landing run – landing touchdown to stopped recorded.

21. Landing with take-off flap setting. (a) Flap position at specified angle, (b) approach speed autopilot approximate average, and (c) measure landing run – landing touchdown to stopped recorded.

22. Landing with zero flap (a) Approach speed autopilot approximate average and (b) measure landing run – landing touchdown to stopped recorded.

Compare the predicted results with the theoretical. Are there discrepancies? Given the above results, is it safe to continue to next flight test?

20.2.2   UAS CoG Flight Test (MANUAL Mode)

The effects of CoG changes are tested.

1. Ground handling. (a) Aircraft speed and direction controllable on ground, (b) UAV can idle and remain stationary on a grass surface suitable for takeoff without intervention, (c) control can be maintained in a fast taxi situation, (d) undercarriage is serviceable, and (e) no loss of aircraft control signal.
2. Aircraft handling and performance at takeoff, low-speed flight, cruise, high-speed flight, and landing to a satisfactory level and as per previous flight test.
3. Move CoG to its maximum forward position. (a) CoG correct and predicted at specified fraction of chord from leading edge, and (b) known mass added at specified forward location.
4. Aircraft handling and performance at takeoff, low-speed flight, cruise, high-speed flight, and landing to a satisfactory level and as per previous flight test.
5. Move CoG to its maximum rearward position. (a) CoG correct and predicted at specified fraction of chord from leading edge, and (b) known mass added at specified rearward location.
6. Aircraft handling and performance at takeoff, low-speed flight, cruise, high-speed flight, and landing satisfactory as per previous flight test.

Did the UAV fly suitably and as predicted?

20.2.3   Fuel Consumption Tests

The fuel consumption is measured for standard flight modes.

1. Ground handling. (a) Aircraft speed and direction controllable on ground, (b) UAV can idle and remain stationary on a grass surface suitable for takeoff without intervention, (c) control can be maintained in a fast taxi situation, (d) undercarriage is serviceable, and (e) no loss of aircraft control signal.
2. Aircraft handling and performance at takeoff, low-speed flight, cruise, high-speed flight, and landing to a satisfactory level and as per previous flight test.
3. Standard flight profile. This will consist of manual takeoff, manual circuit pattern, and manual landing. Flight duration typically to be 30 min unless limited by regulatory requirements or tank size. Throttle back to low speed. Land, checking that engine gives normal throttle response at all times. Repeat process as needed, refueling and recharging batteries when necessary. Inspect airframe as required in accordance with maintenance schedule.

20.2.4   Engine Failure, Idle, and Throttle Change Tests

Engine reliability is checked for a range of throttle settings and operations.

1. Ground handling. (a) Aircraft speed and direction controllable on ground, (b) UAV can idle and remain stationary on a grass surface suitable for take-off without intervention,

(c) control can be maintained in a fast taxi situation, (d) undercarriage is serviceable, and (e) no loss of aircraft control signal.

2. Aircraft handling and performance at takeoff, low-speed flight, cruise, high-speed flight, and landing to a satisfactory level and as per previous flight test.

3. Standard flight profile. This will consist of manual takeoff, manual circuit pattern, and manual landing. Flight duration to be at least 15 min to ensure engine conditions reach full operating temperature. Throttle back to low speed. Land, checking that engine gives normal throttle response at all times. Repeat the process three times to ensure consistency, refueling, and recharging batteries as needed.

4. Simulated power loss in level flight. The test should not be attempted unless it is safe to do so and can be done in any order. The type of power loss, full or partial, should be stated at the discretion of an authorized Flight Test Director in the form of "Full Power Loss." The UAV pilot will then cut the throttle to the UAV. (a) Port engine partial completed safely, (b) port engine full completed safely, (c) starboard engine partial completed safely, and (d) starboard engine full completed safely.

5. Simulated power loss after takeoff. This should be carried out at three heights. The tests should not be attempted unless it is safe to do so and can be carried out in any order. The type of power loss full or partial should be stated at the discretion of an authorized Flight Test Director in the form of "Full Power Loss." The UAV pilot will then cut the throttle to the UAV. (a) Height 1 should require the UAV to land ahead within a 45° arc. This should be demonstrated and a go-around commenced. (i) Port engine partial completed safely, (ii) port engine full completed safely, (iii) starboard engine partial completed safely, and (iv) starboard engine full completed safely. (b) Height 2 should require the UAV to turn back to the airfield and land downwind. This should not be attempted in wind conditions where the pilot deems this to be unsafe. This should be demonstrated and a go-around commenced. (i) Port engine partial completed safely, (ii) port engine full completed safely, (iii) starboard engine partial completed safely, and (iv) starboard engine full completed safely. (c) Height 3 should require the UAV to complete a mini-circuit and land. (i) Port engine partial completed safely, (ii) port engine full completed safely, (iii) starboard engine partial completed safely, and (iv) starboard engine full completed safely.

20.2.5 Autonomous Flight Control

The ability of the autopilot to control the aircraft satisfactorily is tested next. REMOTE refers to an aspect of GSC having been passed to the conventional pilot transmitter which can be in MANUAL or AUTO mode where AUTO refers to control by the GCS. Note that in AUTO mode unselected functions can still be manually controlled by the pilot's transmitter by setting it to REMOTE mode. Also the autopilot we use offers an ASSISTED mode where stick actions on the pilot's transmitter are interpreted as directional and speed commands rather than as servo settings (effectively, a "fly-by-wire" approach): not all autopilots offer this functionality.

1. Check UAV failsafe positions are correct and receiver's failsafe to AUTO MODE.
2. Check, with the transmitter off, that the GCS can still control the UAV.
3. GCS in REMOTE mode.
4. Transmitter in MANUAL mode.
5. Takeoff.

6. Climb to suitable height.

7. Transmitter to AUTO mode.

8. Bank. (a) Input bank command to wings level while in straight and level flight, (b) bank to GCS AUTO. Remaining controls by REMOTE, (c) pilot to maintain straight and level flight, (d) determine correct coefficients to ensure correct, stable, and non-oscillatory aircraft response: this will be achieved by setting various bank angles to be maintained and entered up to 30° and observing aircraft response, (i) Kd established and (ii) Ki established. (e) Return bank to REMOTE control.

9. Elevation. (a) Input elevation angle of zero while in straight and level flight, (b) elevation to GCS AUTO. Remaining controls by REMOTE, (c) pilot to maintain straight and level flight, (d) determine correct coefficients to ensure correct, stable and non-oscillatory aircraft response: this will be achieved by setting various pitch angles to be maintained and entered up to 20° and observing aircraft response, (i) Kd established and (ii) Ki established. (e) Return elevation to REMOTE control.

10. Altitude. (a) Input altitude to current height while in straight and level flight (at say 95 m), (b) altitude to GCS AUTO. Remaining controls by REMOTE, (c) pilot to maintain straight and level flight, (d) determine correct coefficients to ensure correct, stable and non-oscillatory aircraft response: this will be achieved by setting various altitudes to be maintained and observing aircraft response, (i) Kd established and (ii) Ki established. (e) Return altitude to REMOTE control.

11. Airspeed. (a) Input airspeed to cruise airspeed while in straight and level flight (at given cruise speed), (b) airspeed to GCS AUTO. Remaining controls by REMOTE, (c) pilot to maintain straight and level flight, (d) determine correct coefficients to ensure correct, stable, and non-oscillatory aircraft response: this will be done by setting various airspeeds within the safe operating range for the aircraft setup to be maintained and observing aircraft response, (i) Kd established and (ii) Ki established. (e) Return airspeed to REMOTE control.

12. Coefficient check by orbiting a specified waypoint. Observe aircraft response and modify coefficients as required including navigation linked coefficients. (a) Set orbit command in GCS. Bank to AUTO. Remainder in REMOTE. UAV to orbit waypoint via bank only. Check suitable aircraft response. Vary bank angles. (b) Set airspeed to maintain in GCS. Airspeed to AUTO. Remainder in REMOTE. UAV to orbit waypoint via bank and airspeed. Check suitable aircraft response. Vary airspeeds. (c) Set altitude to maintain in GCS. Altitude to AUTO. Fully autonomous control. Vary heights.

13. Maintaining autonomous control. Set a new waypoint to navigate to with a large separation distance with cruise airspeed and safe height. Check aircraft exhibits correct response.

14. Set a waypoint circuit pattern to check coefficients and turn anticipation. Check aircraft exhibits correct response.

15. Test-ASSISTED mode. UAV pilot to take full ASSISTED mode and check correction functionality and response UAV responds correctly.

16. Test of safety radius. Reduce safety radius to 200 m. When in ASSISTED mode, fly deliberately toward this outer radius. UAV should orbit the ERP waypoint when radius exceeded (the same response happens in fully auto mode). (a) UAV responds correctly, and manual control is regained after at least two orbits of the ERP are completed and (b) return safety radius to 500 m.

17. Check the functionality of the autopilot flight envelope. (a) Assisted mode. Envelope tested successfully, and (b) auto mode. Envelope tested successfully.
18. Check communication waypoint return. Over airfield with communication loss time set to 5 s. Ensure emergency waypoint is located correctly. (a) Set UAV to orbit a waypoint away from the emergency waypoint. With RC TX in full AUTO mode, disconnect GCS transceiver on laptop by disconnecting the appropriate cable. (i) After 5 s, UAV moves to emergency waypoint and orbits correctly – on success pilot to override, Reconnect GCS, (b) On successful connection with AP from GCS, fly UAV in ASSISTED mode. (i) After 5 s UAV moves to emergency waypoint and orbits correctly – on success, pilot to override, reconnect GCS.
19. Pilot to land the UAV in manual mode.

Is the UAS flight control system (FCS) set up correctly and did it demonstrate correct control and operation of the UAV?

20.2.6   Auto-Takeoff Test

The ability of the autopilot to carry out auto-takeoff is tested.

1. On each occasion, the autopilot start-up mode should be completed with all control surfaces deflecting and autopilot control initialized (check on first flight and when subsequently powering on the UAV).
2. GCS operator to set waypoints such that the aircraft will take off and climb out into wind.
3. UAV pilot to perform manual takeoff along planned path to confirm correct handling of aircraft. Land.
4. Aircraft to be taxied into the take-off position. Autopilot to control one channel as agreed between UAV pilot and GCS operator. Partially auto take-off and climb out followed by manual landing.
5. If satisfied, this is to be repeated with the autopilot progressively being given more control in an order agreed between the UAV pilot and GCS operator.
6. If satisfied, a fully auto-takeoff is to be completed followed by a manual landing.

20.2.7   Auto-Landing Test

The ability of the autopilot to carry out auto landing is tested.

1. On each occasion, the autopilot start-up mode should be completed with all control surfaces deflecting and autopilot control initialized (check on first flight and when subsequently powering on the UAV).
2. GCS operator to set waypoints such that aircraft will approach and land into wind.
3. UAV pilot to perform manual takeoff and landing along planned path to confirm correct handling of aircraft.
4. GCS operator to set landing script to perform a series of "simulated approaches," whereby the aircraft will approach and flare at a specified altitude directly above the intended touch-down waypoint. UAV pilot to override control upon autopilot flare or sooner if aircraft

deviates from the intended flight path. To be repeated as many times as necessary, with altitude being incrementally decreased as agreed by UAV pilot and GCS operator. On each occasion, check that the altitude control is suitably accurate. Completed successfully.

5. Auto landing to be performed if satisfied. UAV pilot to take control upon touchdown, or sooner if aircraft deviates from the intended approach.

20.2.8 Operational and Safety Flight Scenarios

The use of the GCS to command the aircraft to carry out various flight missions is tested.

1. Standard flight profile. This will consist of manual takeoff, manual circuit pattern, and manual landing. The orbit points shall move a minimum of two times to demonstrate effective operation. Land.
2. Standard flight profile. This will consist of manual takeoff, assisted FCS control circuit pattern, and manual landing. The orbit points shall move a minimum of two times to demonstrate effective operation. Land.
3. Standard flight profile. This will consist of manual takeoff, autonomous circuit pattern, and manual landing. The orbit points shall move a minimum of two times to demonstrate effective operation. Land.
4. Simulated loss of GCS control (laptop crash) while the UAV is in autonomous mode. Simulated by USB to aerial being disconnected.
5. Emergency landing. In the event of a transmitter failure, the UAV should be able to be landed using only the FCS. This should be demonstrated until it is clear that the UAV will land in the designated area. Manual control will be taken, and aircraft will perform a go-around. Control will be taken by the UAV pilot using the autopilot override switch on his discretion of this being demonstrated. No persons should be endangered. This should be fully understood before commencing test. Persons should be kept to safe area. (a) Fly UAV at height in AUTO mode with REMOTE control by pilot, (b) on close communication between pilot and GCS operator, (i) pilot to stop inputs to transmitter, (ii) wait 2 s to simulate a reaction time, and (iii) GCS to recover the aircraft and land the aircraft under the close supervision and authorization of the pilot. (c) Pilot to take manual override control when he is happy that the aircraft would land in the specified field and commence a go-around.
6. Landing site becomes unavailable. A nearby different location is to be chosen and an approach to land commenced followed by a go-around.

20.3 Maintenance

UAS maintenance is of vital importance to their safe operation. This broadly falls into two areas: the actual hardware that flies, and any software used both on and off the airframe in order for it to operate. Specific documented action should be taken to ensure that both hardware and software are properly maintained. For hardware, it will be obvious that the flying airframe operates in a harsh environment subject to large acceleration and deceleration forces, changes in temperature and humidity, and significant vibrations. In addition, various fuels and oils will be in use and may be spilt onto the airframe or its systems. For software, the chief dangers are undetected bugs in flight critical software either being present from the outset or being inadvertently introduced either by firmware upgrades or during operator-based programming of those functions that are designed to be set by the flight team.[2]

20.3.1 Overall Airframe Maintenance

It should go without saying that it is good practice when dealing with any airframe to be neat and tidy and to keep the airframe clean and to ensure that no small items are accidentally lost inside fuselages, nacelles, wings, and so on. Before flying, the appropriate logs should be checked to see if any items on the airframe are nearing the end of a service interval, which they may reach during the planned operations. All structural elements should be inspected and main spars gently flexed to ensure that all parts remain correctly attached and aligned. All flight control surfaces should be inspected for damage and moved through their full range of movement using the hand-held pilots transmitter (and not by manually forcing the surfaces – this can damage both the surface and the control servos that operates it). After every flight, the aircraft should be wiped down and any oil or fuel residues removed from the outer surfaces. If flying is completed for the day, fuel tanks should be drained before the airframe is transported.

20.3.2 Time and Flight Expired Items

A number of items on the airframe will have either time or number of flight limits on them before they should be serviced or replaced.

Engine Servicing

Most obviously, any internal combustion (IC) engine will need regular servicing by appropriately trained staff closely following the manufacturers' manuals. We carry out minor services such as valve clearance adjustments ourselves but prefer to use specialist companies for major overhauls. It is very easy, for example, when using four-stroke IC engines with very small valves to reach a situation where the exhaust valves are not seating correctly. If an engine is flown in such a condition, it is possible for very rapid valve seat wear to occur and the engine to fail in flights as short as 10 min duration. On a single-engine aircraft, this might lead to the total loss of the airframe.

Servos and Control Surfaces

Control surface servos are the next most obvious items that can suffer significant wear during flight. Consequently, we tend to replace these after specified numbers of flying hours. If one is using aero-modeler-grade items, or if the servos are very heavily loaded, they should probably be replaced at intervals of 10 h or less. Certainly, no servo should be used beyond 100 h flying time, even if only very lightly used. The attrition of on-board vibration can of itself lead to the gradual degradation of any complex electromechanical part. Attention should also be paid

---

[2] A classic example is the inadvertent selection of the wrong flight control program from a set available on the pilot's transmitter, all with similar names.

to the linkages and hinge points that are used to locate all control surfaces. It is very easy to accumulate wear in any bearing surfaces that will gradually allow play in the control mechanism, which will make flight control less easy and can induce control surface vibration and flutter.

20.3.3 Batteries

Airframe batteries will either be ground-charged on maintained by on-board generators. For batteries that are being ground-charged, good-quality charging systems should be used that monitor individual cell voltages during charge and control the overall current and voltage values. Charging should be carried out in appropriator locations and logs kept of the amount of charge added to a battery and the time taken to achieve this. Even the best quality LiFe, LiPo, and NiMHy batteries do not last for ever, and so after a given number of charges such items should be replaced. We would advise not recharging any airframe battery more than 50 times before it is recycled. It is also good practice to not allow airframe batteries to become too depleted unless safe recovery during a flight requires it. It is much wiser to stop a flight early or have sufficient excess capacity to ensure that batteries never go below 30% remaining charge.

When using batteries maintained by on-board charging circuits, it is tempting to almost "fit and forget" these items since the intention is that they are never heavily depleted and, moreover, have their status continually monitored by the charging circuits. However, on-board-charged batteries are typically much smaller than those that have to have sufficient capacity for the entire flight at the outset. Then, if there is any interruption to the charging circuit, they can very rapidly discharge. Thus such systems should be used only if ground-based monitoring of the battery charge is possible. Moreover, when not being flown, such batteries need just as careful maintenance as those that are routinely removed for ground-based charging.

All batteries should be periodically subjected to visual inspection to check that there is no damage or distortion to the external casing or connections. Slight bulging of the case is often the first sign of distress in a battery. Any battery with a less than perfect outer case should not be flown.

20.3.4 Flight Control Software

Flight Control Software is split into that on the airframe and that on the GCS. It will consist of firmware that is rarely changed and is largely independent of the aircraft, and operator-set software that may be readily edited and is aircraft-specific. Maintenance of firmware mainly consists of ensuring that the set in use is up to date as confirmed from the manufacturers' Web sites. If such software is updated, great caution should be taken when next flying the aircraft, just in case some unexpected glitch or feature change has been introduced. For systems that are subject to regulatory approvals, such a change may well require a reapplication for permission to fly with suitable justification for the change and evidence that it has been suitably tested.

User-programmable software is generally configured into the GCS before the beginning of each flight session and uploaded to the airframe when the avionics are first powered up. Assuming that the links between the GCS and airframe are functioning as designed, maintenance essentially involves ensuring that the correct set of software is being used and that no unauthorized changes have been made to this since it was last used. An essential part of the first test flights of any new aircraft will be the adjustment of the pilot's hand-held transmitter settings to match the characteristics of the airframe. A good deal of this will be carried out during ground tests, and a safety run through these setting will be needed before every flight to ensure that the main flight control surfaces and propulsion system behave as expected. It is good practice to then keep back-ups and records of these settings so that at the start of flying the transmitter software is in a known condition.

A similar set of parameters will also be need to be established for any autopilot fitted to the airframes. This will primarily involve the controller gain settings but also calibration of pitot tubes and the on-board magnetometer if fitted. Again, once a safe set of parameters has been established, these should be backed up and recorded so that known conditions are established before flying. In all cases, adjustments to transmitter or autopilot settings should be made only by experienced and authorized members of the flight team.

20.3.5 Maintenance Record Keeping

It will be clear from what has already been set out in this chapter that good record keeping is of vital importance to the safe operations of any UAS. This is just as important with regard to maintenance logs as it is to flight logs. The operating team simply has to be sure of the condition of the hardware and software in use to be able to have confidence in it. Such logs should be neat and tidy, readily to hand, and updated after each flight or at the very least after each flight day. Each UAS should have its own dedicated set of logs, and these should be backed up with paper hard copies that are available on the flying field.

21

Lessons Learned

We have been building and flying small fixed-wing unmanned air vehicles (UAVs) since 2008. Figure 21.1 shows our very first student-designed and -built airframe. Since that time, we have built some 30 further aircraft and our methods of working have evolved as we have discovered the best ways of getting our designs into the sky, not always successfully, Figure 21.2.

Perhaps the most important single lesson we have learned is that good-quality detail design is of paramount importance. Even the very best concepts will be spoiled by poor detailing, while even quite poor concepts will often perform acceptably if all the detail design work is of good quality. Initially, this attention to detail also extended to the skills needed to carry out aircraft manufacture. The airframe in Figure 21.1 involved much labor-intensive construction using typical aero-modeler-type approaches with wings covered with aero-modeler film and glued wooden fuselage construction. Now, however, following our commitment to digital manufacture, this mainly concerns detail CAD-based design work backed up with appropriate simulations of the underlying physics and the mechanisms being deployed. It is thus very important that a well-rehearsed approach to the detail design phase and the supporting analyses and tests that go with it be deployed. One of the key reasons for writing this book was to ensure that our student teams be aware of these things from the outset of their project work.

Next, we would point out the importance of experiments and test flights. It is often the case that building a Mark 1 airframe early on in a project, so as to gain information that can inform a Mark 2 version, is a better approach that committing too much effort to analysis that may be of limited fidelity or based on incorrect assumptions. It is also almost impossible to get every thing "right first time," even in the most sophisticated design organizations. Early prototypes can be invaluable in improving detail design work.

Thirdly, we would re-emphasize the critical importance of weight and center of gravity control and their impact on stall speed, pitch stability, and take-off and landing velocities. Once the wings have been sized, any increase in weight leads inexorably to an increase in stall speed and thus worsens take-off and landing performance. Such weight increases over the design weight budget stem from two main areas: first, it is all too easy to neglect vital components when making weight build-up assessments (in particular, wiring and servo linkages are often much heavier in total than one might image); second, there is always the temptation to add things; in particular, when designing the structure, it is easy to add small strengthening and

Small Unmanned Fixed-wing Aircraft Design: A Practical Approach, First Edition.
Andrew J. Keane, András Sóbester and James P. Scanlan.
© 2017 John Wiley & Sons Ltd. Published 2017 by John Wiley & Sons Ltd.

Figure 21.1   Our first student-designed UAV.

Figure 21.2   Not all test flights end successfully!

stress-relieving elements without being fully aware of their total impact on weight. At all times one should be looking for places that are overly strong and removing weight, rather than just worrying about areas that are too weak and adding to them. Provided sensible sized spars are in use and basic stress analysis has been carried out, in our experience small UAV airframes are generally too strong. In all our work, we have only ever had one flight-critical structural failure and that was due to poor manufacturing and not to poor design or analysis. It is also wise to allow for some internal heavy items such as batteries to be moved forward and aft in the airframe without major redesign. This allows fine-tuning of the longitudinal center of gravity and thus control of the static margin. If there is great uncertainty about the longitudinal center of gravity position during the early stages of design, it is sensible to design the fuselage in a modular manner so that its length can be changed. We have even built aircraft where the fuselage length can be changed depending on the payload being deployed, see Figure 21.3. On smaller aircraft, it is sometimes possible to arrange for the main wing attachment point to be adjustable to achieve the same effect.

Fourthly, we note that if a very conventional type of airframe is to be produced, then simply following previous satisfactory design concepts will likely ensure a successful outcome. Perhaps the best place to start is with the choice of payload, powerplant, and overall airframe topology. It is only when the designs become more radical or extreme performance is required that the full gamut of well-established aeroengineering calculations become critical. That is not to say that such calculations should not be carried out for all projects; rather it is to point out that by following a previous well-designed example, much of the knowledge accumulated over the last 100 years of flight will implicitly be embedded in the new design. However, when it comes to making any calculation, we would emphasize that the care that goes into the calculation should be related to how important the result is, not how technically difficult the calculation is. So, such simple but crucial calculations as the choice of wing area should be checked and double-checked before they are accepted by the design team. One does not want to get to the airfield before it dawns on one that the wings are out of proportion to the rest of the airframe (see below)!

Good team working is, of course, also vital to any project undertaken by more than a single individual. Good teams need a mix of both personal and technical skills. Not everyone can be the team leader and many engineers are not good at such roles. Equally, it takes a particular

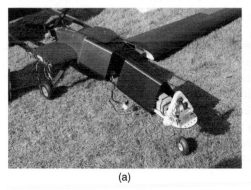
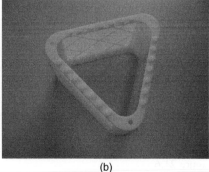

(a)   (b)

Figure 21.3   Aircraft with variable length fuselage. (a) Fuselage split open. (b) Spare fuselage section.

sort of mindset to carefully and relentlessly control weight growth in a project. It is also not wise to form a team comprised of 10 aerodynamicists, just one structural engineer, and no avionics experts. As in all aspects of aircraft design and manufacture, balance is crucial.

Finally, and by no means least importantly, always carry out initial flight trials at a properly controlled test airfield and with an experienced test pilot. As already noted, we have for many years worked with Paul Heckles of the Paul Heckles Flight Centre.[1] If your test pilot is not of the first grade, then any unexpected behavior is likely going to lead to a major crash and total loss of the airframe.

21.1 Things that Have Gone Wrong and Why

We have learned these lessons the hard way: over the years we have had a number of "uncontrolled descents," mostly of student-designed and -built aircraft (but not always). We try not to intervene too strongly into our student groups during their project work since we find they tend to learn more if allowed to make their own mistakes. What follows is a list of some the things that have gone wrong and why – it is a bit of a rogue's gallery but we would not pretend that we only ever make perfect aircraft at the University of Southampton:

- A failure to ensure a positive static margin led to an aircraft taking off and immediately stalling during a fast taxi test (this occurred when a test was being carried out without Paul being present).
- A failure to ensure dynamic stability led one aircraft to exhibit uncontrollable Dutch roll. This was a flying boat and the very large area of hull forward of the aerodynamic center needed a far larger vertical tail volume than is normally required on conventional aircraft, see Figure 21.4 (note that an XFLR5 analysis would not have revealed this since XFLR5 does not model fuselage side forces).
- A failure to ensure that a set of laser-cut plywood spar parts had been correctly placed in the clamping jig during manufacture led to a main spar failure during extreme maneuvers. A simple spar load or ground vibration test would have revealed this, but neither was carried out prior to flight.
- An engine failure on a single-engine aircraft meant that the aircraft did not have sufficient height to glide back to our test runway, leading to a controlled ditching in the adjacent field. The low-speed screw on the carburetor was poorly set, and when opening the throttle rapidly from idle, the engine stalled.
- The adoption of a radical, all-steerable and split elevator on a research aircraft had a highly nonlinear impact on controllability, causing the aircraft to suddenly nose-dive on a steep turn into toward final approach, see Figure 21.5. The aircraft had previously appeared to have benign characteristics. One should always fully explore the envelope with sufficient altitude to maximize the chances of recovery from such mishaps. The rebuilt aircraft has a fence at the elevator split, which mitigates this problem.
- Engine vibrations caused an autopilot to fail. We now extensively test the mountings of our autopilots in the lab prior to operations, using signals measured on the target aircraft, see Figure 21.6.

---

[1] http://www.paulhecklesrc.co.uk/

Figure 21.4   Student-designed flying boat with large hull volume forward and insufficient vertical tail volume aft.

(a)   (b)

Figure 21.5   Aircraft with split all-moving elevator. (a) Without dividing fence. (b) With fence.

- Engine vibrations caused a power isolation switch to momentarily open and thus initiated the engine ignition safety cut-out system, forcing the pilot to glide the unpowered aircraft back to the landing strip.
- A sizing error during concept design led to an aircraft being built with wings that were too small given the installed power; this was not revealed until its first catapult launch, Figure 21.7.

Figure 21.6   Autopilot on vibration test.

Figure 21.7   Student UAV with undersized wings. The open payload bay also added to stability issues.

Figure 21.8   2SEAS aircraft after failure of main wheel axle.

- Repeated autonomous landings on a concrete runway led to low-cycle fatigue of a main wheel axle due to the significant impact loads experienced. Subsequent use of laser height finding mitigated these issues on automated landings, Figure 21.8.

Despite this apparent litany of failures, our aircraft fly successfully much more often than not; so we would encourage anyone setting out to design and build his/her own unmanned air system (UAS) to go right ahead. It is not nearly as daunting as one might initially think (even given the size of this book) and it can be a great deal of fun. Certainly, our student teams enjoy the experience enormously and clearly get a great thrill when "their" aircraft takes wing for the first time. Even those whose aircraft do not perform as desired learn a great deal from actually being directly involved in the whole process from first ideas to final flight.

Part V

Appendices, Bibliography, and Index

Appendix A

Generic Aircraft Design Flowchart

Figure A.1 is based on the work of Karuzic [28] as set out in his M.Sc. thesis. It summarizes the entire design process of a typical aircraft, whether manned or unmanned. As such, it forms a useful checklist of the stages in the design process.

Small Unmanned Fixed-wing Aircraft Design: A Practical Approach, First Edition.
Andrew J. Keane, András Sóbester and James P. Scanlan.
© 2017 John Wiley & Sons Ltd. Published 2017 by John Wiley & Sons Ltd.

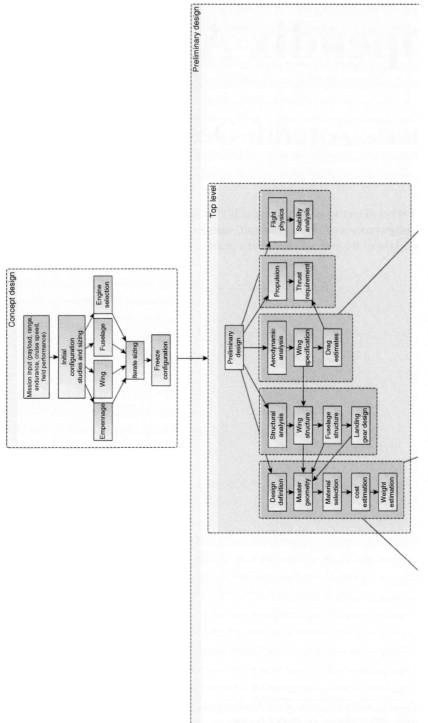

Figure A.1 Generic aircraft design flowchart.

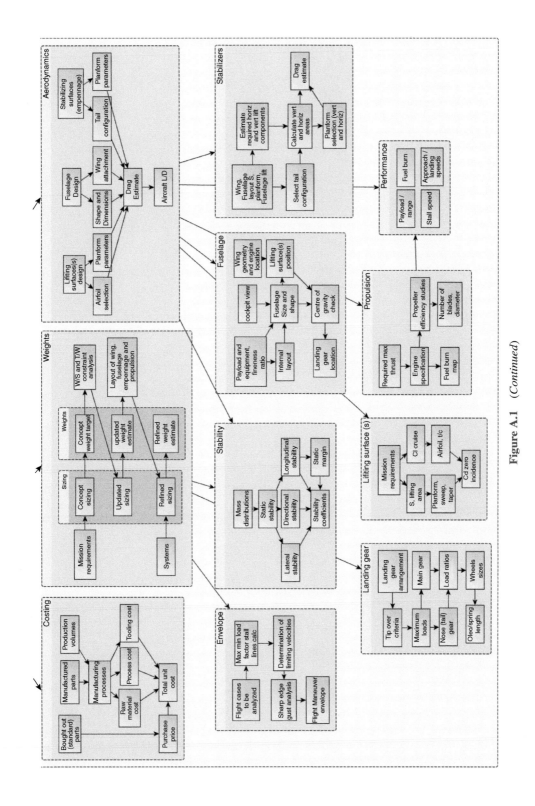

Figure A.1 (*Continued*)

Appendix B

Example AirCONICS Code for Decode-1

The Decode-1 airframe model used in the chapters on preliminary design was created using the AirCONICS system[1] that drives the Rhino CAD platform.[2] The specific code used to create the aiframe is given as follows. It may also be downloaded from https://aircraftgeometrycodes .wordpress.com/airconics-uav/.

[1] https://aircraftgeometrycodes.wordpress.com/airconics/.
[2] http://www.rhino3d.com/.

Small Unmanned Fixed-wing Aircraft Design: A Practical Approach, First Edition.
Andrew J. Keane, András Sóbester and James P. Scanlan.
© 2017 John Wiley & Sons Ltd. Published 2017 by John Wiley & Sons Ltd.

```
# ===============================================================
# Example Python script for building a twin boom UAV geometry, to be run in the
# Rhinoceros 5.0 CAD engine.
# ===============================================================
# Provided as part of Keane, A. J., Sobester, A., Scanlan, J. P., "Small
# Unmanned Fixed-wing Aircraft Design: A Practical Approach", John Wiley & Sons,
# 2017.
#
# This script uses the AirCONICS aircraft geometry toolbox (included):
# Aircraft CONfiguration through Integrated Cross-disciplinary Scripting
# version 0.1.1b
# Andras Sobester, 2014.
#
# It also uses the UIUC library of airfoil coordinates, downloaded in 2016 from
# the following URL: http://m-selig.ae.illinois.edu/ads/coord_database.html
# ===============================================================
#
# Before running this script, please edit airconics_setup.py, where you can
# specify the path where the code provided is installed, as well as the location
# of the airfoil coordinate library.
# ===============================================================
#
# This program, including the AirCONICS aicraft geometry toolbox, are free
# software: you can redistribute it and/or modify it under the terms of the GNU
# Lesser General Public License as published by the Free Software Foundation,
# either version 3 of the License, or any later version.
#
# This program is distributed in the hope that it will be useful, but
# WITHOUT ANY WARRANTY; without even the implied warranty of
# MERCHANTABILITY or FITNESS FOR A PARTICULAR PURPOSE. See the GNU Lesser
# General Public License for more details.
#
# You should have received a copy of the GNU General Public License and GNU
# Lesser General Public License along with this program. If not, see
# <http://www.gnu.org/licenses/>.

# ===============================================================
# PREAMBLE
# ===============================================================
```

```python
from __future__ import division
import math, rhinoscriptsyntax as rs
import primitives
import airconics_setup
import liftingsurface
import AirCONICStools as act
import HighLiftDevices as HLD
import landinggear as LG
import uavauxtools as UAT
import payload

global BoomInner, BoomOuter
global SettingAngle, TaperRatio
global fo

rs.UnitAbsoluteTolerance(0.0001)
rs.UnitAngleTolerance(0.1)
rs.UnitRelativeTolerance(0.1)

#================================================================
# GEOMETRY PARAMETERS
#================================================================

# === Airframe geometry parameters taken directly from spreadsheet ==============

# Name        Value          # Long Name / Definition
Awing    =    965830.1       #      total wing area
AR       =    9              # aspect  ratio (span^2 / area)
Span     =    2948.3         # total wing span (rect wing)
Chord    =    327.6          # aerodynamic mean chord
Dprop    =    435.0          # propellor diameter
Atail    =    158945.7       #      tailplane area
ARtail   =    4.0            # tailplane aspect ratio (span^2 / area)
Height_fin  =  293.0         #    fin height (or semi-span) for two fins
Chord_fin   =  195.3         #    fin mean chord
Depth_Fuse  =  200.0         #    fuselage depth
```

```
Width_Fuse     = 150.0     # fuselage width
Len_Nose       = 150.0     # nose length (forward of front bulkhead)
Dia_Wheels     = 100.0     # diam of main undercarriage wheels
Len_Engine     = 125.0     # length of engine
x_fnt_bkhd     = 640.3     # long position of front bulkhead
x_tail_spar    = -995.3    # long position of tailplane spar
x_rear_bkhd    = -200.0    # long position of rear bulkhead
x_mid_bkhd     = 220.2     # long position of middle bulkhead
z_fuse_base    = -110.0    # vert position of base of fuselage
z_tail_boom    = 0.0       # vert position of tailboom
z_engine       = 60.0      # vert position of engine
z_uncarriage   = -300.0    # vert position of centre of main undercarriage wheels
y_tail_boom    = 239.2     # horizontal position of tail booms
x_main_spar    = 0.0       # long position of main spar

FinAngle = 90
TaperRatio = Span / Chord

SWEEP_QTR_deg = 0
SWEEP_QTR_rad = math.radians(SWEEP_QTR_deg)
SWEEP_LE_rad = math.atan(math.tan(SWEEP_QTR_rad)+\
(1/AR)*(1-TaperRatio)/(1+TaperRatio))
SWEEP_LE_deg = math.degrees(SWEEP_LE_rad)

# Aileron geometry
ASpanStart = 0.6
ASpanEnd = 0.98
AChord = 0.3
ATaper = 1.0
ADeflectionAngle = 20.0

# Elevator geometry
ESpanStart = 0.01
ESpanEnd = 0.98
EChord = 0.3
ETaper = 1.0
EDeflectionAngle = 0.0
```

```
# Rudder geometry
RSpanStart = 0.0555
RSpanEnd = 0.98
RChord = 0.3
RTaper = 1.0
RDeflectionAngle = 0.0

# Use these flags to turn components on and off

VTail = False
Pod = True
NoseGear = False
Ailerons = True
Elevators = True
Rudder = True

# The geometry of the internal structure
# (all dimensions in m)
RibThick = 0.005
CutterSize = 5
MainSparODia = 0.02
MainSparIDia = 0.016
BoomODia = 0.02
BoomIDia = 0.016
TailSparODia = 0.016
TailSparIDia = 0.013
FinSparODia = 0.01
FinSparIDia = 0.008
HingeSparODia = 0.005
HingeSparIDia = 0.003
SparClearance = 0.0

# cut up the structure if needed
CutStructure = True

# =========== Variables to sweep ==================================
# EXAMPLE: 'Decode 1' pusher UAV

PayloadPodLength = 0.4
```

```
#-x_fnt_bkhd/1000.0, z_uncarriage/1000 - nose/tail gear length
# NoseGear - if False tail gear is drawn
# ================================================================================

Span = math.sqrt(AR*Awing)
Chord = Awing/Span

Span_tail = math.sqrt(ARtail*Atail)
Chord_tail = Atail/Span_tail

# ==== 3D wing main wing definition ======================================
# ==== through definition of spanwise parameter variations ==============

def myDihedralFunctionPusher(Epsilon):
    # User-defined function describing the variation of dihedral as a function
    # of the leading edge coordinate
    return 0

def myTwistFunctionPusher(Epsilon):
    # User-defined function describing the variation of twist as a function
    # of the leading edge coordinate
    RootTwist = 0
    TipTwist = 2 # typical reduction in twist to give washout
    TipTwist = 0
    return RootTwist + Epsilon*(TipTwist-RootTwist)

def myChordFunctionPusher(Epsilon):
    # User-defined function describing the variation of chord as a function of
    # the leading edge coordinate

    if Epsilon < 0.2:
        return 1
    else:
        ChordLengths = [1, TaperRatio]
        EpsArray = [0.2, 1]
        f = act.linear_interpolation(EpsArray, ChordLengths)
        return f(Epsilon)
```

```python
def myAirfoilFunctionPusher(Epsilon, LEPoint, ChordFunct, ChordFactor, \
        DihedralFunct, TwistFunct):
    # Defines the variation of cross section as a function of Epsilon

    AirfoilChordLength = (ChordFactor*ChordFunct(Epsilon))/math.cos \
        (math.radians(TwistFunct(Epsilon)))

    # Instantiate class to set up a generic airfoil with these basic parameters
    Af = primitives.Airfoil(LEPoint, AirfoilChordLength, DihedralFunct(Epsilon)\
        , TwistFunct(Epsilon),
    EnforceSharpTE = True)

    SmoothingPasses = 1

    # Add airfoil curve to document and retrieve handles to it and its chord
    # - in this case NACA23012, with DesignLiftCoefficient = 0.3,
    # MaxCamberLocFracChord = 0.15 and MaxThicknessPercChord = 15
    Airf, Chrd = primitives.Airfoil.AddNACA5(Af, 0.3, 0.15, 15, SmoothingPasses)

    # A possible alternative - a NACA 4-digit section (e.g., NACA2212)
    # Airf, Chrd = primitives.Airfoil.AddNACA4(Af, 2, 2, 12, SmoothingPasses)

    return Airf, Chrd

def mySweepAngleFunctionPusher(Epsilon):
    # User-defined function describing the variation of sweep angle as a
    # function of the leading edge coordinate
    return SWEEP_LE_deg

# ==== 3D tailplane definition ============================================
# ==== through definition of spanwise parameter variations ================

def myTwistFunctionTailplane(Epsilon):
    # User-defined function describing the variation of twist as a function
    # of the leading edge coordinate
```

```python
    # 2.85 gives the correct trim at 30 m/s with zero main wing setting angle
    # according to XFLR5
    #RootTwist = 2.85
    #TipTwist = 2.85
    # 0.34 gives the correct trim at 30 m/s with 2.53 main wing setting angle
    # according to XFLR5
    RootTwist = 0.34
    TipTwist = 0.34
    return RootTwist + Epsilon*(TipTwist-RootTwist)

def myChordFunctionTailplane(Epsilon):
    # User-defined function describing the variation of chord as a function of
    # the leading edge coordinate

    return 1

def myAirfoilFunctionTailplane(Epsilon, LEPoint, ChordFunct, ChordFactor, \
DihedralFunct, TwistFunct):
    # Defines the variation of cross section as a function of Epsilon

    AirfoilChordLength = (ChordFactor*ChordFunct(Epsilon))/math.cos\
(math.radians(TwistFunct(Epsilon)))

    # Instantiate class to set up a generic airfoil with these basic parameters
    Af = primitives.Airfoil(LEPoint, AirfoilChordLength, DihedralFunct(Epsilon)\
, TwistFunct(Epsilon),
EnforceSharpTE = True)

    SmoothingPasses = 1

    # Add NACA0012 airfoil curve to document and retrieve handles to it and
    # its chord
    Airf, Chrd = primitives.Airfoil.AddNACA4(Af, 0, 0, 12, SmoothingPasses)

    return Airf, Chrd

def mySweepAngleFunctionTailplane(Epsilon):
    # User-defined function describing the variation of sweep angle as a funct.
    # of the leading edge coordinate
```

```python
    return 0

def myDihedralFunctionTailplane(Epsilon):
    # User-defined function describing the variation of dihedral as a function
    # of the leading edge coordinate
    return 0

#=================================================================
# ==== end of tailplane definition ===============================


# ==== 3D tailfin definition =====================================
# ==== through definition of spanwise parameter variations =======


def myTwistFunctionFin(Epsilon):
    # User-defined function describing the variation of twist as a function
    # of the leading edge coordinate
    return 0

def myChordFunctionFin(Epsilon):
    # User-defined function describing the variation of chord as a function of
    # the leading edge coordinate
    return 1

def myAirfoilFunctionFin(Epsilon, LEPoint, ChordFunct, ChordFactor, \
    DihedralFunct, TwistFunct):
    # Defines the variation of cross section as a function of Epsilon

    AirfoilChordLength = (ChordFactor*ChordFunct(Epsilon))/math.cos(\
    math.radians(TwistFunct(Epsilon)))

    # Instantiate class to set up a generic airfoil with these basic parameters
    Af = primitives.Airfoil(LEPoint, AirfoilChordLength, DihedralFunct(Epsilon)\
    , TwistFunct(Epsilon),
    EnforceSharpTE = True)

    SmoothingPasses = 1

    # Add NACA0012 airfoil curve to document and retrieve handles to it and
    # its chord
```

```python
    Airf , Chrd = primitives . Airfoil . AddNACA4( Af ,  0 ,  0 ,  12 ,  SmoothingPasses )

    return Airf ,  Chrd

def mySweepAngleFunctionFin ( Epsilon ):
    # User-defined function describing the variation of sweep angle as a fn
    # of the leading edge coordinate
    return 0

def myDihedralFunctionFin ( Epsilon ):
    # User-defined function describing the variation of dihedral as a function
    # of the leading edge coordinate
    return 0

#======== end of tailfin definition ==========================================
#======== END GEOMETRY PARAMETER DEFINITION ==================================


#=============================================================================
# AIRFRAME GEOMETRY GENERATION
#=============================================================================

rs . EnableRedraw ( False )

# Wing apex location
P = (0 ,0 ,0)

LooseSurf = 3 # Tightly fit surface (loft_type arhument in AddLoftSrf)
SegmentNo = 2 # Number of airfoil sections to loft wing surface over

SettingAngle = 2.53 # as computed with XFLR5 to give Cl of 0.28 at 30 m/s

TipRequired = 1 # Flat wingtip

SectionsRequired = False # when optimizing this currently needs to be False but
                         # when creating XFLR5 files it must be True
```

```python
Wing = liftingsurface.LiftingSurface(P, mySweepAngleFunctionPusher, \
myDihedralFunctionPusher, myTwistFunctionPusher, myChordFunctionPusher, \
myAirfoilFunctionPusher, LooseSurf, SegmentNo, TipRequired, SectionsRequired)

# The wing thus defined, we now scale it to the target aspect ratio and area
Wing.TargetAspectRatio = AR
Wing.TargetArea = Awing/1000000.0
Wing.wTargetAspectRatio = 1
Wing.wTargetArea = 1

# Initial iterate
ChordFactor = 0.26
ScaleFactor = 1.47

OptimizeChordScale=1

SLS, ActualSemiSpan, LSP_area, RootChord, AR, SWingTip, Sections =\
Wing.GenerateLiftingSurface(ChordFactor, ScaleFactor, OptimizeChordScale)


# Rotate the wing to the specified setting angle
TipX = math.tan( SWEEP_LE_rad ) * ActualSemiSpan
RotVec = rs.VectorCreate((0,0,0),(0,1,0))
SLS = rs.RotateObject(SLS, (0,0,0), -SettingAngle, axis = RotVec)
SWingTip = rs.RotateObject(SWingTip, (0,0,0), -SettingAngle, axis = RotVec)
SLS_TE = UAT.ExtractTrailingEdge(SLS)
# shift wing up to allow for setting angle about LE and
# not 1/4 chord point where we attach it
SLS = rs.MoveObject(SLS, \
(0.0, 0.0, 0.25*RootChord*math.sin(math.radians(SettingAngle))))
SWingTip = rs.MoveObject(SWingTip, \
(0.0, 0.0, 0.25*RootChord*math.sin(math.radians(SettingAngle))))
SLS_TE = rs.MoveObject(SLS_TE, \
(0.0, 0.0, 0.25*RootChord*math.sin(math.radians(SettingAngle))))
PWingTip = act.MirrorObjectXZ(SWingTip)
rs.DeleteObject(SLS_TE)

# Adding ailerons to each wing
```

```
if Ailerons:

    SLScopy=rs.CopyObject(SLS)
    Aflapconfig1 = \
    ['simpleflap', ASpanStart, ASpanEnd, AChord, ATaper, ADeflectionAngle]
    Devices = [Aflapconfig1]
    SAileron, Area, Cutter, Cutter1, Cutter2, SAileronCutBrick = \
    HLD.AddHighLiftDevices(SLS, Devices)
    rs.DeleteObjects([Cutter, Cutter1, Cutter2])
    PLS = act.MirrorObjectXZ(SLS)
    PAileron = act.MirrorObjectXZ(SAileron)

else:
    PLS = act.MirrorObjectXZ(SLS)


# ===== Booms ================================================================

    PodLength = 1.5*Chord/1000.0
    BoomLength = -x_tail_spar/1000.0
    BoomY = y_tail_boom/1000.0
    BoomZ = z_tail_boom/1000.0
    [SPodSurf] = payload.pod([0.025*PodLength,BoomY,BoomZ],PodLength,11, 0.1)
    if CutStructure:
        SPodSurf = UAT. CutCylinder(RootChord/4.0,\
        BoomY,BoomZ, BoomLength,BoomY,BoomZ,BoomODia, SPodSurf)

        SBoom = UAT. CreateSpar(RootChord/4.0+MainSparODia/2.0,BoomY,BoomZ,\
        BoomLength-TailSparODia/2.0,BoomY,BoomZ, BoomODia-SparClearance, BoomIDia)
    else:
        SBoom = \
        rs.AddCylinder([RootChord/4.0,BoomY,BoomZ], [BoomLength,BoomY,BoomZ], 0.01)

    PPodSurf = act.MirrorObjectXZ(SPodSurf)
    PBoom = act.MirrorObjectXZ(SBoom)
    if CutStructure: # cut booms to enable biased meshing along their lengths away
#                      from where the pod ends
        SBoom, SBoom1 = UAT. BooleanSplitX(0.875*PodLength, CutterSize, SBoom)
        PBoom, PBoom1 = UAT. BooleanSplitX(0.875*PodLength, CutterSize, PBoom)
```

```
# ==== Tailplane ==================================================
TailChord = Chord_tail/1000.0
FinX = -x_tail_spar/1000.0-TailChord/4.0
P_TP = [FinX,0,0]
LooseSurf_TP = 1
SegmentNo_TP = 2

TipRequired_TP = 1
SectionsRequired_TP = False

Tailplane = liftingsurface.LiftingSurface(P_TP, mySweepAngleFunctionTailplane ,\
myDihedralFunctionTailplane , myTwistFunctionTailplane , myChordFunctionTailplane ,\
myAirfoilFunctionTailplane , LooseSurf_TP, SegmentNo_TP , TipRequired_TP ,\
SectionsRequired_TP)

if VTail: # we will use a pair of tails rotated to make the V tail so
#            allow for this in area and aspect ratio
    CantAngle = math.atan((2*Height_fin)/Span_tail)
    TailSpan = Span_tail/1000.0/math.cos(CantAngle)/2.0
    Tailplane.TargetAspectRatio = ARtail/math.cos(CantAngle)/2.0
    Tailplane.TargetArea = Atail/1000000.0/math.cos(CantAngle)/2.0
else:
    CantAngle=0.0
    TailSpan = Span_tail/1000.0
    Tailplane.TargetAspectRatio = ARtail
    Tailplane.TargetArea = Atail/1000000.0

Tailplane.wTargetAspectRatio = 1
Tailplane.wTargetArea = 1
ChordFactor_TP = 0.5
ScaleFactor_TP = 0.4

OptimizeChordScale_TP = True

SLS_TP, ActualSemiSpan_TP , LSP_area_TP ,  \
RootChord_TP , AR_TP, SWingTip_TP , Sections_TP =\
```

```
Tailplane.GenerateLiftingSurface(ChordFactor_TP, ScaleFactor_TP, OptimizeChordScale_TP)

SLS_TP_TE = UAT.ExtractTrailingEdge(SLS_TP)

if Elevators:
    Eflapconfig = \
    ['simpleflap', ESpanStart, ESpanEnd, EChord, ETaper, EDeflectionAngle]
    Devices = [Eflapconfig]
    SElevator, Area, Cutter, Cutter1, Cutter2, CutBrick = \
    HLD.AddHighLiftDevices(SLS_TP, Devices)
    rs.DeleteObjects([Cutter, Cutter1, Cutter2])
    SLS_TP = rs.BooleanDifference(SLS_TP, CutBrick)
    # shift elevator down to allow for setting angle about LE and not 1/4 chord
    # point where we attach it
    SElevator = rs.MoveObject(SElevator,(0.0, 0.0, -0.25*RootChord_TP*math.sin \
    (math.radians(myTwistFunctionTailplane(0.0)))))
    PElevator = act.MirrorObjectXZ(SElevator)
    STail = rs.JoinSurfaces([SLS_TP, SWingTip_TP])
    rs.DeleteObjects([SLS_TP, SWingTip_TP])
    PTail = act.MirrorObjectXZ(STail)
    Tail = rs.JoinSurfaces([STail, PTail])
    rs.DeleteObjects([STail, PTail])

# shift tail down to allow for setting angle about LE and
# not 1/4 chord point where we attach it

Tail = rs.MoveObject(Tail,(0.0, 0.0, -0.25*RootChord_TP*math.sin \
(math.radians(myTwistFunctionTailplane(0.0)))))
SLS_TP_TE = rs.MoveObject(SLS_TP_TE,(0.0, 0.0, -0.25*RootChord_TP*math.sin \
(math.radians(myTwistFunctionTailplane(0.0)))))

if VTail:
    RotVec = rs.VectorCreate([FinX,y_tail_boom/1000.0,0],[FinX+1,y_tail_boom/1000.0,0])
    RotCent = [FinX,y_tail_boom/1000.0,0]
    if CutStructure:
        #Tail = rs.BooleanUnion([Tail, SFin, PFin])
        Tail = UAT.CutCylinder(BoomLength,-ActualSemiSpan_TP, BoomZ, BoomLength, \
        ActualSemiSpan_TP, BoomZ, TailSparODia, Tail)
        TailSpar = UAT.CreateSpar(BoomLength,-ActualSemiSpan_TP, BoomZ, BoomLength, \
```

```python
ActualSemiSpan_TP ,BoomZ,  TailSparODia−SparClearance ,  TailSparIDia )

Tail , TailS1  = UAT. BooleanSplitY (BoomY−(y_tail_boom/1000.0*(1.0−1.0/math.cos\
(CantAngle))+ActualSemiSpan_TP)+0.05*Chord/1000.0, CutterSize , Tail)
Tail , TailS2  = UAT. BooleanSplitY (BoomY−(y_tail_boom/1000.0*(1.0−1.0/math.cos\
(CantAngle))+ActualSemiSpan_TP)−0.05*Chord/1000.0, CutterSize , Tail)
Tail , TailS3  = UAT. BooleanSplitY (ESpanStart*ActualSemiSpan_TP+0.00175,CutterSize , Tail)
TailS1 , TailS4  = UAT. BooleanSplitY (ESpanEnd*ActualSemiSpan_TP−0.00075,CutterSize , TailS1)
Tail , TailS5  = UAT. BooleanSplitY (−ESpanStart*ActualSemiSpan_TP−0.00175,CutterSize , Tail)
Tail , TailS6  = UAT. BooleanSplitY (−ESpanEnd*ActualSemiSpan_TP+0.00075,CutterSize , Tail)

SLS_TP_TE_P  =  act .MirrorObjectXZ (SLS_TP_TE)
SLS_TP_TE_S  =  rs .CopyObject (SLS_TP_TE)
rs .DeleteObjects ([SLS_TP_TE])
SLS_TP_TE = rs .JoinCurves ([SLS_TP_TE_S,  SLS_TP_TE_P])
rs .DeleteObjects ([SLS_TP_TE_S,  SLS_TP_TE_P])

Tail = rs .MoveObject (Tail,  [0,y_tail_boom/1000.0*(1.0−1.0/math.cos(CantAngle))+ActualSemiSpan_TP ,0])
SLS_TP_TE = rs .MoveObject (SLS_TP_TE,  [0,y_tail_boom/1000.0*\
(1.0−1.0/math.cos(CantAngle))+ActualSemiSpan_TP ,0])
Tail = rs .RotateObject (Tail ,RotCent ,math.degrees(CantAngle),  axis  =  RotVec)
SLS_TP_TE = rs .RotateObject (SLS_TP_TE,RotCent ,math.degrees(CantAngle),  axis  =  RotVec)
TailP  =  act .MirrorObjectXZ (Tail)
rs .DeleteObject (SLS_TP_TE)
CentrePod  =  payload .pod ([FinX,  0,  y_tail_boom/1000.0*math.tan(CantAngle)],RootChord_TP,12,  0.0)

if  CutStructure :
    ypos1=y_tail_boom/1000.0*(1.0−1.0/math.cos(CantAngle))+ActualSemiSpan_TP
    TailS1 = rs .MoveObject (TailS1 ,  [0,ypos1 ,0])
    TailS2 = rs .MoveObject (TailS2 ,  [0,ypos1 ,0])
    TailS3 = rs .MoveObject (TailS3 ,  [0,ypos1 ,0])
    TailS4 = rs .MoveObject (TailS4 ,  [0,ypos1 ,0])
    TailS5 = rs .MoveObject (TailS5 ,  [0,ypos1 ,0])
    TailS6 = rs .MoveObject (TailS6 ,  [0,ypos1 ,0])
    TailSpar = rs .MoveObject (TailSpar,  [0,ypos1 ,0])
    TailS1 = rs .RotateObject (TailS1 ,RotCent ,math.degrees(CantAngle),  axis  =  RotVec)
    TailS2 = rs .RotateObject (TailS2 ,RotCent ,math.degrees(CantAngle),  axis  =  RotVec)
    TailS3 = rs .RotateObject (TailS3 ,RotCent ,math.degrees(CantAngle),  axis  =  RotVec)
    TailS4 = rs .RotateObject (TailS4 ,RotCent ,math.degrees(CantAngle),  axis  =  RotVec)
```

```
TailS5  =  rs.RotateObject(TailS5,RotCent,math.degrees(CantAngle),  axis  =  RotVec)
TailS6  =  rs.RotateObject(TailS6,RotCent,math.degrees(CantAngle),  axis  =  RotVec)
TailSpar  =  rs.RotateObject(TailSpar,RotCent,math.degrees(CantAngle),  axis  =  RotVec)
TailS2  =  UAT.CutCylinder(0.25*Chord/1000.0,BoomY,BoomZ,  BoomLength,BoomY,BoomZ,BoomODia,TailS2)
TailP1  =  act.MirrorObjectXZ(TailS1)
TailP2  =  act.MirrorObjectXZ(TailS2)
TailP3  =  act.MirrorObjectXZ(TailS3)
TailP4  =  act.MirrorObjectXZ(TailS4)
TailP5  =  act.MirrorObjectXZ(TailS5)
TailP6  =  act.MirrorObjectXZ(TailS6)
TailPSpar  =  act.MirrorObjectXZ(TailSpar)
Tail  =  rs.BooleanUnion([Tail,  TailP])

if  Elevators:
    SElevator  =  rs.MoveObject(SElevator,  [0,y_tail_boom/1000.0*\
    (1.0−1.0/math.cos(CantAngle))+ActualSemiSpan_TP,0])
    PElevator  =  rs.MoveObject(PElevator,  [0,y_tail_boom/1000.0*\
    (1.0−1.0/math.cos(CantAngle))+ActualSemiSpan_TP,0])
    SElevator  =  rs.RotateObject(SElevator,RotCent,math.degrees(CantAngle),  axis  =  RotVec)
    PElevator  =  rs.RotateObject(PElevator,RotCent,math.degrees(CantAngle),  axis  =  RotVec)
    SElevatorP  =  act.MirrorObjectXZ(SElevator)
    PElevatorP  =  act.MirrorObjectXZ(PElevator)

# =====  Tailfin  ==================================================================

if  VTail  ==  False:
    P_Fin  =  [FinX+RootChord_TP/30.0,BoomY,0.0]
    FinChord  =  Chord_fin/1000.0
    FinSpan  =  Height_fin/1000.0
    FinTC  =  12
    LooseSurf_Fin  =  1
    SegmentNo_Fin=  100

    TipRequired_Fin  =  1
    SectionsRequired_Fin  =  True
    Fin  =  liftingsurface.LiftingSurface(P_Fin,  mySweepAngleFunctionFin,  myDihedralFunctionFin,\
    myTwistFunctionFin,  myChordFunctionFin,  myAirfoilFunctionFin,  LooseSurf_Fin,  SegmentNo_Fin,\
    TipRequired_Fin,  SectionsRequired_Fin)
```

```python
# Specify the desired aspect ratio and span
Fin.TargetAspectRatio = Height_fin*2/Chord_fin
Fin.TargetArea = Chord_fin*Height_fin/10000000.0
Fin.wTargetAspectRatio = 1
Fin.wTargetArea = 1
ChordFactor_Fin = 0.658
ScaleFactor_Fin = 0.208
OptimizeChordScale_Fin=0

SLS_Fin, ActualSemiSpan_Fin, LSP_area_Fin, RootChord_Fin, AR_Fin, SWingTip_Fin, Sections_Fin =\
Fin.GenerateLiftingSurface(ChordFactor_Fin, ScaleFactor_Fin, OptimizeChordScale_Fin)

if Rudder:
    Rflapconfig1 = ['simpleflap', RSpanStart, RSpanEnd, RChord, RTaper, RDeflectionAngle]
    Devices = [Rflapconfig1]
    SRudder, Area, Cutter, Cutter1, Cutter2, CutBrick = HLD.AddHighLiftDevices(SLS_Fin, Devices)
    rs.DeleteObjects([Cutter, Cutter1, Cutter2])
    SLS_Fin = rs.BooleanDifference(SLS_Fin, CutBrick)

SFin = rs.JoinSurfaces([SLS_Fin, SWingTip_Fin])
P_FinAft = [P_Fin[0]+1, P_Fin[1], P_Fin[2]]
RotVec = rs.VectorCreate(P_Fin, P_FinAft)
SFin = rs.RotateObject(SFin, P_Fin, -FinAngle, axis = RotVec)

if Rudder:
    SRudder = rs.RotateObject(SRudder, P_Fin, -FinAngle, axis = RotVec)
    # move fins up to prevent them poking through the bottom of an elevator with setting angle
    SRudder = rs.MoveObject(SRudder, [0.0, 0.0, 0.25*RootChord_TP*math.sin(math.radians\
    (myTwistFunctionTailplane(0.0)))])
    PRudder = act.MirrorObjectXZ(SRudder)

# move fins up to prevent them poking through the bottom of an elevator with setting angle
SFin = rs.MoveObject(SFin, [0.0, 0.0, 0.25*RootChord_TP*math.sin(math.radians\
(myTwistFunctionTailplane(0.0)))])
rs.CapPlanarHoles(SFin)
PFin = act.MirrorObjectXZ(SFin)

rs.DeleteObjects([SLS_Fin, SWingTip_Fin])
```

```python
# =====    CentrePod    ============================================================

if Pod:

    CentrePod = payload.pod([-0.05, 0, z_uncarriage/1000.0/2],\
    PayloadPodLength,14, 0.3)

    P_Fin = [0.05,0,0]
    FinChord = Chord/1000.0
    FinSpan = -z_uncarriage/1000.0/4.0
    FinTC = 12
    LooseSurf_Fin = 1
    SegmentNo_Fin= 2
    TipRequired_Fin = 1
    SectionsRequired_Fin = False

    PodFin = liftingsurface.LiftingSurface(P_Fin,\
    mySweepAngleFunctionTailplane,\
    myDihedralFunctionTailplane, myTwistFunctionTailplane,\
    myChordFunctionTailplane,\
    myAirfoilFunctionTailplane, LooseSurf_Fin, SegmentNo_Fin, TipRequired_Fin,\
    SectionsRequired_Fin)

    # Specify the desired aspect ratio and span
    PodFin.TargetAspectRatio = -z_uncarriage/Chord*2.5
    PodFin.TargetArea = (-z_uncarriage/1000.0/2)*(Chord/1000.0)
    PodFin.wTargetAspectRatio = 1
    PodFin.wTargetArea = 1
    ScaleFactor_Fin = 0.175
    ChordFactor_Fin = 1.0231386102
    OptimizeChordScale_Fin=0

    CLS_Pod, ActualSemiSpan_PodFin, LSP_area_PodFin,  RootChord_PodFin, AR_PodFin,\
    WingTip_PodFin, Sections_PodFin = PodFin.GenerateLiftingSurface(ChordFactor_Fin,\
    ScaleFactor_Fin, OptimizeChordScale_Fin)

    PodFin = rs.JoinSurfaces([CLS_Pod,WingTip_PodFin])
```

```
rs.DeleteObjects([CLS_Pod,WingTip_PodFin])
P_FinAft = [P_Fin[0]+1, P_Fin[1], P_Fin[2]]
RotVec = rs.VectorCreate(P_Fin, P_FinAft)
PodFin = rs.RotateObject(PodFin, P_Fin, 90, axis = RotVec)
CentrePod = rs.BooleanUnion([CentrePod, PodFin])
rs.DeleteObjects(PodFin)


# ===== Fuselage =================================================

FuselageLength = (Len_Nose+2*Len_Engine+x_fnt_bkhd-x_rear_bkhd)/1000.0

FuselageFinennesInPercent = ((Width_Fuse/1000.0)/FuselageLength)*100.0

MainEngineNacelle = payload.pod([-(Len_Nose+x_fnt_bkhd)/1000.0, 0, 0],FuselageLength ,\
FuselageFinennesInPercent, 0.6)

PropDia = Dprop/1000.0
PropX = (2*Len_Engine-x_rear_bkhd-Width_Fuse/4.0)/1000.0

MainPropP1 = rs.AddPoint([PropX,0,PropDia/2])
MainPropP2 = rs.AddPoint([PropX,PropDia/2,0])
MainPropP3 = rs.AddPoint([PropX,0,-PropDia/2])
MainProp = rs.AddCircle3Pt(MainPropP1,MainPropP2,MainPropP3)

PropDisk = rs.AddPlanarSrf(MainProp)

rs.DeleteObjects([MainPropP1,MainPropP2,MainPropP3,MainProp])


# ===== Landing gear =============================================

if NoseGear:
    # Nose gear
    NoseWheelScaleFactor = 1 # with respect to main gear
```

```
        NoseWheelX = -x_fnt_bkhd/1000.0
        NoseWheelCentre = [NoseWheelX, 0, z_uncarriage/1000.0]
        NoseWheelRadius = 0.5*NoseWheelScaleFactor*Dia_Wheels/1000.0
        NoseStrutLength = -z_uncarriage/1000
        CompleteNoseGear = LG.LandingGear(NoseWheelCentre, NoseWheelRadius, NoseStrutLength)
else:
        # Tail gear
        TailWheelScaleFactor = 0.5 # with respect to main gear
        TailWheelX = -x_tail_spar/1000.0
        STailWheelCentre = [TailWheelX, BoomY, z_uncarriage/1000.0/2]
        TailWheelRadius = 0.5*TailWheelScaleFactor*Dia_Wheels/1000.0
        TailStrutLength = -z_uncarriage/1000/2
        TailGearStbd = LG.LandingGear(STailWheelCentre, TailWheelRadius, TailStrutLength)
        TailGearPort = act.MirrorObjectXZ(TailGearStbd)

# MainGear
MainWheelX = -x_rear_bkhd/1000.0
MainWheelY = y_tail_boom/1000.0
MainWheelCentre = [MainWheelX, MainWheelY, z_uncarriage/1000.0]
MainWheelRadius = 0.5*Dia_Wheels/1000.0
MainStrutLength = -z_uncarriage/1000
MainGearPort = LG.LandingGear(MainWheelCentre, MainWheelRadius, MainStrutLength)
MainGearStbd = act.MirrorObjectXZ(MainGearPort)

# ===== Assembly ==============================================
# now union all the bits together to create the airframe
#
TWing = rs.JoinSurfaces([SLS,PLS,SWingTip, PWingTip])
rs.DeleteObjects([SLS,PLS,SWingTip, PWingTip])

if CutStructure:
        TWing = UAT.CutCylinder(RootChord/4.0, -ActualSemiSpan,0.0, RootChord/4.0, ActualSemiSpan, \
        0.0, MainSparODia, TWing)
        Spar = UAT.CreateSpar(RootChord/4.0, -ActualSemiSpan,0.0, RootChord/4.0, ActualSemiSpan, \
        0.0, MainSparODia-SparClearance, MainSparIDia)
        TWing, TWingS1 =      UAT.BooleanSplitY(BoomY+0.1*Chord/1000.0, CutterSize, TWing)
        TWingP1, TWing =      UAT.BooleanSplitY(-BoomY-0.1*Chord/1000.0, CutterSize, TWing)
        TWing, TWingS8 =      UAT.BooleanSplitY(BoomY-0.1*Chord/1000.0, CutterSize, TWing)
        TWingP8, TWing =      UAT.BooleanSplitY(-BoomY+0.1*Chord/1000.0, CutterSize, TWing)
```

```python
TWingS2, TWingS3 = UAT.BooleanSplitY(0.9945*ASpanStart*Span/2000.0-RibThick, CutterSize, TWingS1)
TWingS4, TWingS5 = UAT.BooleanSplitY(0.9945*ASpanStart*Span/2000.0, CutterSize, TWingS3)
TWingS6, TWingS7 = UAT.BooleanSplitY(0.9991*ASpanEnd*Span/2000.0, CutterSize, TWingS5)
TWingP3, TWingP2 = UAT.BooleanSplitY(-0.9945*ASpanStart*Span/2000.0+RibThick, CutterSize, TWingP1)
TWingP5, TWingP4 = UAT.BooleanSplitY(-0.9945*ASpanStart*Span/2000.0, CutterSize, TWingP3)
TWingP7, TWingP6 = UAT.BooleanSplitY(-0.9991*ASpanEnd*Span/2000.0, CutterSize, TWingP5)

SPodSurf = UAT.CutCylinder(RootChord/4.0, -ActualSemiSpan, 0.0, RootChord/4.0, ActualSemiSpan, 0.0, \
MainSparODia, SPodSurf)
PPodSurf = UAT.CutCylinder(RootChord/4.0, -ActualSemiSpan, 0.0, RootChord/4.0, ActualSemiSpan, 0.0, \
MainSparODia, PPodSurf)

TWingS8 = UAT.CutCylinder(RootChord/4.0, BoomY, BoomZ,  BoomLength, BoomY, BoomZ, BoomODia, TWingS8)
TWingP8 = UAT.CutCylinder(RootChord/4.0, -BoomY, BoomZ,  BoomLength, -BoomY, BoomZ, BoomODia, TWingP8)
SPodSurf = rs.BooleanUnion([SPodSurf, TWingS8])
PPodSurf = rs.BooleanUnion([PPodSurf, TWingP8])

# insert aileron control surface hinge spars
if Ailerons:
    print('Aileron hinges OFF')
    xpos1=((1.0-4.9*AChord/6.0)*myChordFunctionPusher(ASpanStart))\
    *RootChord*math.cos(math.radians(SettingAngle))
    ypos1a=0.994*ASpanStart*ActualSemiSpan-RibThick
    ypos1b=-0.994*ASpanStart*ActualSemiSpan+RibThick
    zpos1=-((1.0-4.9*AChord/6.0)*myChordFunctionPusher(ASpanStart)-0.25)\
    *RootChord*math.sin(math.radians(SettingAngle))
    xpos2=((1.0-5.0*AChord/6.0)*myChordFunctionPusher(ASpanEnd))\
    *RootChord*math.cos(math.radians(SettingAngle))
    ypos2=1.001*ActualSemiSpan
    zpos2=-((1.0-5.0*AChord/6.0)*myChordFunctionPusher(ASpanEnd)-0.25)\
    *RootChord*math.sin(math.radians(SettingAngle))
    TWingS4 = UAT.CutCylinder(xpos1, ypos1a, zpos1, xpos1, ypos1a, zpos2, HingeSparODia, TWingS4)
    TWingS7 = UAT.CutCylinder(xpos1, ypos1a, zpos1, xpos1, ypos1a, zpos2, HingeSparODia, TWingS7)
    SAileronSpar = UAT.CreateSpar(xpos1, ypos1a, zpos1, xpos1, ypos1a, zpos2, HingeSparODia\
    -SparClearance, HingeSparIDia)
    TWingP4 = UAT.CutCylinder(xpos1, ypos1b, zpos1, xpos1, ypos1b, zpos2, -ypos2, zpos2, HingeSparODia, TWingP4)
    TWingP7 = UAT.CutCylinder(xpos1, ypos1b, zpos1, xpos1, ypos1b, zpos2, -ypos2, zpos2, HingeSparODia, TWingP7)
    PAileronSpar = UAT.CreateSpar(xpos1, ypos1b, zpos1, xpos1, ypos1b, zpos2, -ypos2, zpos2, HingeSparODia\
    -SparClearance, HingeSparIDia)
```

```
PAileronCutBrick = act.MirrorObjectXZ(SAileronCutBrick)
PAileronCutBrick_cpy = rs.CopyObject(PAileronCutBrick,[0,-RibThick,0])
SAileronCutBrick_cpy = rs.CopyObject(SAileronCutBrick,[0,RibThick,0])

TWingP6 = rs.BooleanDifference(TWingP6,PAileronCutBrick)
TWingP6 = rs.BooleanDifference(TWingP6,PAileronCutBrick_cpy)

TWingS6 = rs.BooleanDifference(TWingS6,SAileronCutBrick)
TWingS6 = rs.BooleanDifference(TWingS6,SAileronCutBrick_cpy)

if VTail == False:
    SFin = UAT.CutCylinder(FinX+RootChord_TP/30.0+RootChord_Fin/4.0,BoomY,BoomZ,\
    FinX+RootChord_TP/30.0+RootChord_Fin/4.0,BoomY,BoomZ+CutterSize, FinSparODia,SFin)
    SFinSpar = UAT.CreateSpar(FinX+RootChord_TP/30.0+RootChord_Fin/4.0,BoomY,BoomZ+BoomODia/2,\
    FinX+RootChord_TP/30.0+RootChord_Fin/4.0,BoomY,BoomZ+ActualSemiSpan_Fin, FinSparODia\
    -SparClearance, FinSparIDia)
    PFin = UAT.CutCylinder(FinX+RootChord_TP/30.0+RootChord_Fin/4.0,-BoomY,BoomZ,\
    FinX+RootChord_TP/30.0+RootChord_Fin/4.0,-BoomY,BoomZ+CutterSize, FinSparODia,PFin)
    PFinSpar = UAT.CreateSpar(FinX+RootChord_TP/30.0+RootChord_Fin/4.0,-BoomY,BoomZ+BoomODia/2\
    ,FinX+RootChord_TP/30.0+RootChord_Fin/4.0,-BoomY,BoomZ+ActualSemiSpan_Fin, FinSparODia\
    -SparClearance, FinSparIDia)

    SFin, SFin1 = UAT.BooleanSplitZ(BoomZ+0.0128,CutterSize, SFin)
    PFin, PFin1 = UAT.BooleanSplitZ(BoomZ+0.0128,CutterSize, PFin)
    SFin1, SFin2 = UAT.BooleanSplitZ(BoomZ+RSpanEnd*ActualSemiSpan_Fin-0.00075,CutterSize, SFin1)
    PFin1, PFin2 = UAT.BooleanSplitZ(BoomZ+RSpanEnd*ActualSemiSpan_Fin-0.00075,CutterSize, PFin1)

    # insert rudder control surface hinge spars
    if Rudder:
        xpos1=FinX+RootChord_TP/30.0+RootChord_Fin/4.0+(0.75-4.9*RChord/6.0)\
        *myChordFunctionFin(RSpanStart)*RootChord_Fin*math.cos(math.radians\
        (myTwistFunctionFin(RSpanStart)))
        xpos2=FinX+RootChord_TP/30.0+RootChord_Fin/4.0+(0.75-4.9*RChord/6.0)\
        *myChordFunctionFin(RSpanEnd)*RootChord_Fin*math.cos(math.radians\
        (myTwistFunctionFin(RSpanEnd)))
        SFin = UAT.CutCylinder(\
        xpos1\
        ,BoomY,BoomZ,\
```

```
xpos2\
,BoomY,BoomZ+ActualSemiSpan_Fin ,HingeSparODia ,SFin)
SRudder = UAT.CutCylinder(xpos1 ,BoomY ,BoomZ ,xpos2 ,BoomY ,BoomZ+ActualSemiSpan_Fin ,\
HingeSparODia ,SRudder)
SFin2 = UAT.CutCylinder(xpos1 ,BoomY,BoomZ, xpos2 ,BoomY ,BoomZ+ActualSemiSpan_Fin +0.001, \
HingeSparODia ,SFin2)
SRudderSpar = UAT.CreateSpar(xpos1 ,BoomY ,BoomZ, xpos2 ,BoomY ,BoomZ+ActualSemiSpan_Fin +0.001,\
HingeSparODia-SparClearance , HingeSparIDia)
PFin = UAT.CutCylinder(xpos1 ,-BoomY,BoomZ, xpos2 ,-BoomY ,BoomZ+ActualSemiSpan_Fin ,\
HingeSparODia, PFin)
PRudder = UAT.CutCylinder(xpos1 ,-BoomY ,BoomZ, xpos2 ,-BoomY ,BoomZ+ActualSemiSpan_Fin , \
HingeSparODia , PRudder)
PFin2 = UAT.CutCylinder(xpos1 ,-BoomY ,BoomZ, xpos2 ,-BoomY ,BoomZ+ActualSemiSpan_Fin +0.001, \
HingeSparODia , PFin2)
PRudderSpar = UAT.CreateSpar(xpos1 ,-BoomY ,BoomZ, xpos2 ,-BoomY ,BoomZ+ActualSemiSpan_Fin +0.001, \
HingeSparODia-SparClearance , HingeSparIDia)

Tail = rs.BooleanUnion([Tail , SFin , PFin])
Tail = UAT.CutCylinder(BoomLength,-ActualSemiSpan_TP ,BoomZ, BoomLength ,\
ActualSemiSpan_TP ,BoomZ, TailSparODia , Tail)
Tail = UAT.CutCylinder(FinX+RootChord_TP/30.0+RootChord_Fin/4.0 ,BoomY,BoomZ,\
FinX+RootChord_TP/30.0+RootChord_Fin/4.0 ,BoomY ,BoomZ+ActualSemiSpan_Fin , FinSparODia , Tail)
Tail = UAT.CutCylinder(FinX+RootChord_TP/30.0+RootChord_Fin/4.0 ,-BoomY ,BoomZ,\
FinX+RootChord_TP/30.0+RootChord_Fin/4.0 ,-BoomY ,BoomZ+ActualSemiSpan_Fin , FinSparODia , Tail)
TailSpar = UAT.CreateSpar(BoomLength,-ActualSemiSpan_TP ,BoomZ, BoomLength , BoomLength ,\
ActualSemiSpan_TP ,BoomZ, TailSparODia-SparClearance , TailSparIDia)

Tail , TailS1 = UAT.BooleanSplitY(BoomY+0.05*Chord/1000.0 ,CutterSize , Tail)
TailP1 , Tail = UAT.BooleanSplitY(-BoomY-0.05*Chord/1000.0 ,CutterSize , Tail)
Tail , TailS2 = UAT.BooleanSplitY(BoomY-0.05*Chord/1000.0 ,CutterSize , Tail)
TailP2 , Tail = UAT.BooleanSplitY(-BoomY+0.05*Chord/1000.0 ,CutterSize , Tail)
Tail , TailS3 = UAT.BooleanSplitY(ESpanStart*ActualSemiSpan_TP+0.00175 ,CutterSize , Tail)
TailP3 , Tail = UAT.BooleanSplitY(-ESpanStart*ActualSemiSpan_TP-0.00175 ,CutterSize , Tail)
TailS1 , TailS4 = UAT.BooleanSplitY(ESpanEnd*ActualSemiSpan_TP-0.002 ,CutterSize , TailS1)
TailP4 , TailP1 = UAT.BooleanSplitY(-ESpanEnd*ActualSemiSpan_TP+0.002 ,CutterSize , TailP1)

TailS2 = UAT.CutCylinder(0.25*Chord/1000.0 ,BoomY ,BoomZ, BoomLength ,BoomY ,BoomZ, BoomODia , TailS2)
TailP2 = UAT.CutCylinder(0.25*Chord/1000.0 ,-BoomY ,BoomZ, BoomLength ,-BoomY ,BoomZ, BoomODia , TailP2)
```

```python
# insert elevator control surface hinge spars
if Elevators:
    xpos1=BoomLength+(0.75 -4.9*EChord/6.0)*myChordFunctionTailplane(ESpanStart)\
        *RootChord_TP*math.cos(math.radians(myTwistFunctionTailplane(ESpanStart)))
    zpos1=(0.75 -4.9*EChord/6.0)*myChordFunctionTailplane(ESpanStart)\
        *RootChord_TP*math.sin(math.radians(myTwistFunctionTailplane(ESpanStart)))
    xpos2=BoomLength+(0.75*(1.0 -EChord))*myChordFunctionTailplane(ESpanEnd)\
        *RootChord_TP*math.cos(math.radians(myTwistFunctionTailplane(ESpanEnd))\
    zpos2=(0.75*(1.0 -EChord))*myChordFunctionTailplane(ESpanEnd)\
        *RootChord_TP*math.sin(math.radians(myTwistFunctionTailplane(ESpanEnd)))
    Tail = UAT.CutCylinder(xpos1,-ActualSemiSpan_TP, zpos1, xpos2, ActualSemiSpan_TP, zpos2, \
        HingeSparODia, Tail)
    TailS4 = UAT.CutCylinder(xpos1, 0.0, zpos1, xpos2, ActualSemiSpan_TP, zpos2, \
        HingeSparODia, TailS4)
    SElevator = UAT.CutCylinder(xpos1, 0.0, zpos1, xpos2, ActualSemiSpan_TP, zpos2, \
        HingeSparODia, SElevator)
    SElevatorSpar = UAT.CreateSpar(xpos1, 0.0, zpos1, xpos1, xpos2, ActualSemiSpan_TP, zpos2, \
        HingeSparODia-SparClearance, HingeSparIDia)
    TailP4 = UAT.CutCylinder(xpos1, 0.0, zpos1, xpos2,-ActualSemiSpan_TP, zpos2, \
        HingeSparODia, TailP4)
    PElevator = UAT.CutCylinder(xpos1, 0.0, zpos1, xpos2,-ActualSemiSpan_TP, zpos2, \
        HingeSparODia, PElevator)
    PElevatorSpar = UAT.CreateSpar(xpos1, 0.0, zpos1, xpos1, xpos2,-ActualSemiSpan_TP, zpos2, \
        HingeSparODia-SparClearance, HingeSparIDia)


if VTail:
    if CutStructure:
        AirFrame = rs.BooleanUnion([SPodSurf, TWing, MainEngineNacelle, PPodSurf])
    else:
        AirFrame= rs.BooleanUnion([SBoom, SPodSurf, PBoom, TWing, MainEngineNacelle, \
            PPodSurf, Tail, TailP, CompleteNoseGear, MainGearPort, MainGearStbd])
else:
    if CutStructure:
        AirFrame = rs.BooleanUnion([TWing, MainEngineNacelle])
    else:
        AirFrame= rs.BooleanUnion([SBoom, SPodSurf, PBoom, TWing, MainEngineNacelle, \
            PPodSurf, Tail, SFin, PFin, CompleteNoseGear, MainGearPort, MainGearStbd])

# ===== end of geometry construction ==================================================
```

```
# remove unwanted construction entities

rs.DeleteObjects(Sections)
rs.DeleteObjects(Sections_TP)
rs.DeleteObject(SLScopy)

if VTail == False:
    rs.DeleteObjects(Sections_Fin)

rs.DeleteObject(SLS)
```

Appendix C

Worked (Manned Aircraft) Detail Design Example

This example shows the generation of parametric preliminary design geometry based loosely around the Vans RV7 aircraft[1] shown in Figure C.1. Although this is a manned aircraft, the procedure is fundamentally no different than that used for the development of unmanned aircraft.

C.1 Stage 1: Concept Sketches

At the concept stage of an aircraft design, the team needs to make sure that they understand all the constraints and preferences of the proposed customers. The team will have considered many different configurations and selected the best configuration (and record here the logical arguments to justify this). This configuration must then be turned into an initial sketch to illustrate the topology and initial sizing of the aircraft (hand sketches are fine at this stage). Such sketches are shown in Figure C.2 and can be used as the basis for creating a CAD-based geometry model.

The sketch is imported into Solidworks part using the "Tools/Sketch Tools/Sketch Picture" commands as explained in Chapter 17. This is shown in Figure C.3, where a "side elevation" has been imported and accurately scaled to be exactly 6.2 m within the part.

Hand sketches of the plan view and front elevation are also required to capture the configuration and initial sizing. These are also imported and scaled. Each view is placed on the relevant orthogonal plane, as shown in Figure C.4. These hand sketches can now be used to capture the essential geometry and parameters to generate a parametric model of the whole aircraft.

First, a Solidworks assembly is created and the hand sketch images are imported into this assembly as a part. This allows these hand sketch images to be switched off later in the design process. To begin, a sketch is traced on top of the hand sketch images. This must capture all

[1] The full manufacturing drawings and assembly instructions can be downloaded from http://smilinpete.com/docs/RV-7_Preview_Drawings_Searchable.pdf

Small Unmanned Fixed-wing Aircraft Design: A Practical Approach, First Edition.
Andrew J. Keane, András Sóbester and James P. Scanlan.
© 2017 John Wiley & Sons Ltd. Published 2017 by John Wiley & Sons Ltd.

Figure C.1   Vans RV7 Aircraft. Cropped image courtesy Daniel Betts https://creativecommons.org/licenses/by-sa/2.0/ – no copyright is asserted by the inclusion of this image.

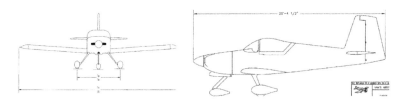

Figure C.2   Concept sketches of an aircraft.

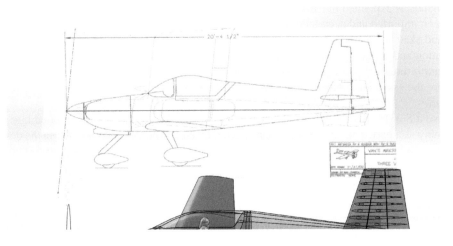

Figure C.3   Side elevation hand sketch imported and scaled.

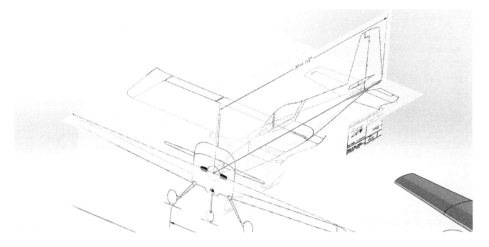

Figure C.4   Plan, side, and front view imported and scaled.

the essential geometric features of the hand sketch. The level of fidelity is left to the user. However, it is better to keep things as simple as possible, as the complexity of the model can become overwhelming. Here, for example, the essential geometry sketch of the side elevation shown in Figure C.5 uses only straight lines and does not attempt to follow the hand sketch contours of things such as the canopy and fairings. This detail is best left out at this stage, as it can be added later.

These Solidworks parametric geometry sketches are created as "assembly level" sketches and do not belong to any one particular part. This is very important, as all the subsequent

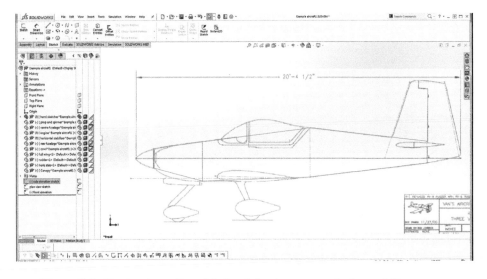

Figure C.5   Tracing the outline of the hand sketch to capture the "essential" geometry.

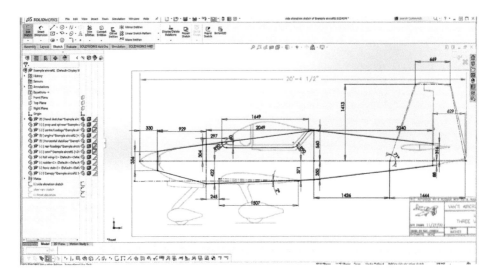

Figure C.6   Dimensioned parametric geometry sketch.

aircraft parts need to be connected to these sketches and the assembly level sketches therefore need to be at the top of the hierarchy. The parametric geometry sketches now need to be
dimensioned so that they can be referenced by the various parts in the assembly. This is shown
in Figure C.6.

Similar dimensioned parametric geometry sketches need to be created for the plan and front
views. We can see in Figure C.7 that this can get very complex. It is important to be very
selective in capturing only the essential geometry at this stage, see Figure C.8. Too much

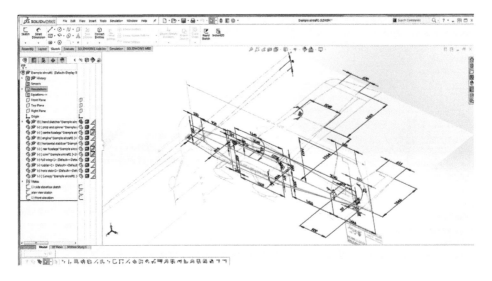

Figure C.7   View of all three of the dimensioned parametric geometry sketches.

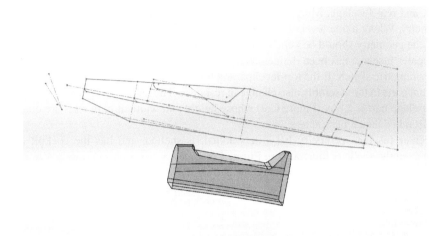

Figure C.8   Center fuselage part, with side elevation parametric geometry sketch in the background.

detail can be overwhelming. It is also crucial that one becomes familiar with how to filter out clutter in a view by switching off dimensions, annotations, and other parts.

C.2   Stage 2: Part Definition

Parts can now be defined, constructed, and tied in with the driving top-level parametric geometry sketches. Notice that the solid parts that are being realized are displaced to the side of the parametric geometry. This declutters the view and allows the emerging parts and driving sketch to be displayed next to each other.

At this stage, decisions have to be made concerning the breakdown of the aircraft into sections. For the fuselage, it has been decided to break it into three sections: nose, center

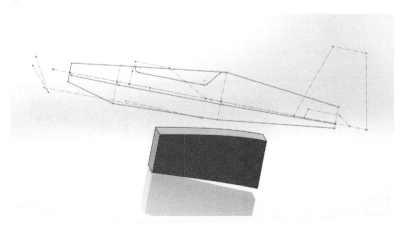

Figure C.9   Underlying geometry for the center fuselage.

fuselage, and rear fuselage. More sections could be used, but this adds to the complexity of the geometry. Where a fine level of resolution is required, this might be necessary, but where possible the geometry should be kept as simple as possible.

This center fuselage has been defined such that all the key dimensions are linked with the driving parametric sketch. If the top-level parametric sketch dimensions are changed, then all the relevant parts in the assembly also change in response. The underlying geometry behind the center fuselage part is a box (Figure C.9) whose height and width at both ends are controlled by the assembly sketch parameters.

The center fuselage geometry has fillets added, is shelled, and has the cockpit cut-out removed from the shell, as illustrated in Figure C.10.

Next, the rear fuselage has been added, as shown in Figure C.11. Note that there is some shared geometry between these two parts. Hence at the interface between them they share the

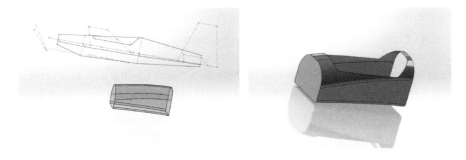

Figure C.10   Completion of center fuselage.

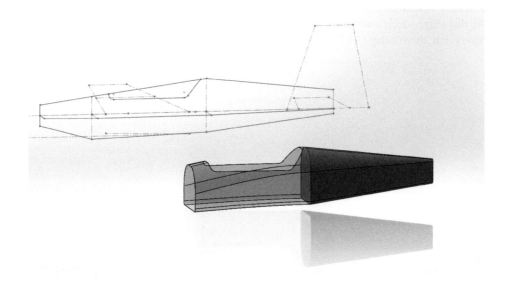

Figure C.11   Rear fuselage synchronized with center fuselage at shared interface.

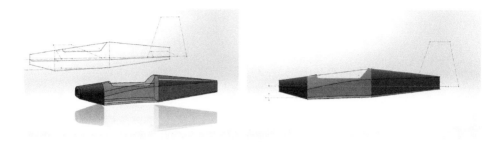

Figure C.12   Fully realized fuselage geometry.

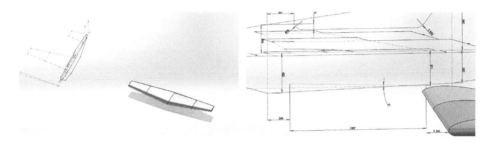

Figure C.13   Two-panel wing and wing incidence and location line in side elevation.

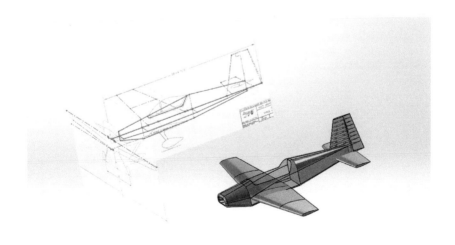

Figure C.14   All the major airframe surface parts added.

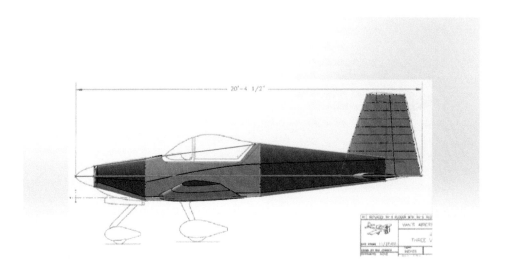

Figure C.15   Checking against original sketch.

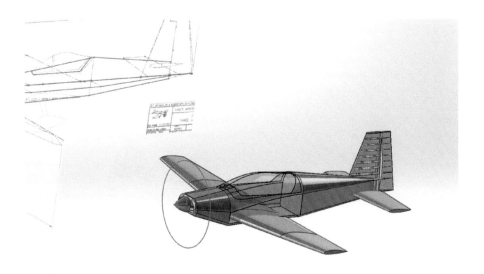

Figure C.16   Addition of propeller disk and spinner so that ground clearance can be checked.

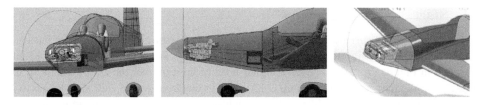

Figure C.17 Engine installation checking cowling clearance and cooling (note: lightweight decal engine geometry).

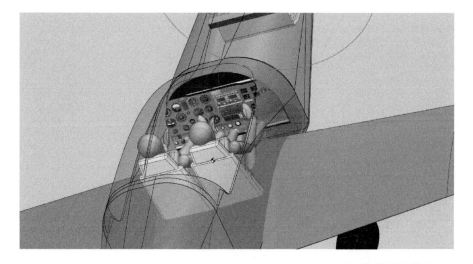

Figure C.18 Checking the instrument panel fit (again use of decal for instruments).

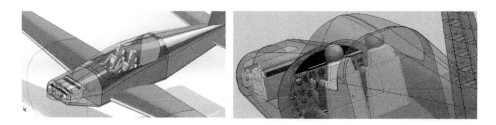

Figure C.19 Checking the ergonomics of crew seating and canopy clearance/view.

same fuselage width, height, and radii of curvature of the upper and lower fillets. Leaving these independent causes a nonsense geometry to be created and adds to the number of (apparently) independent driving parameters and hence complexity.

Finally, the nose section of the fuselage is put in place. Note in Figure C.12 how the emerging aircraft is synchronized with the background driving sketch.

C.3   Stage 3: "Flying Surfaces"

Next come the 'flying surfaces' of wings and empennage. Starting with the wings, it has been decided to model these in the form of two panels per wing, as discussed in Chapter 17 and illustrated in Figure 17.32. This gives a higher level of control over the configuration of the wings. The wing part is shown in Figure C.13 linked to the driving plan view sketch next to it. An important detail from the side elevation sketch is the location of the wing in the fuselage as well as the angle of incidence of the whole wing. The dihedral is defined by a parameter in the front elevation sketch.

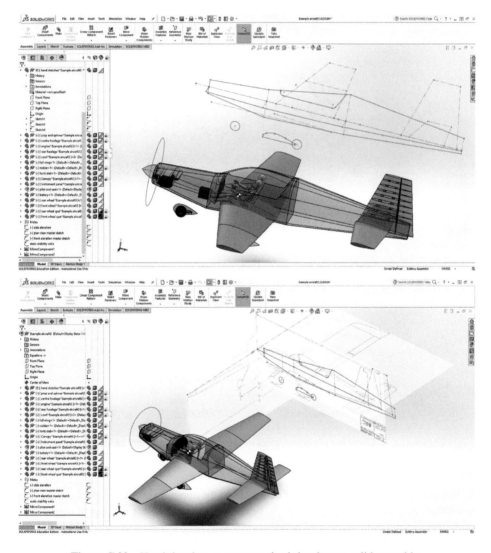

Figure C.20   Hand sketches–to parameterized sketches–to solid assembly.

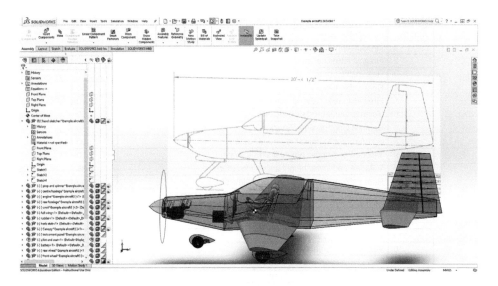

Figure C.20   (*Continued*)

Similarly, the horizontal and vertical stabilizers are added to the fuselage. This gives the basis to the whole aircraft to which detail can now be added. At this stage, all of the major items can be added to the aircraft and the mass distribution and center of gravity can be investigated. Also, the fit of the major items can be checked. In particular, the cockpit size and ergonomics can be checked with a 90 percentile mannequin. Further details are added, as shown in Figures C.14–C.19. The whole process is illustrated in Figure C.20.

C.4   Stage 4: Other Items

Because the sketches are defined as belonging to the assembly (rather than any one particular part), it is now easy to create a design table for integration with design calculations. It is worth noting that even this "simple" and somewhat crude aircraft model needs nearly *50 independent variables* to control it (Figure C.21). However, the assembly sketches can be used to do simple geometrical calculations to allow, for example, the static margin to be calculated as a "live" output by using the center of gravity feature as in Figure C.22.

C.5   Stage 5: Detail Definition

In the final detail design stage, the geometry is refined to the point where all geometry is fully defined as shown in Figure C.23. The geometry shown is still not sufficiently detailed for manufacture in this case, however. Figure C.24 shows a view of the manufacturer's full CAD model of the aircraft. Current CAD tools are simply not sophisticated enough to allow a fully *parameterized* detail assembly to be constructed and maintained without very considerable effort. Even if this becomes possible, the model will tend to be very fragile and capable of

Number	Parameter name	Value
1	Firewall upper height@side elevation	304
2	Firewall lower height@side elevation	422
3	Rear cabin upper height@side elevation	700
4	Rear cabin lower height@side elevation	350
5	Firewall to rear cabin@side elevation	2049
6	Spinner diameter@side elevation	356
7	Spinner length@side elevation	330
8	Spinner rear to firewall@side elevation	1500
9	Tailcone end upper height@side elevation	316
10	Tailcone end lower height@side elevation	88
11	Rear cabin to end tailcone@side elevation	3000
12	Dist or wing root TE from mid fuselage@side elevation	371
13	Wing root chord@side elevation	1500
14	Wing incidence@side elevation	4
15	Vert stab root chord@side elevation	1444
16	Vert stab tip chord@side elevation	700
17	Vert stab sweep angle@side elevation	71
18	Vert stab height@side elevation	1600
19	Rear cabin to vert stab root LE@side elevation	1426
20	Firewall to wing root LE@side elevation	245
21	Dist vert stab TE to end tailcone@side elevation	131
22	Canopy to firewall@side elevation	297
23	Rad of cockpit cut out@side elevation	90
24	Rad cockpit cutout@side elevation	34
25	Rim height cockpit@side elevation	60
26	Canopy length@side elevation	1649
27	Mainwheel horiz from wingTE@side elevation	800
28	Mainwheel vert from HFD@side elevation	900
29	Front wheel dist from HFD@side elevation	830
30	Front wheel dist from firewall@side elevation	671
31	Nose half width@plan view master sketch	377
32	Firewall half width@plan view master sketch	496
33	Rear cabin half width@plan view master sketch	436
34	Tail end half width@plan view master sketch	57
35	Panel 1 span@plan view master sketch	1964
36	Panel 1 sweep@plan view master sketch	80
37	Panel 1 tip chord@plan view master sketch	1049
38	Panel 2 span@plan view master sketch	1561
39	Panel 2 tip chord@plan view master sketch	670
40	Panel 2 sweep@plan view master sketch	80
41	Tailplane tip chord@plan view master sketch	615
42	Tailplane sweep@plan view master sketch	82
43	Tailplane semi span@plan view master sketch	1346
44	D16@plan view master sketch	470
45	Canopy cut out length@plan view master sketch	1204
46	Dist cockpit cutout from firewall@plan view master sketch	656
47	Main wheel semi track@front elevation master sketch	1465

Figure C.21 Whole aircraft parametric variables.

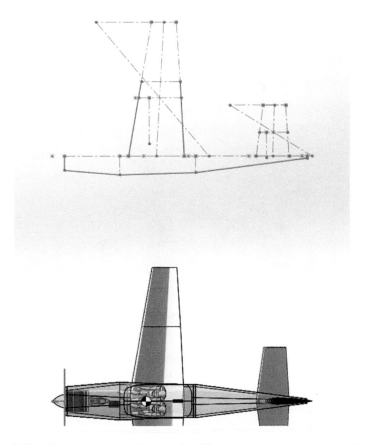

Figure C.22 Wing geometry used to calculate lift centers for static margin calculations.

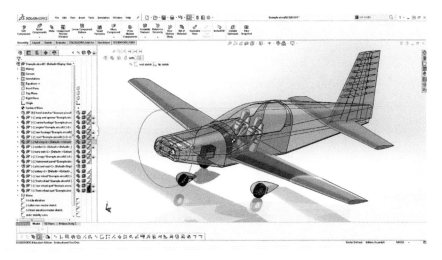

Figure C.23 Final parametric aircraft design with all major masses added.

Figure C.24   Final detailed geometry. Courtesy of Vans Aircraft, Inc.

only small adjustments to geometry. Hence it is important to do as much analysis and design checking at the preliminary design stage so that these design decisions can be "frozen" and the next level of detail created.

Bibliography

[1] Torenbeek, E. (1982) *Synthesis of Subsonic Airplane Design*, Kluwer Academic Publishers, Dordrecht, ISBN: 90-247-2724-3.

[2] van Schaik, J.R., Scanlan, J.P., Keane, A.J., Takeda, K., and Gorissen, D. (2011) Improving design rationale capture during embodiment design, in *Proceedings of International Conference on Engineering Design, ICED11* (eds T. Howard, K. Mougaard, T. McAloone, and C. Hansen), The Design Society, pp. 436–445, ISBN: 978-1-904670-32-2.

[3] Gorissen, D., Quaranta, E., Ferraro, M., Scanlan, J., Keane, A., and Takeda, K. (2012) Architecting a Decision Environment for Complex Design Evaluation. Proceedings of the 53rd AIAA/ASME/ASCE/AHS/ASC Structures, Structural Dynamics and Materials Conference, Honolulu, AIAA-2012-1806, doi: 10.2514/6.2012-1806.

[4] Gorissen, D., Quaranta, E., Ferraro, M., Schumann, B., van Schaik, J., Gisbert, M.B.I., Keane, A., and Scanlan, J. (2014) Value-based decision environment: vision and application. *J. Aircr.*, **51** (5), 1360–1372.

[5] Stinton, D. (1983) *The Design of the Aeroplane*, Blackwell Science, Oxford, ISBN: 0-632-01877-1.

[6] Cousin, J. and Metcalf, M. (1990) The BAE (Commercial Aircraft) LTD Transport Aircraft Synthesis and Optimization Program. AIAA/AHS/ASEE Aircraft Design, Systems and Operations Conference, AIAA 90-3295, Dayton.

[7] Keane, A.J. (2003) Wing optimization using design of experiment, response surface, and data fusion methods. *J. Aircr.*, **40** (4), 741–750.

[8] Walsh, J., Townsend, J., Salas, A., Samareh, J., Mukhopadhyay, V., and Barthelemy, J.-F. (2000) Multidisciplinary High-Fidelity Analysis and Optimization of Aerospace Vehicles, Part 1: Formulation. Proceedings of the 38th Aerospace Sciences Meeting and Exhibit. AIAA 2000-0418, Reno, NV.

[9] Hoerner, S.F. (1992) *Fluid-Dynamic Drag*, Hoerner Fluid Dynamics, Bakersfield, CA.

[10] Hoerner, S.F. (1992) *Fluid-Dynamic Lift*, Hoerner Fluid Dynamics, Bakersfield, CA.

[11] Raymer, D. (2012) *Aircraft Design: A Conceptual Approach*, AIAA Education Series, 3rd edn, AIAA.

[12] Brandt, J.B. and Selig, M.S. (2011) Propeller Performance Data at Low Reynolds Numbers. Proceedings of the 49th AIAA Aerospace Sciences Meeting, Orlando, AIAA-2011-1255.

[13] Falk, M. and Logan, G. (2002) *Systems Engineering and Open Systems, SYS-301*, Defense Acquisition University.

[14] ACSYNT (1995) *ACSYNT: Installation Manual and User Guide, Version 3.0*, ACSYNT Institute, Virginia Polytechnic Institute and State University, Blacksburg, VA 24061-0002.

Small Unmanned Fixed-wing Aircraft Design: A Practical Approach, First Edition.
Andrew J. Keane, András Sóbester and James P. Scanlan.
© 2017 John Wiley & Sons Ltd. Published 2017 by John Wiley & Sons Ltd.

[15] Gudmundsson, S. (2014) *General Aviation Aircraft Design*, Butterworth-Heinemann, Oxford.

[16] Koza, J. (1992) *Genetic Programming: On the Programming of Computers by Means of Natural Selection*, MIT Press, Cambridge, MA.

[17] Sobester, A. (2014) Four Suggestions for Better Parametric Geometries. 10th AIAA Multidisciplinary Design Optimization Conference, National Harbor, MD.

[18] Gundlach, J. (2012) *Designing Unmanned Aircraft Systems: A Comprehensive Approach*, AIAA, Reston, VA.

[19] Anderson, J.D. Jr. (2011) *Introduction to Flight*, 7th edn, McGraw Hill, New York.

[20] Abbott, I.H. and von Doenhoff, A.E. (1959) *Theory of Wing Sections: Including a Summary of Airfoil Data*, Dover Publications, New York.

[21] Sivells, J.C. and Spooner, S.H. (1949) Investigation in the Langley 19-Foot Pressure Tunnel of Two Wings of NACA 65-210 and 64-210 Airfoil Sections with Various Type Flaps. Technical report 942, NACA.

[22] Gentry, G.L., Bezos, G.M., Dunham, R.E. Jr., and Melson, W.E. Jr. Wind Tunnel Aerodynamic Characteristics of a Transport-Type Airfoil in a Simulated Heavy Rain Environment. Nasa-cr-181119, NASA, 1992.

[23] Etkin, B. and Reid, L.D. (1996) *Dynamics of Flight Stability and Control*, 3rd edn, John Wiley & Sons, Inc., New York.

[24] Phillips, W. (2000) Improved closed-form approximation for Dutch roll. *Journal of Aircr.*, **37**, 484–490.

[25] Phillips, W. (2000) Phugoid approximation for conventional airplanes. *J. Aircr.*, **37**, 30–36.

[26] Mader, C.A. and Martins, J. (2011) Computation of aircraft stability derivatives using an automatic differentiation adjoint approach. *AIAA J.*, **49**, 2737–2750.

[27] Golland, L. (1952) Aileron Reversal of Straight and Swept Wings at High Subsonic Speeds. Technical report 52-231, Wright Air Development Center.

[28] Karuzic, F.G. (2007) Conceptual design of silent and green airliners. Msc thesis. Cranfield.

Index

2SEAS, 391
3D printing, 313, 332

Abaqus, 240, 247
accelerometer, 297
acceptance tests, 352, 358, 368
Access Hatches, 54
Accident List, 364
Accident Sequences and Mitigation, 366
ACIS, 191, 242
ACSYNT, 123
ADS, 123
aero-elasticity, 265, 297
aerocalc, 147
aerodynamics, 125
    airframe, 214
    codes, 125
    lifting surfaces, 21
    RANS, 228
    simple wing theory, 33
ailerons, 22, 40, 80, 221, 269
air traffic, 353, 368
AirCONICS, 189
    designing Decode-1, 192
    example code for Decode-1, 399
    structural modeling, 240
airfoil, 191
    three-dimensional analysis, 210
    two-dimensional analysis, 208
Airframe Load Tests, 290

airworthiness, 103, 349, 375
Anaconda, 147
Ancillaries, 88
Ansys Fluent, 196, 200
AnyLogic, 125
Apache, 126
approach speed, 145, 149, 180, 376
approval, *see* Regulatory Approval
Arduino, 86
aspect ratio, 21, 34, 162, 167, 170, 206, 266
Assembly Mechanisms, 54, 342
auto landing, *see* Flight Tests
auto take-off, *see* Flight Tests
autonomous flight, 4, 357, 378
autopilot, 73, 84, 86
    data capture, 368
    failure, 108
    radio channels, 357
    testing, 378
Aviation Authority Requirements, 349
avionics, 73
    diagram, 73
    placement, 47
    power supplies, 76
    System Description, 356
    testing, 300
    trays, 50

batteries, 71, 383
bayonets, 345

Small Unmanned Fixed-wing Aircraft Design: A Practical Approach, First Edition.
Andrew J. Keane, András Sóbester and James P. Scanlan.
© 2017 John Wiley & Sons Ltd. Published 2017 by John Wiley & Sons Ltd.

beam theory, 239, 243
beyond line of sight, *see* BLOS
blockage, *see* wind tunnels tests
BLOS, 108, 359
bolts, conventional, 346
Boxer, 204
Breguet range equation, 125, 169
bulk modulus, 334

CAD Codes, 120
camera mountings, 51
carbon fiber, 332
Cases, Storage and Transport, 347
cell height
    boundary layer, 203
Centaur, 204
Center of Gravity, *see* CoG
CFD, 11, 119, 195
CFRP, 332
chopped strand, 340
chord, 177, 319
Clamps, 346
climb performance, 148, 167, 174, 224, 375
coefficient
    flap hinge moment, 221
    of drag, 35, 151, 167, 202, 216, 285
    of friction, 167
    of lift, 33, 35, 41, 124, 149, 167, 176,
        198, 202, 230, 268
    of parasitic drag, 167, 232
    of pitching moment, 35, 221, 223, 268
    of roll, 223
    of thrust, 174
    of yaw, 223
    tail volume, 221
CoG, 124, 176, 273, 310
    control, 279
    Estimation, 170
    typical table, 273
    UAS CoG flight test, 377
compliance testing, 113
components, externally sourced, 4, 331
composites, 340
Computational Fluid Dynamics, *see* CFD
constraint analysis, 144
    the constraint space, 146

Constraints, 101
contingency
    planning, 371
    weight, 278
control reversal, 265, 297
    estimating onset speeds, 266
control surface, 22, 82, 262, 272, 290, 352,
        382
    failure, 108
    inputs, 370
control system, 24, 73, 356
    software management, 368
costs, 125
    management, 114
covers, 37
cruise performance, 148

data-bases, 126
database, 352
databases, 363
decision making, 7
decision support, 123, 125
DECODE, 6
Decode-1
    AirCONICS, 192
    AirCONICS code, 399
    analysis with Fluent, 228
    analysis with XFLR5
        Aerodynamics, 215
        Control Surfaces, 221
        Stability, 223
    control surfaces, 221
    detail design, 313
    FEA, 245
    Fluent, 228
    ground vibration test, 297
    lifting surfaces, 230
    spar deflections, 244
    spar layout, 246
    spreadsheet, 174
    stability, 223
    Vn diagram, 238
    wind tunnel testing, 284
    XFLR5, 215
Decode-2
    spreadsheet, 177

design
  algorithm, 146
  checklist, 117
  concept, 8, 123
    initial constraint analysis, 127
    managing the process, 144
  constraints, 129
    computational implementation, 145
    constraint analysis report, 146
  detail, 11
    Decode-1, 313
    fuselage, 306
    hand sketches, 311
    master sketches, 311
    wings, 318
  drivers, 29
  in-service and de-commissioning, 13
  manufacturing, 12
  preliminary, 10
    aerodynamic and stability analysis,
      195
    geometry, 189
    structural analysis, 237
  requirements, 145
  responsibility allocation, 112
  the brief, 113, 127
  the process of, 101
  the stages of, 6
  topology, 130
detail design, see design
Digital DatCom, 229
dihedral, 191
divergence, 265, 297
  estimating onset speeds, 266
Documentation, 349
drag
  induced, 42
Dutch roll, 223

electric motors
  see motors, 66
elevators, 22, 80, 139, 221, 269, 287
  structure, 240
  XFLR5, 200
Emergency Procedures, 360
empennage, 45

endurance, 15, 62, 72, 77, 89, 102, 129,
  166, 355
engines
  control, 70
  failure, 107
  glow-plug IC, 62
  IC Liquid Fueled, 59
  mountings, 48
  servicing, 382
  spark ignition gasoline IC, 62
  testing, 65
environment, 128
epoxy, 340
ESDU, 229
Euler-Bernoulli beam theory, 243
EVLOS, 359
Experimental Testing and Validation, 281
Expired Items
  Time and Flight, 382
extended visual line of sight, see EVLOS

failure modes, 104
  aerodynamic and control, 105
  autopilot, 108
  control surface, 108
  effects, 362
  engine, 107
  motor, 107
  primary Tx/Rx, 108
  safety case, 350
  structural, 106
FDM, 335
  ABS, 335
FEA, 11, 245
  analysis of 3D printed parts, 255
  analysis of fiber or Mylar clad foam
    parts, 255
  complete spar and boom model,
    250
  model preparation, 246
Finite Element Analysis, see FEA
flaps, 22, 41, 221, 269
Flight Control Software, 383
flight envelope, 369
Flight Planning Manual, 368
flight simulators, 227

flight tests
    auto landing, 380
    auto take-off, 380
    Autonomous Flight Control, 378
    Engine Failure, Idle and Throttle
        Change, 377
    Fuel Consumption, 377
    Operational and Safety Flight Scenarios,
        381
    safety, 361
    UAS CoG (MANUAL mode), 377
    UAS performance (MANUAL mode),
        375
Fluent, see Ansys Fluent
flutter, 265, 297
    estimating onset speeds, 266
foam, 337
    cladding
        fiber, 339
        Mylar, 339
    hot-wire cutting, 337
Fuel Systems, 70
fuel tanks, 24
    integral, 44, 52
Fused Deposition Modeling, see FDM
fuselage, 23, 45
    detail design, 306

generators, 71
Geometry Codes, 120
glass fiber, 340
glow-plug, see engines
Goals, 101
GPS, 25
Gridgen, 204
ground control system, see control
        system
ground vibration test, see structural
        testing
gust load, 238

Hazards
    Operational, 363
health monitoring, 17, 90
hinge moment, 221
horseshoe vortex, 198

IGES, 191, 242
induced drag, see drag
inertia
    stability, 226
    structural dynamics, 265
    XFLR5, 198
Instrumentation and recording of flight test
        data, 370
interface definitions, 112
iron bird, 342
ISA, 128

Javaprop, 68, 166
Jupyter, 145, 146

k-ω , 205
k-ω SST, 200

landing gear, 27, 191
Laser Cutting, 339
laser sintering
        see SLS, 332
Lessons Learned, 385
lifting line, 198
Lifting Surfaces, 21
LiPo see batteries, 71
locking pins, 345
Longitudinal Stability, 170

Maintenance, 381
    Overall Airframe, 382
    Record Keeping, 384
    Schedule, 360
maneuver load, 238
Manufacture, 331
Manufacturing Methods, 5
Maximum Take-Off Weight, see MTOW
Meshing, 204
mission, 128
Mission Planning, 125
mode shape, 269, 296
morphology, 7
motors
    brushless, 66, 77
    control, 70
    failure, 107
    mountings, 48

MTOW, 15, 60, 146, 166, 247, 310
Mylar, 339

NACA airfoil, 197, 230, 320
nacelle structure, 45
NASA, 134, 196
natural frequency, 269, 296

OpenFoam, 196
OpenVSP, 123
Operational Simulation, 125
Operations Manual, 358
    Brief Technical Description, 359
    Maintenance Schedule, 360
    Operating Limits, Conditions, and
        Control, 359
    Operational and Emergency Procedures,
        360
    Operational Area and Flight Plans, 360
    Organization, 359
    Team Roles, 359
Oswald span efficiency, 153, 170

PaceLab, 123
panel method, 196, 198, 262
    XFoil and XFLR5, 196
parasitic drag, 226
Parasolid, 191, 242
payload, 20, 27, 51
Payload Communications Systems, 87
phugoid, 223
pitch, 172
Pitot static, 87, 369, 384
Pixhawk, 87
Plane Maker, 228
Poisson's ratio, 248, 334
polyester, 340
polyurethane, 337
Powering, 171
preliminary design, see design
Primary Control Transmitter and Receivers,
    73
Primary Tx/Rx Failure, 108
propeller, 20, 68
    actuator disc, 193, 233
    data, 166

diameter, 172, 177
    sizing, 171
Propulsion, 24, 59
Python, 147, 190

radio links, 73, 87
range, 15, 89, 102, 129, 169
RANS, 196
    Solvers – Fluent, 200
receiver, 24, 358
    GPS, 90
    primary control, 73
    video, 53
Record Keeping
    Maintenance, 384
Regulatory Approval, 349, 368, 383
reliability, 65, 107, 377
requirements, see design
requirements flowdown, 112
Resilience and Redundancy, 90
Rhinoceros CAD package, 190
ribs, 38
ring vortex, 198
Risk Assessment Process, 362
roll control, 40
roll damping, 223
rudders, 22, 139, 221, 269, 290

safety case, 361
Schrenk approximation, 240
screws, conventional, 346
sealing, components, 335
Selective Laser Sintering, see SLS
sense and avoid, 27
servos, 78
    testing, 302
shear modulus, 334
short period modes, 223
SIMPLE solution method, 200
simulators
    see flight simulators
SLS, 332
    metal, 334
    nylon, 334
    sealing, 335
SolidWorks, 120, 303

Spalart-Allmaras, 200, 205
spar deflections, 244, 250
spars, 21, 37, 190, 221
  FEA, 250
  sizing, 169
  testing, 293
spats, 95
spiral mode, 223
SPOTTER, 20
  avionics diagram, 73
  design brief, 184
  engines, 64
  fuel tank sensors, 71
  geometry, 187
  integral fuel tank, 53
  nacelle, 46
  payload pod, 51
  spreadsheet, 182
  wing, 37
Spreadsheet Based Concept Design, 165
stability derivatives, 224
stall, 34, 195, 281, 355
Star-CD, 196
static margin, 170
Steering, 95
STEP, 191, 242
structural analysis, 119, 125
  codes, 125
  preliminary, 237
  using simple beam theory, 243
structural dynamics, 265
structural failure, 106
structural loading calculations, 169
structural mounting and loading, 293
structural testing
  dynamic, 296
  instruments, 290
  static, 294
structure
  internal, 23
SULSA, 51, 93
Suspension, 95
SVN, 126
sweep, 191, 319
synthesis, 7

System Description, 351
  Airframe, 352
  Flight Data, Acceptance, 358
  Ground Control System, 356
  Performance, 355
systems engineering, 110

tail, 45, 57
take-off, 68
  auto, 380
  performance, 148, 356
  run, 375
taxonomy, 7
telemetry, 25, 88
Test Flight
  examples, 375
  planning, 369
Tgrid, 204
Three-Dimensional Printing, see 3d printing
thrust, 124, 172
thrust to weight ratio, 146, 154, 176
Tool Selection, 119
topologies, 133
TortoiseSVN, 126
transmitter, 358
  payload, 87
  primary control, 73
transponders, 27
TRIZ, 136
turbulence model, 196, 200
  choice, 204
turn performance, 148
twist, 191, 319

UAVs
  A Brief Taxonomy, 15
  morphology, 19
UIUC database, 59, 166, 198
undercarriage, 27, 93
  attachment, 55
  load test, 295
  retractable
    wing housed, 42
  retractable systems, 97

V-tail, 58, 139
validation, 281
value driven design, 6
viscous sublayer, 209
visual line of sight, *see* VLOS
VLOS, 359
Vn diagram, 238, 352

weight, 125
  control, 273
  estimation, 170
  management, 114
Wheels, 93
wind tunnels tests, 282
  blockage effects, 284
  calibrating the test, 284
  elevator effectiveness, 287
  mounting the model, 282
  rudder effectiveness, 290
  typical results, 287

wing, 33
  attachment, 47
  covers, 37
  divergence, *see* divergence
  fuselage attachments, 38
  loading, 146, 154, 177
  morphing, 15, 40, 137
  ribs, 38
  span, 177, 319
  tips, 42
Wiring Looms, 342
Wiring, Buses and Boards, 82
work-breakdown structure, 110
woven rovings, 340
X-Plane, 227
XFLR5, 195
XFoil, 195

$y+$, 203
Young's modulus, 244, 334

Printed and bound by CPI Group (UK) Ltd, Croydon, CR0 4YY

16/04/2025

14658382-0002